Recent Advances in Iron Catalysis

Editor

Hans-Joachim Knölker

MDPI • Basel • Beijing • Wuhan • Barcelona • Belgrade • Manchester • Tokyo • Cluj • Tianjin

Editor
Hans-Joachim Knölker
Technische Universität Dresden
Germany

Editorial Office
MDPI
St. Alban-Anlage 66
4052 Basel, Switzerland

This is a reprint of articles from the Special Issue published online in the open access journal *Molecules* (ISSN 1420-3049) (available at: https://www.mdpi.com/journal/molecules/special_issues/iron_catalysis).

For citation purposes, cite each article independently as indicated on the article page online and as indicated below:

LastName, A.A.; LastName, B.B.; LastName, C.C. Article Title. *Journal Name* **Year**, *Article Number*, Page Range.

ISBN 978-3-03943-118-2 (Hbk)
ISBN 978-3-03943-119-9 (PDF)

Contents

About the Editor

Hans-Joachim Knölker, Professor Dr. rer. nat. habil., received his Ph.D. in 1985 at the University of Hannover, Germany. After postdoctoral studies in 1986 at the University of California in Berkeley, he completed his habilitation in 1990 at the University of Hannover. In 1991, he was appointed full professor of organic chemistry at the University of Karlsruhe, and in 2001, he moved to the Technische Universität Dresden. In 2000, he was a visiting scientist in India as a fellow of the Indian National Science Academy. He received the JSPS award twice and spent time as a visiting scientist in Japan as a fellow of the Japan Society for the Promotion of Science (JSPS), in 1998 at the University of Tsukuba and in 2007 at Kyushu University. In 2006, Professor Knölker was elected as an Ordinary Member of the Saxon Academy of Sciences, and he became a fellow of the Royal Society of Chemistry. Since 2011, he has been the editor-in-chief of the series *The Alkaloids* (Academic Press, London). His research areas include the development of novel synthetic methodologies, organometallic chemistry, natural product synthesis, biomolecular chemistry, and medicinal chemistry.

Preface to "Recent Advances in Iron Catalysis"

Transition metal-catalyzed reactions play a key role in many transformations which are applied in synthetic organic chemistry. For most of these reactions, noble metals have been used as catalysts with palladium being in the focus (for example, the Heck coupling, a broad range of palladium(0)-catalyzed cross-coupling reactions of organometallic reagents, the Wacker oxidation, and many others). Over the last two decades, we have witnessed a development of using more and more first row transition metals as catalysts for organic reactions, with iron taking the center stage. The driving forces behind this development are not only the high costs for the noble metals but also their toxicity. Iron is the most abundant transition metal in the Earth's crust, and thus, it is considerably cheaper than the precious noble metals often used in catalysis. Moreover, iron compounds have been involved in many biological processes early on during evolution, and in consequence, iron exhibits a relatively low toxicity. Because of this low toxicity, iron-catalyzed reactions have become an integral part of environmentally benign sustainable chemistry. However, iron catalysts are not only investigated to replace noble metals; they also offer many applications in synthesis beyond those of classical noble metal catalysts. Several articles of the present book emphasize the complementarity of iron-catalyzed reactions as compared to reactions catalyzed by noble metals. The book shows intriguing recent developments and the current standing of iron-catalyzed reactions as well as applications to organic synthesis.

Hans-Joachim Knölker
Editor

Review

On the Use of Iron in Organic Chemistry

Arnar Guðmundsson [1] and Jan-E. Bäckvall [1,2,*]

[1]   Department of Organic Chemistry, Arrhenius Laboratory, Stockholm University, SE-10691 Stockholm, Sweden; arnar.gudmundsson@su.se
[2]   Department of Natural Sciences, Mid Sweden University, Holmgatan 10, 85179 Sundsvall, Sweden
*   Correspondence: jeb@organ.su.se; Tel.: +46-08-674-71-78

Received: 2 March 2020; Accepted: 10 March 2020; Published: 16 March 2020

Abstract: Transition metal catalysis in modern organic synthesis has largely focused on noble transition metals like palladium, platinum and ruthenium. The toxicity and low abundance of these metals, however, has led to a rising focus on the development of the more sustainable base metals like iron, copper and nickel for use in catalysis. Iron is a particularly good candidate for this purpose due to its abundance, wide redox potential range, and the ease with which its properties can be tuned through the exploitation of its multiple oxidation states, electron spin states and redox potential. This is a fact made clear by all life on Earth, where iron is used as a cornerstone in the chemistry of living processes. In this mini review, we report on the general advancements in the field of iron catalysis in organic chemistry covering addition reactions, C-H activation, cross-coupling reactions, cycloadditions, isomerization and redox reactions.

Keywords: iron; organic synthesis; C-H activation; C-C coupling

---

1. Introduction

Iron is the most abundant element on Earth by mass and is used ubiquitously by living organisms [1]. The ability of iron to assume many oxidation states (from −2 up to +6) coupled with its low toxicity makes it an attractive, versatile and useful catalyst in organic synthesis. The wide range of oxidation states available for iron and its ability to promote single electron transfer (SET) allows it to cover a wide range of transformations. In low oxidation states, iron becomes nucleophilic in character and takes part in reactions such as nucleophilic substitutions, reductions and cycloisomerizations. At higher oxidation states, iron behaves as a Lewis acid, activating unsaturated bonds and at very high oxidation states (+3 to +5) iron complexes can take part in C-H activation. Due to iron's central position in the periodic table, it can have the property of both an "early" and "late" transition metal and with the many oxidation states available, any type of reaction is, in principle, within reach. Iron cations also bind strongly to many N- and O-based ligands, and these ligands can replace phosphine ligands in iron chemistry.

As the atmosphere on Earth changed following the Great Oxidation Event about 2.4 billion years ago, ferrous (+2) iron complexes became less stable and ferric (+3) complexes became predominant [2]. Iron in its ferric oxidation state typically forms complexes that are water-insoluble like hematite (Fe_2O_3) or magnetite (Fe_3O_4), especially under basic conditions when exposed to air. This propensity of iron complexes to precipitate can be a hindrance for catalysis, although, despite the fact that ferric complexes are mostly water-insoluble at biological pH, iron is still the most common transition metal in living organisms and is indispensable for the chemical processes of life—oxygen binding, electron transport, DNA synthesis, and cell proliferation, to name only a few.

Modern catalysis has been dominated by noble transition metals, such as palladium, platinum, ruthenium and iridium, and these metals have been used in a wide range of reactions. The main advantage that noble transition metals have over their base counterparts is their preference for undergoing two-electron processes. They do, however, have significant drawbacks: high cost, non-renewable supply and/or precarious toxicological and ecological properties. These factors may not pose too much of a problem for academic research, but have profound implications for industry and for the future of sustainable chemistry. It is for these reasons that attention has shifted towards base transition metals like iron, copper and nickel. Iron is particularly well suited as it is the second most abundant metal in the Earth's crust after aluminum and is consequently attractive economically and ecologically. Despite these advantages of iron, organoiron catalysts tend to suffer from serious drawbacks such as difficult synthetic pathways, lack of robustness, poor atom economy and low activity or enantioselectivity. Although circumventing these limitations will be necessary for iron catalysis to reach its full potential, base metal catalysis will no doubt gain importance in the future and it is reasonable to think that base metals such as iron will eventually supplant the traditional dominance of noble transition metals as the field matures. In recent times, the area of iron catalysis has exploded and Beller in 2008 and Bolm in 2009 declared that the age of iron has begun [3,4]. An intriguing outlook on the future of homogeneous iron catalysis was published in 2016 by Fürstner [5].

This review focuses on the recent advancements in iron catalysis pertaining to organic chemistry from 2016 to February 2020. An excellent and comprehensive review from 2015 by Knölker on all the types of iron-catalyzed reactions discussed in this review in addition to others can be consulted for the interested reader [6]. Other more specialized reviews may be found in each respective subsection. It should be mentioned that metathesis reactions are omitted from this review.

2. Iron in Organic Synthesis

Organoiron chemistry began in 1891 with the discovery of iron pentacarbonyl by Mond and Berthelot [7,8]. It was used sixty years later industrially in the Reppe process of hydroformylation of ethylene to form propionaldehyde and 1-propanol in basic solutions [9]. An important event was the discovery of ferrocene in 1951 [10], the structure of which was determined in 1952 [11,12] and led to the Nobel prize being awarded to Wilkinson and Fischer in 1973. The discovery of the Haber–Bosch process was an additional milestone in iron chemistry. The latter process uses an inorganic iron catalyst for the production of ammonia and sparked an agricultural revolution [13,14]. In modern organoiron chemistry, iron is used in a great number of diverse reactions, as will be apparent from this review, though perhaps, just as in the chemistry of life, its most ubiquitous role is in redox chemistry.

2.1. Addition Reactions

The first example of an iron-catalyzed racemization of alcohols was reported in 2016 with the use of an iron pincer catalyst [15]. Between 2016 and 2017, the groups of Bäckvall [16] and Rueping [17] independently reported the dynamic kinetic resolution (DKR) of sec-alcohols using a combination of iron catalysis for racemization and a lipase for resolution, which demonstrates a useful combination of enzyme and transition metal catalysis (Scheme 1). In one study, Knölkers complex (**II**) was used directly [17], and in the other study, a bench-stable precursor to Knölkers complex (**II**), iron tricarbonyl complex **I**, was used, which was activated through oxidative decarbonylation with TMANO to form coordinatively unsaturated iron complex **I′** [16]. In the latter study, various benzylic and aliphatic esters could be produced in good to excellent yields with excellent *ee*. Two different enzymes, Candida antarctica lipase B (CalB/Novozyme 435) and Burkholderia cepasia (PS-C), could be used and the procedure could be reproduced on gram-scale. The group of Zhou also reported a related work using hexanoate as the acyl donor [18]. Rodriguez published a review on the synthesis, properties and reactivity of this interesting class of iron catalysts in 2015 [19].

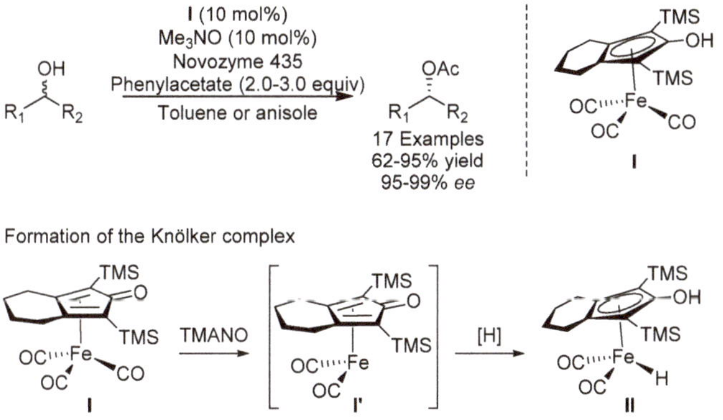

Formation of the Knölker complex

Scheme 1. Bäckvall's and Rueping's DKR of *sec*-alcohols.

The indole ring is a ubiquitous heterocyclic motif in natural products and methods for constructing chiral polycyclic systems with indole skeletons has attracted considerable attention. In 2017, the group of Zhou reported the first intramolecular enantioselective cyclopropanation of indoles that was catalyzed either by iron or copper in the presence of a chiral ligand (Scheme 2) [20]. Many functional groups were tolerated and various cyclopropanated indoles were prepared in high to excellent yields and in almost all cases with excellent ee. The mechanism of the enhancement of the enantioselectivity is currently unknown, although the R_2 group was found to be important and had to be different from hydrogen for the reaction to proceed.

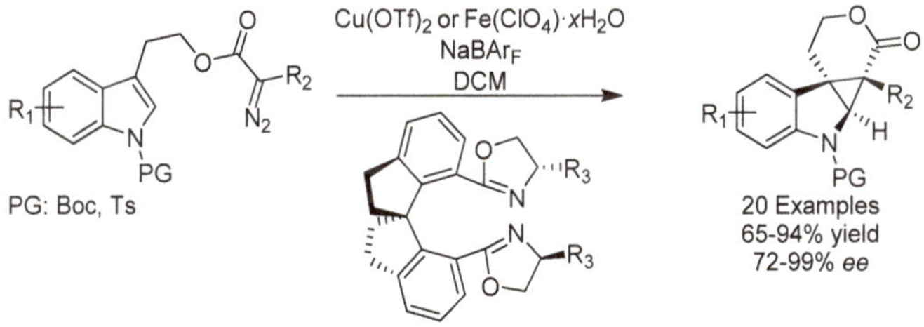

Scheme 2. Zhou's enantioselective cyclopropanation.

Hydroamination of alkenes is an atom-economic approach and the amines produced are some of the most common functionalities found in fine chemicals and pharmaceuticals. Hydroamination of terminal alkenes typically gives the Markovnikov product selectivity, but in 2019 the group of Wang reported the first iron-catalyzed anti-Markovnikov addition of allylic alcohols [21]. For this purpose, an iron-PNP pincer complex was used. The reaction proceeds through a hydrogen-borrowing strategy where the iron complex temporarily activates the alcohol by dehydrogenation to the α,β-unsaturated carbonyl compound. The latter compound reacts with an amine to form an iminium ion, which undergoes conjugate additon at the β-position with another amine followed by hydrolysis and reduction to give the product. Various amines were produced in good yields with this method. Interestingly, hydroamidation could also be performed (Scheme 3).

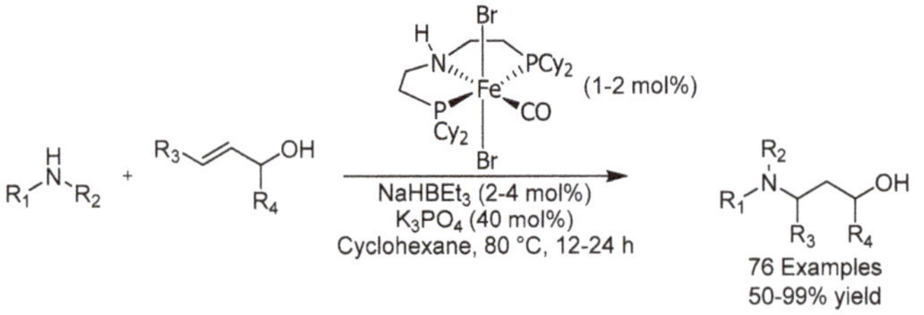

Scheme 3. Wang's iron-catalyzed anti-Markovnikov hydroamination.

C-C and C-N bonds are important bonds in organic chemistry and one of the most effective ways of creating these bonds simultaneously is through the carboamination of olefins. In 2017, the group of Bao reported the diastereoselective construction of amines and disubstituted β-amino acids through the carboamination of olefins (Scheme 4) [22]. Aliphatic acids were used as an alkyl source and nitriles as a nitrogen source. The protocol could be performed on a gram-scale and various carboamination products were obtained in good yields and excellent diastereoselectivity. The choice of acid was found to have a strong effect on the diastereoselectivity. TsOH was used for the carboamination of olefins, but for the carboamination of esters, binary and ternary acids had a more positive effect over monoacids, with H_2SO_4 giving the best result. The addition of the nitrile group was found, through Density Functional Theory (DFT) calculations, to be diastereoselectivity-determining, and hyperconjugation was proposed to account for the anti-selectivity.

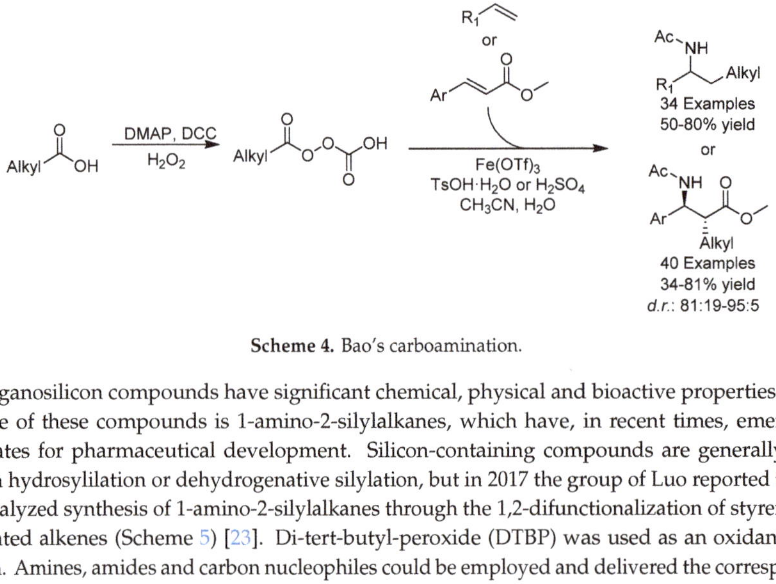

Scheme 4. Bao's carboamination.

Organosilicon compounds have significant chemical, physical and bioactive properties and an example of these compounds is 1-amino-2-silylalkanes, which have, in recent times, emerged as candidates for pharmaceutical development. Silicon-containing compounds are generally made through hydrosylilation or dehydrogenative silylation, but in 2017 the group of Luo reported the first iron-catalyzed synthesis of 1-amino-2-silylalkanes through the 1,2-difunctionalization of styrenes and conjugated alkenes (Scheme 5) [23]. Di-tert-butyl-peroxide (DTBP) was used as an oxidant in the reaction. Amines, amides and carbon nucleophiles could be employed and delivered the corresponding products in mostly good yields. The reaction was proposed to proceed via a silicon-centered radical from oxidative cleavage of the Si-H bond followed by addition across the C=C bond and a N-H oxidative functionalization cascade.

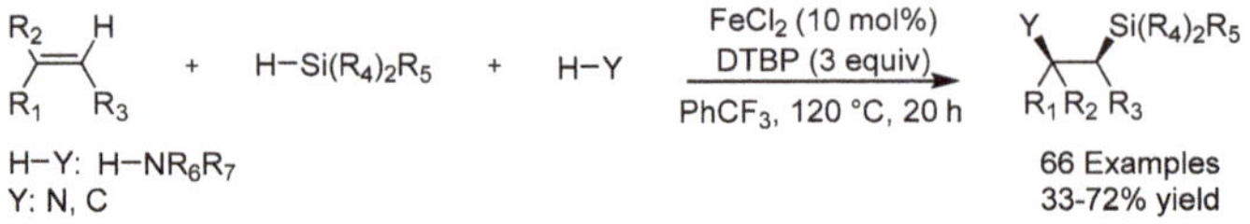

Scheme 5. Luo's silylation.

2.2. C-H Bond Activation

The first example of a C-H activation was a Friedel–Crafts reaction reported by Dimroth in 1902, where benzene reacted with mercury (II) acetate to give phenylmercury (II) acetate [24]. Later, in 1955, Murahashi reported the cobalt-catalyzed chelation-assisted C-H functionalization of (E)-N-1-diphenylmethaneimine to 2-phenylisoindolin-1-one [25]. A great advance in the field occurred in 1966 when Shilov reported that K_2PtCl_4 could induce isotope scrambling between methane and heavy water [26,27]. Shilov's discovery led to the so called "Shilov system", which remains to this day as one of the few catalytic systems that can accomplish selective alkene functionalizations under mild conditions. In 2008, the synthetic power of C-H activation was expanded to include organoiron catalysis by Nakamura in his arylation of benzoquinolines (Scheme 6) [28]. An excellent review on the subject of iron in C-H activation reactions by Nakamura was published in 2017 [29] and a review on oxidative C-H activation was published by Li in 2014 [30].

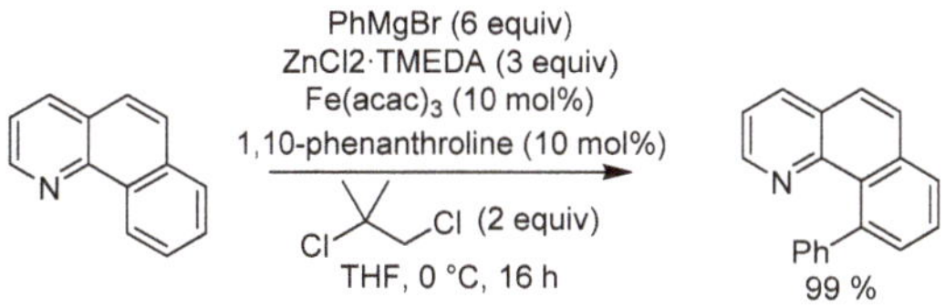

Scheme 6. Nakamura's C-H activation.

The group of Arnold has in the past used engineered cytochrome P450, which is a type of enzyme that uses a heme cofactor, to enantioselectively α-hydroxylate arylacetic acid derivatives via C-H activation [31]. In 2017, they reported the directed evolution of cytochrome P450 monooxygenase, for enantioselective C-H activation to give C-N bonds (Scheme 7) [32]. It uses a variant of P411 based on the P450 monooxygenase which has an axial serine ligand on the haem iron instead of the natural cysteine. The method utilizes a tosyl azide as a nitrene source which generates an iron nitrenoid that subsequently reacts with an alkane to deliver the C-H amination product. The P411 variant has a turnover number (TON) of 1300, which is considerably higher than the best reported, to our knowledge, for traditional chiral transition metal complexes, which is a chiral manganese porphyrin with a turnover number of 85 [33]. A variety of benzylic tosylamines could be produced with excellent ees.

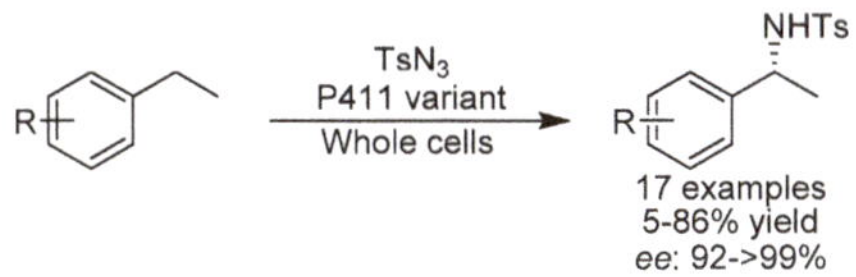

Scheme 7. Arnold's C-H amination.

In 2019, Arnold and coworkers extended their methodology using another variant of P411 in C-H alkylation using diazoesters (Scheme 8) [34]. The diazo substrate scope could be extended beyond ester-based reagents to Weinreb amides and diazoketones and gave the corresponding products with excellent *ees* and with total turnover numbers (TTN) of up to 2330. These studies together show the potential for generating C-H alkylation enzymes that can emulate the scope and selectivity of Natures C-H oxygenation catalysts.

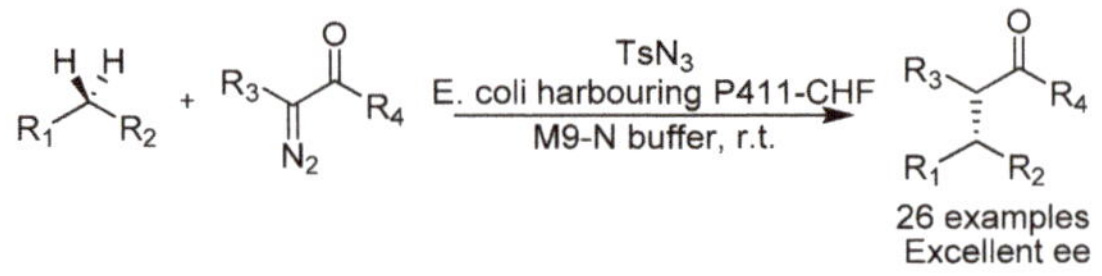

Scheme 8. Arnold's sp^3 C-H activation.

C(sp^3)-H alkylation via an isoelectronic iron carbene intermediate was first reported in 2017 by the group of White using an iron phthalocyanine (Scheme 9) [35]. Iron carbenes generally prefer cyclopropanation over C-H oxidation, but, in this case, allylic and benzylic C(sp^3)-H bonds could be alkylated with a broad scope. Mechanistic studies indicated that an electrophilic iron carbene was mediating homolytic C-H cleavage followed by recombination with the resulting alkyl radical to form the new C-C bond. The C-H cleavage was found to be partially rate determining.

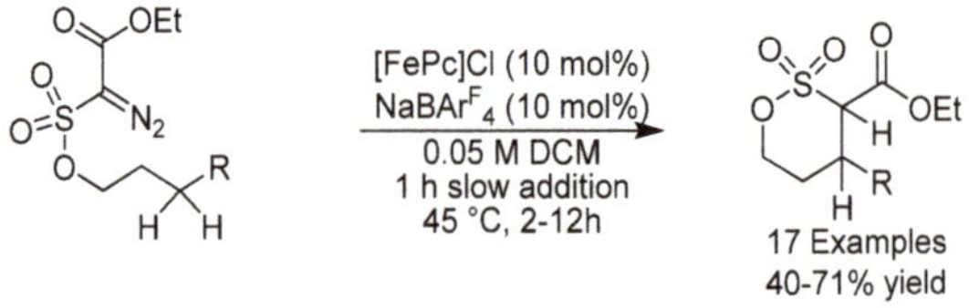

Scheme 9. White's isoelectronic carbene C(sp^3)-H oxidation.

In 2018, the group of Ackermann reported on an allene annulation through an iron-catalyzed C-H/N-H/C-O/C-H functionalization sequence (Scheme 10) [36]. The mechanism was shown to involve an unprecedented 1,4-iron migration C-H activation manifold. Alkyl chlorides were tolerated under these reaction conditions, with no cross-coupling being observed. Various dihydroisoquinolones could be produced through the use of this method in excellent yields and the modular nature of the triazole group allowed for the synthesis of exo-methylene isoquinolones as well.

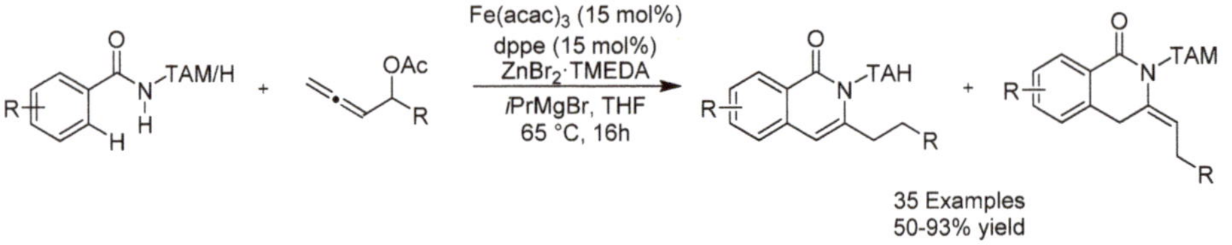

Scheme 10. Ackermann's allene annulation.

In late 2019, the group of Wang demonstrated an iron-catalyzed α-C-H functionalization of π-bonds in the hydroxyalkylation of alkynes and olefins (Scheme 11) [37]. Propargylic and allylic C-H bonds were functionalized with this method and a wide variety of homopropargylic and homoallylic alcohols could be produced in excellent yields, although with modest stereoselectivity. The key to the success of this approach is the fact that coordination of the iron catalyst to the unsaturated bond is known to lower the pKa of a propargylic or allylic proton from ≈38 and ≈43, respectively, to <10 [38]. An (α-allenyl)iron or (π-allyl)iron complex for propargylic or allylic complexes, respectively, is then formed in the presence of a base, which is utilized as the coupling partner.

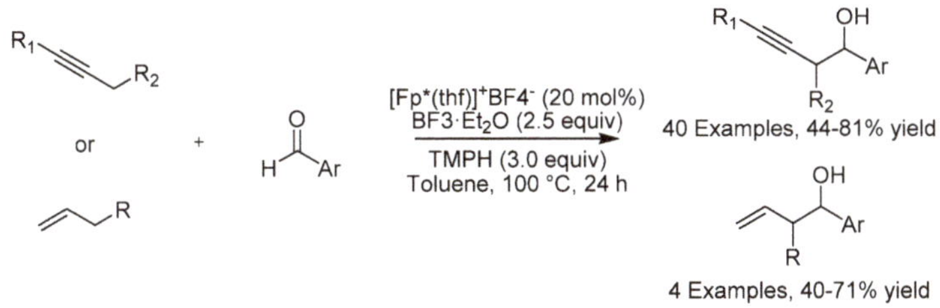

Scheme 11. Wang's α-C-H functionalization.

In 2018, the group of Liu reported the unprecedented iron(II)-catalyzed fluorination of C(sp³)-H bonds using alkoxyl radicals (Scheme 12) [39]. The procedure was applied to a wide range of substrates and it was found that a range of functional groups were tolerated, including halide and hydroxyl groups. N-fluorobenzenesulfonamide (NFSI) was used as the fluoride source and the substrate scope could be extended from fluorination to chlorination, amination and alkylation. The authors also demonstrated a one-pot application of their protocol starting from a simple alkane.

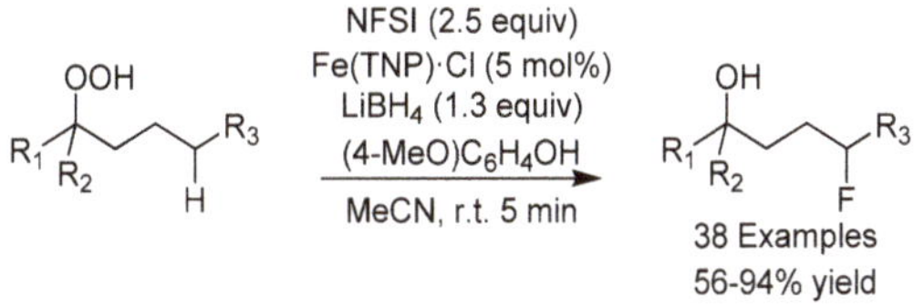

Scheme 12. Liu's C-H fluorination.

2.3. Cross-Coupling Reactions

Transition-metal-catalyzed cross coupling protocols have become an important tool in the organic chemist's arsenal. This area has been important in chemistry for about five decades, and in 2010 it received formal recognition when Richard Heck, Akira Suzuki and Ei-ichi Negishi received the Nobel prize for palladium-catalyzed cross-couplings in organic synthesis. Although a powerful technique, its applications have been dominated by the use of expensive palladium- and nickel-based catalysts, which are often toxic. The most common types of cross coupling reactions using iron are those involving Grignard reagents as the transmetalating nucleophile. The first example of an alkenylation of alkyl Grignard reagents with organic halides using iron(III) chloride was reported in 1971 by Kochi and Tamura (Scheme 13) [40]. A review on the subject of iron-catalyzed cross-coupling reactions with a focus on mechanistic studies was published in 2016 by Byers [41]. A more focused review on the use of iron-catalyzed cross-coupling for the synthesis of pharmaceuticals was released in 2018 by Szostak [42].

Scheme 13. Kochi's and Tamura's original alkenylation.

In 2016, Bäckvall and coworkers reported the coupling of propargyl carboxylates and Grignard reagents using the environmentally benign Fe(acac)₃ to synthesize substituted allenes and protected α-allenols (Scheme 14) [43,44]. The mild reaction conditions tolerate a broad range of functional groups (silyl ethers, carbamates and acetals) and could be applied to more complex molecules such as steroids. Tri and tetra substituted allenes were obtained in excellent yields, whereas the yield was found to drop for less substituted allenes. A variety of alkyl and aryl Grignard reagents could be applied and it was demonstrated that the protocol can be readily performed on a gram-scale.

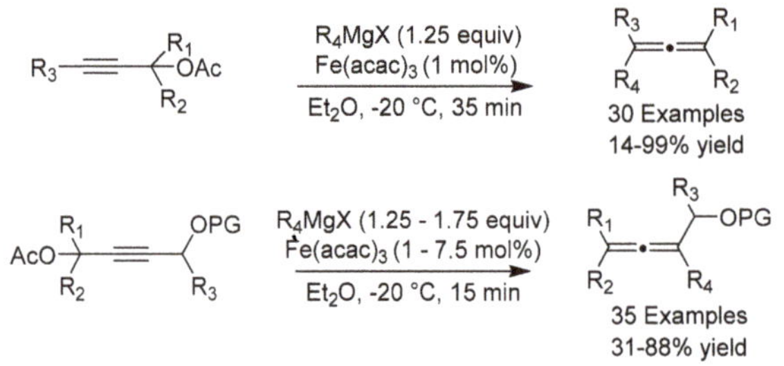

Scheme 14. Bäckvall's synthesis of substituted allenes and protected α-allenols from carboxylates.

In 2016, the group of Frantz reported on a highly stereoselective iron-catalyzed cross coupling using $FeCl_3$ to couple Grignard reagents and enol carbamates (Scheme 15) [45]. Many functional groups, such as ethers, silanes, primary bromides, alkynes and alkenes, were tolerated. In almost all cases, the yield and E/Z selectivity was excellent, with (E)-carbamates leading to (E)-acrylates and (Z)-carbamates leading to (Z)-acrylates. This study constitutes the only example so far of an iron-catalyzed cross-coupling, where an oxygen-based electrophile is favored over a vinylic halide (a Cl group at R_2 in Scheme 15).

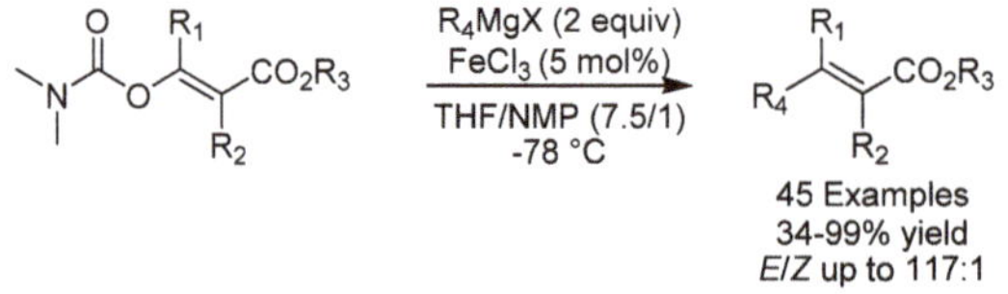

Scheme 15. Frantz' stereoselective synthesis of acrylates.

Aryl C-glycosides are interesting pharmaceutical candidates because of their biological activities and resistance to metabolic degradation. In 2017, the group of Nakamura developed a highly diastereoselective iron-catalyzed cross-coupling of glycosyl halides and aryl metal reagents to form these compounds using $FeCl_2$ in conjunction with a SciOPP ligand (Scheme 16) [46]. A variety of aryl, heteroaryl and vinyl metal reagents based on magnesium, zinc, boron and aluminium could be applied. The reaction was found to proceed through the generation and stereoselective trapping of glycosyl radical intermediates and represents a rare example of a highly stereoselective carbon-carbon bond formation based on iron catalysis.

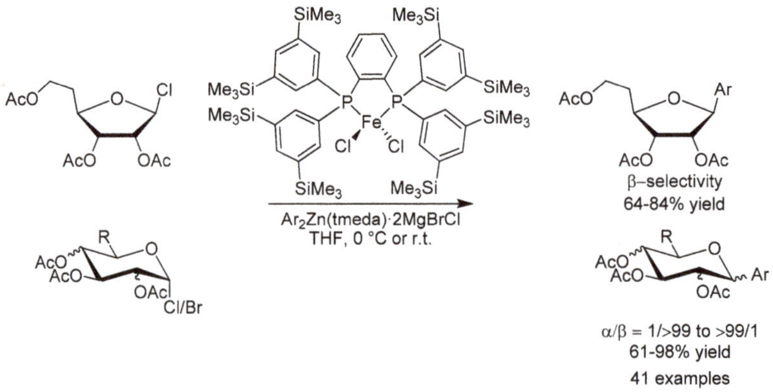

Scheme 16. Nakamura's diastereoselective synthesis of aryl C-glycosides using (Sciopp) $FeCl_2$.

Heterocyclic motifs are common in biologically active compounds and the presence of heteroatoms arranged around a quaternary carbon center often endows a certain spacial definition that can be useful, for example, in enhancing drug-binding. These types of spirocyclic motifs are most often generated through [2 + 3] cycloadditions, but, in 2017, the group of Sweeney developed an elegant cross-coupling cascade reaction to generate these motifs with inexpensive Fe(acac)$_3$ as the catalyst directly from feedstocks chemicals directly available from plant sources (Scheme 17) [47]. The protocol delivered diastereomerically enriched nitrogen- and oxygen-containing cis-heterospirocycles and was applicable to substrates with typically sensitive functionalities like esters and aryl chlorides.

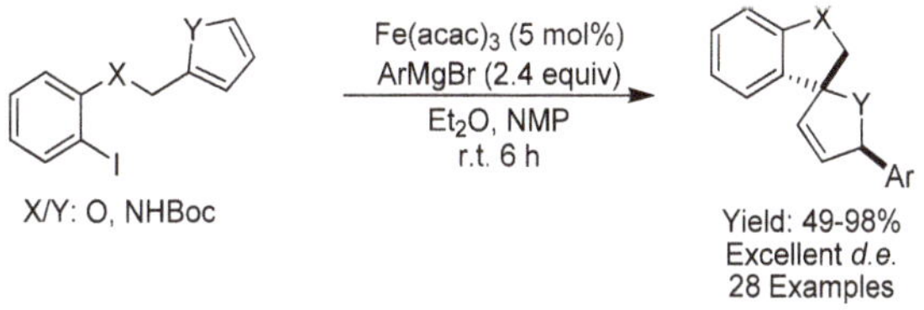

Scheme 17. Sweeney's cyclization.

Palladium catalysts have traditionally been ubiquitous in cross-coupling reactions. Utilizing iron for the Suzuki coupling to provide simple biaryl compounds remained elusive until recently, when the group of Bedford showed that simple π-coordinating N-pyrrole amides on the aryl halide substrate could facilitate activation of the relatively unreactive C-X bond (Scheme 18) [48]. The use of an NHC ligand with the appropriate steric bulk proved crucial. A variety of biaryl products were obtained in excellent yields. This study demonstrates that iron-catalyzed Suzuki couplings to give biaryls are achievable, though to make the procedure general, efforts will have to be made to develop non-directed halide bond activation.

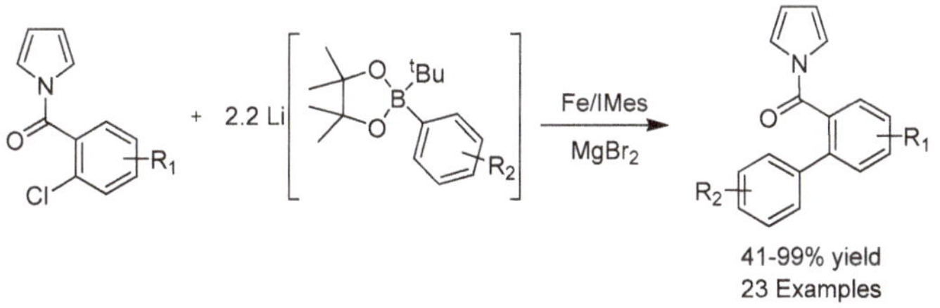

Scheme 18. Bedford's cross-coupling.

2.4. Cycloadditions

Medium-sized rings (7–11 membered) are important structural motifs with applications in the synthesis of polymers, fragrances and other specialty chemicals. Among these cyclic compounds, eight-membered rings are particularly useful for the synthesis of polyethylene derivatives. Metal-catalyzed [4 + 4]-cycloaddition of two butadienes to produce 1,5-cyclooctadiene is well established, although the use of 1,3-dienes is challenging owing to unwanted side-products including linear oligomers, [2 + 2]- and [4 + 2]-cycloadducts, as well as regio- and stereoisomeric [4 + 4] products. Catalyst design must be guided with these concerns in mind. Recently, the group of Chirik developed an iron-catalyzed [4 + 4]-cycloaddition of 1,3-dienes to access these compounds using (imino)pyridine iron bis-olefin and α-diimine iron complexes (Scheme 19) [49]. A wide variety of 1,5-cyclooctadienes could be produced in good to excellent yields and with controlled chemo- and regioselectivity. Kinetic analysis and Mößbauer spectroscopy provided evidence for a mechanism in which oxidative cyclization of the two dienes determines the regio- and diastereoselectivity.

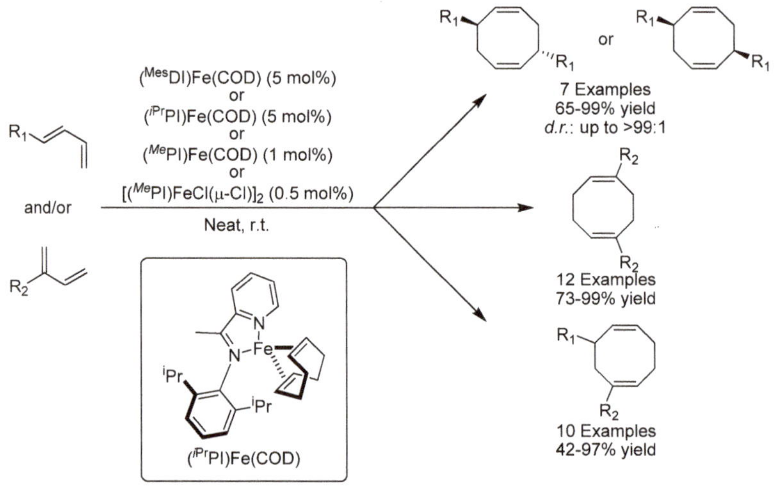

Scheme 19. Chirik's iron-catalyzed [4 + 4]-cycloaddition.

Indolines are pharmaceutically significant aza-heterocycles and the first example of a synthesis of these compounds via an iron-catalyzed decarboxylative [4 + 1]-cycloaddition was described in 2016 by the group of Xiao. A wide range of functionalized indolines could be prepared from vinyl benzoxazinanones and sulfur ylides in good to excellent yields in high diastereoselectivity using the nucleophilic $Bu_4N[Fe(CO)_3(NO)]$ (TBAFe) catalyst (Scheme 20) [50]. The authors postulated that the reaction proceeds via the formation of an electrophilic (π-allyl) iron complex, which would subsequently react with the sulfur ylide and undergo an intramolecular S_N2 displacement of dimethyl sulfide by the tosyl amide anion. This study is noteworthy and constitutes the first example of the exploitation of an interesting reverse-electron-demand (π-allyl) iron species.

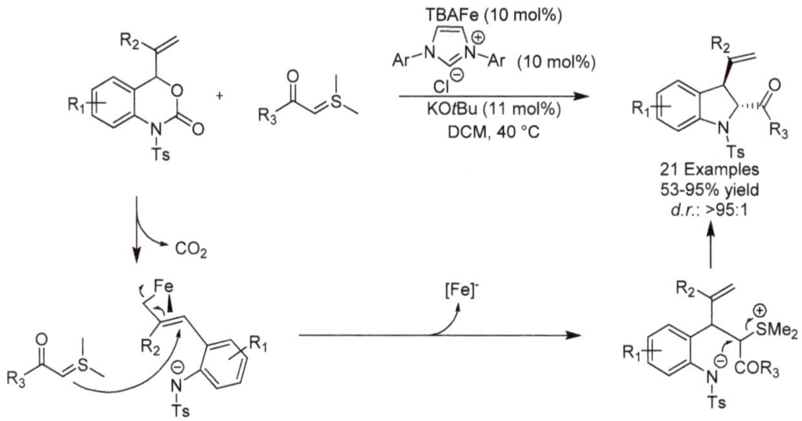

Scheme 20. Xiaos' [4 + 1]-cycloaddition.

Oxidative [4 + 2] annulations of imines and alkynes for the synthesis of isoquinolines are well established [51] but an interesting redox-neutral variant was reported in 2016 by the group of Wang involving an iron-carbonyl-catalyzed C-H transformation of an arene (Scheme 21) [52]. A wide variety of isoquinolines were isolated in good to excellent yield and only the cis-stereoisomer was observed.

Mechanistic studies revealed that there was an essential synergy of dinuclear irons in the oxidative C-H addition and turnover-limiting H-transfer to the alkyne in the catalytic cycle.

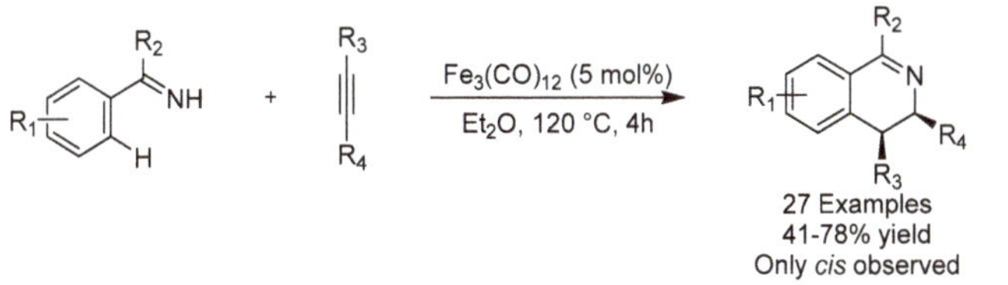

Scheme 21. Wang's [4 + 2] annulation/C-H activation.

Cyclotrimerization of alkynes mediated by transition metals is a key approach towards the synthesis of substituted aromatic compounds. In 2017, the group of Jacobi von Wangelin reported the iron-catalyzed trimerization of terminal alkynes in the absence of a reducing agent (Scheme 22) [53]. The approach was based on having a simple Fe (II) precatalyst and an internal bulky basic ligand which could mimic the effect of an external reductant. Terminal alkynes readily reacted, whereas internal ones did not, which is in accordance with the mechanistic postulate that the reaction is initiated through alkyne deprotonation. Good regioselectivity was observed with almost exclusive formation of the 1,2,4-substituted product over the 1,3,5-substituted and a wide variety of arenes were produced in mostly excellent yields.

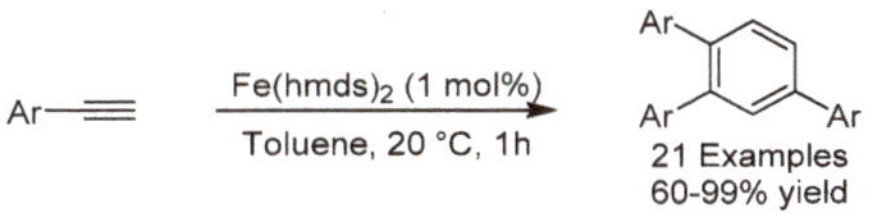

Scheme 22. Jacobi van Wangelin's cyclotrimerization.

2.5. Isomerizations

Iron-catalyzed isomerization of unsaturated alcohols to carbonyls was first reported by Emerson and Pettit in the 1960s [54]. In this study, butadiene-(tricarbonyl)iron complexes were exposed to strong acids to form their corresponding π-allyl iron complexes. When these were reacted with water, η^2-(allyl alcohol)iron complexes formed, which isomerized to the enol-iron complex. Upon decomposition, butanone was obtained.

Between 2017 and 2019, the groups of Bäckvall [55,56] and Rueping [57,58] independently reported the iron-catalyzed cycloisomerization of allenols and allenic sulfonamides using iron tricarbonyl complex **I** (Scheme 23). An analogous ruthenium-catalyzed cyclization of α-allenols had previously been reported by Bäckvall [59]. In these protocols, **I** is activated by TMANO to form the active catalyst **I'**. The protocols have a wide scope and deliver the corresponding heterocycles in excellent yields. Interestingly it was found that through the introduction of a substituent in the R_2 position of the allene, the reaction was made diastereoselective, giving the trans-product predominantly in the case of dihydrofurans or exclusively in the case of dihydropyrroles [54,55]. The origin of the diastereoselectivity, revealed through DFT calculations, was the participation of the non-innocent cyclopentadienyl ligand of the catalyst lowering the transition state, leading to the trans product.

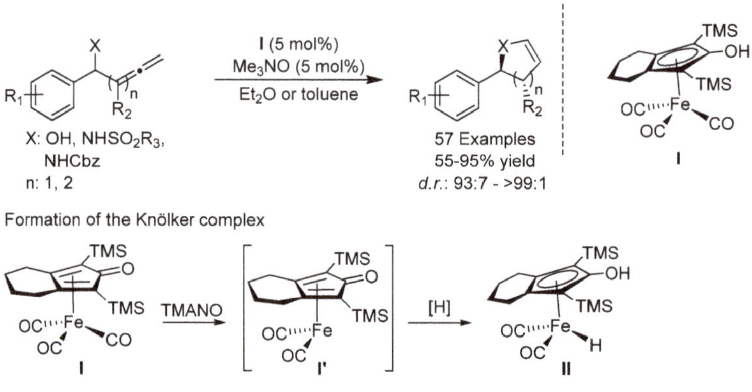

Scheme 23. Bäckvall's and Rueping's cycloisomerizations.

According to the Woodward–Hofmann rules, [2 + 2] cycloadditions of two olefins are photochemically allowed, but thermally forbidden [60]. Due to this restriction, photochemical [2 + 2] cycloaddition between two olefins is the most common strategy to form cyclobutanes. In 2018, the group of Plietker reported an unusual cycloisomerization of enynes using a cationic iron nitrosyl catalyst (Scheme 24) [61]. Their 1,6- and 1,7-enyneacetates were expected to isomerize to allenes, but instead isomerized to their corresponding bicyclo [3.2.0] or [4.2.0] alkylamides in good to excellent yield under mild conditions. The reaction tolerates esters, amides and halides, and afforded the products generally in good to high yields.

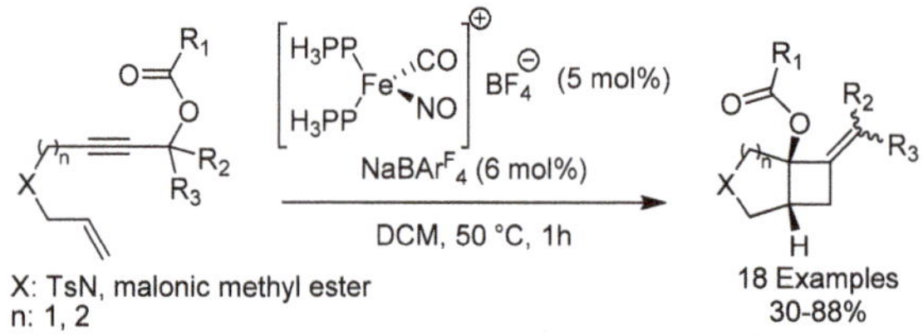

Scheme 24. Plietker's cycloisomerization of enynes.

Iron carbonyl complexes have been used in the isomerization of allylic alcohols, but a major drawback associated with the use of these complexes is the fact that they require UV light for activation and are thus not amenable to industrial synthesis [62]. In 2018, the group of de Vries reported the use of a pincer PNP-iron complex for the isomerization of allylic alcohols and found it to be an excellent racemization catalyst in conjunction with tBuOK (Scheme 25) [63]. Both benzylic and aliphatic allylic and homoallylic alcohols could be isomerized in mostly excellent yields. The sterically hindered trans-sobrerol could be reduced as well, although in a modest yield of 36%. A two-step hydrogen borrowing mechanism involving dehydrogenation-hydrogenation isomerization was proposed on the basis of DFT calculations.

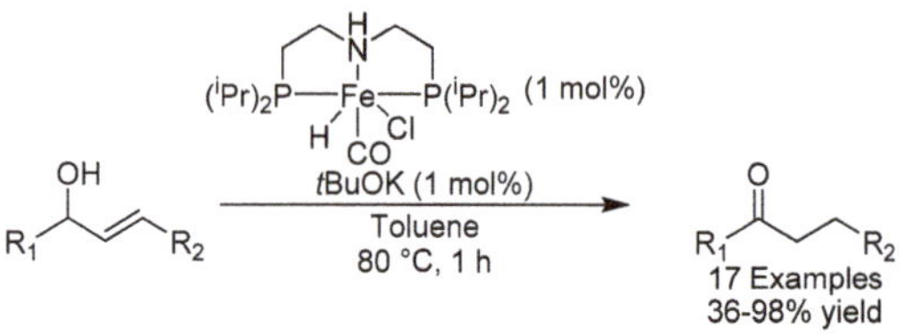

Scheme 25. de Vries' allylic alcohol isomerization.

An elegant cascade reaction involving cross-coupling and cycloisomerization forming two C-C bonds and highly functionalized 1,3-dienes with a stereodefined tetrasubstituted alkene unit, which is difficult to obtain by other means, was developed in 2016 by the group of Fürstner (Scheme 26) [64]. A series of X-ray structures showed that the substituent delivered by the Grignard reagent and the transferred alkenyl substituent were on the same side of the central tetrasubstituted alkene unit. Any loss of stereoselectivity is likely the result of a secondary isomerization process. Many variously substituted 1,3-dienes could be produced with this method in good to excellent yield and enantioselectivity and both primary and secondary Grignard reagents could be used.

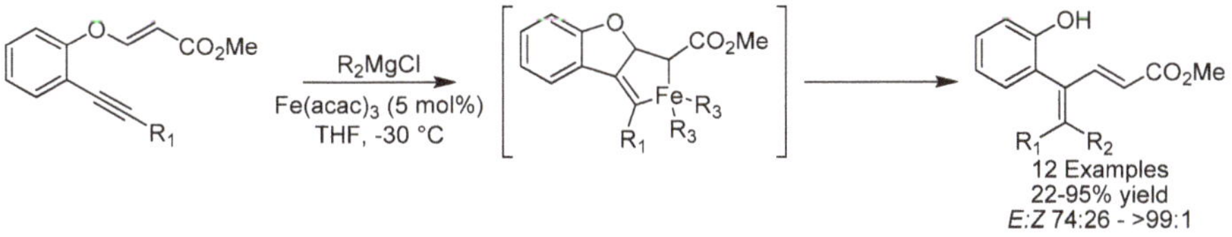

Scheme 26. Fürstner's cycloisomerization/cross-coupling cascade reaction.

2.6. Redox Reactions

Redox chemistry is what iron has traditionally been most known for, due to its multiple possible oxidation states. It is a very important field, not only for organic synthesis but also for biochemistry and industrial applications. A review on the use of iron (and other base metals) in borrowing hydrogen strategies was released by Morrill in 2019 [65]. A review on iron in reduction and hydrometalation was published in 2019 by Darcel [66].

One of the main bottlenecks in synthetic fuel production powered by sunlight or electricity is the oxidation of water. The use of abundant and environmentally benign metals for this purpose is desireable and, to this end, the group of Masaoka, in 2016, reported the use of a pentanuclear iron catalyst $[Fe^{II}_4Fe^{III}(\mu_3\text{-}O)(\mu\text{-}L)_6]^{3+}$ ((L = 3,5-bis(2-pyridyl)pyrazole) [67]. A multinuclear core typically gives rise to a redox flexibility that would be expected to be favourable in a reaction involving multi-electron transfer, such as water oxidation, as is known from enzyme examples in nature such as photosystem II in plants where Mn_4Ca clusters are used. The turnover frequency was found to be 1900 per second, which is about three orders of magnitude greater than that of other iron-based systems, but despite the advantage of this system, it required a $H_2O/MeCN$ mixture and a large overpotential of 0.5 V, preventing it from practical use. These findings do, however, clearly indicate that efficient water oxidation protocols can be developed based on multinuclear iron catalysts.

Oxidation reactions are of fundamental interest in organic chemistry, and an ideal oxidant to use in such processes is O_2. To this end, the group of Bäckvall in 2020 reported the first example of a biomimetic oxidation of alcohols (Scheme 27) [68]. Various primary and secondary alcohols could be oxidized to their corresponding aldehydes and ketones with this method in good to excellent yields. The process was inspired by the electron transport chain used in living organisms and involves coupled redox processes that lead to a low-energy pathway through a stepwise oxidation using electron transfer mediators (ETMs) with O_2 as the terminal oxidant. An iron(II)-catalyst was used for this purpose, which exhibited surprising stability in the presence of O_2, in addition to a benzoquinone-derivative and a Co(salen)-type complex as ETMs.

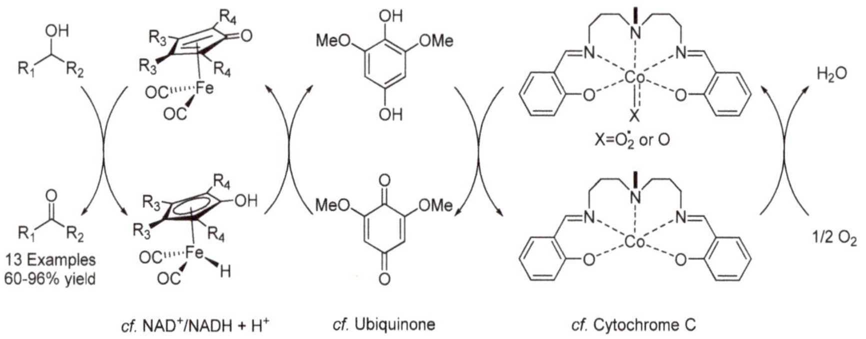

cf. NAD$^+$/NADH + H$^+$ *cf.* Ubiquinone   *cf.* Cytochrome C

Scheme 27. Bäckvall's biomimetic oxidation of alcohols.

The oxidation of alcohols to carboxylic acids typically involves the use of stoichiometric amounts of toxic oxidants, such as $KMnO_4$ or CrO_3. In 2016, the group of Ma developed an iron-catalyzed aerobic oxidation of alcohols to carboxylic acids to circumvent this problem (Scheme 28) [69]. A wide variety of alcohols (and aldehydes) could be oxidized to their corresponding carboxylic acids and synthetically useful groups such as halides, alkynes, olefins, esters and ethers were tolerated. The synthetic application of this protocol was also demonstrated through the first total synthesis of the naturally occurring allene, phlomic acid, and in a 55 g synthesis of palmitic acid.

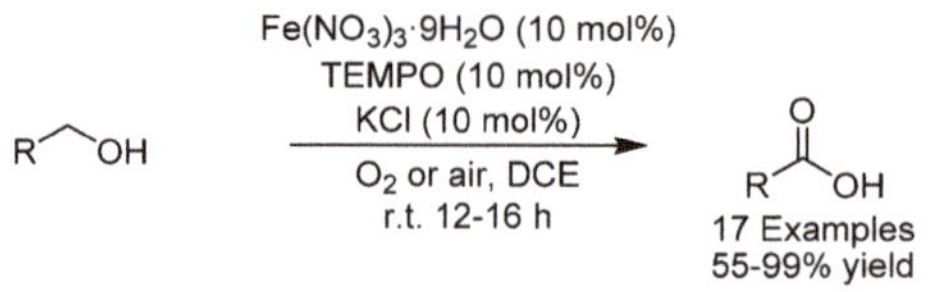

Scheme 28. Ma's iron-catalyzed aerobic oxidation of alcohols.

Cyclic α-epoxide enones are structures found routinely in natural products and are also useful synthons. Producing these compounds through enantioselective epoxidation is notoriously difficult. In 2016, the group of Costas reported the first iron-catalyzed and highly enantioselective epoxidation of these compounds using a bulky tetradentate iron catalyst (Scheme 29) [70]. The reaction was found to proceed via the generation of an electrophilic oxidant, which is unusual for this class of reactions, where nucleophilic oxidants usually account for the reaction through a Weitz–Scheffer-type mechanism. 2-Ethylhexanoic acid (2-eha) was found to be necessary as an additive to help activate H_2O_2. Various epoxides could be produced in mostly excellent yields and excellent *ees*.

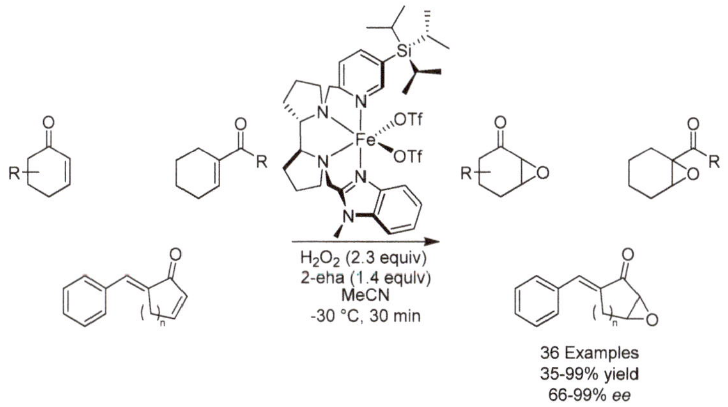

Scheme 29. Costas' enantioselective epoxidation.

A number of pharmaceutically active compounds for psychiatric therapy or cardic arrythmias have optically active alcohols or oxaheterocycle motifs. The standard methods of synthesizing these compounds rely on expensive and partly toxic noble metals and it would thus be advantageous to employ base metals for this purpose. In 2018, the group of Gade developed an enantioselective reduction of functionalized ketones to form these compounds (Scheme 30) [71]. A chiral pincer "boxmi"-type iron catalyst with a low catalyst loading of 0.5 mol% was used which provided access to a broad range of functionalized halohydrins, oxaheterocycles and amino alcohols in excellent yields and mostly excellent *ee*s. The protocol could also be performed on a gram scale to form a chiral halohydrin that forms the basis for the preparation of the antidepressant (R)-fluoxetine.

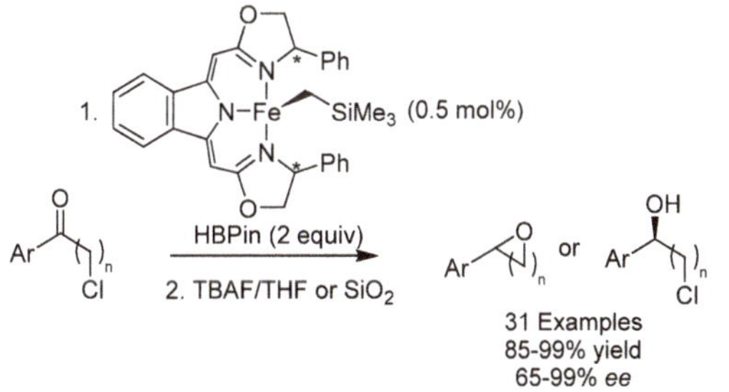

Scheme 30. Gade's enantioselective reduction of functionalized ketones.

Transfer hydrogenation processes are attractive alternatives to classical catalytic hydrogenation since the use of high pressure and expensive reactors can be avoided. The group of de Vries in 2019 reported the base-free transfer hydrogenation of esters using EtOH as the hydrogen source (Scheme 31) [72]. It is interesting to note that EtOH can be used as the reductant, which is usually a problem, since the acetaldehyde formed can potentially deactivate the catalyst by decarbonylation. Various aliphatic and aromatic esters, including lactones, could be reduced to their corresponding alcohols in good to excellent yields. In addition the reaction could be scaled up tenfold. Similar iron-based protocols using a base are known to reduce C=C bonds [73], but in this base-free protocol, esters were selectively reduced. One important topic in green chemistry is the recycling of polymers by conversion to their monomers. This study shows the first transfer-hydrogenation-catalyzed depolymerization of a polyester (Dynacol 7360, made from adipic acid and 1,6-hexanediol), demonstrating the possibility of using iron-catalysis for the recycling of plastic waste.

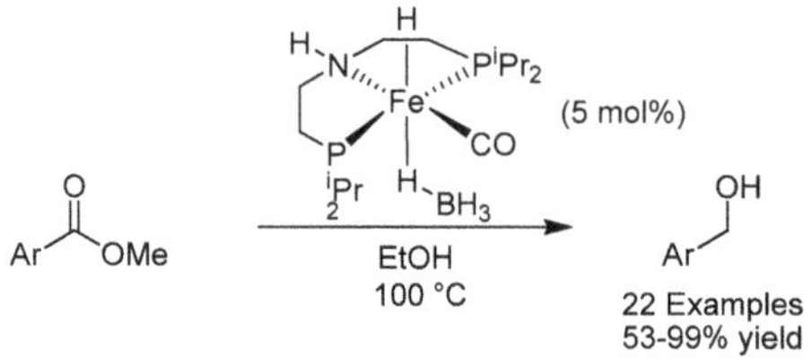

Scheme 31. de Vries' transfer hydrogenation of esters.

Simple carboxylic acids are excellent carbonyl donors for functionalized carboxylic acid derivatives. Generating the enolate, enediolate, of a carboxylic acid is, however, quite challenging, due to the low acidity of the α-proton. The Brønsted acidity of the carboxylic acid is a further complication for the deprotonation of the α-proton, and therefore two equivalents of a strong base such as LDA are normally required for efficient enolization. In 2020, the group of Ohshima reported the α-functionalization of carboxylic acids through an iron-catalyzed $1e^-$ radical process [74]. The use of molecular sieves was found to be crucial in suppressing undesired decarboxylation and the alkali metal salt component of the sieves (sodium or potassium carboxylate) was found to play a crucial role in promoting the reaction. The reaction had wide scope and functional groups typically sensitive to oxidation conditions, such as primary benzylic alcohols and thioethers were tolerated. This study constitutes the first example of the generation of redox-active heterobimetallic enediolates from unprotected carboxylic acids under catalytic conditions (Scheme 32).

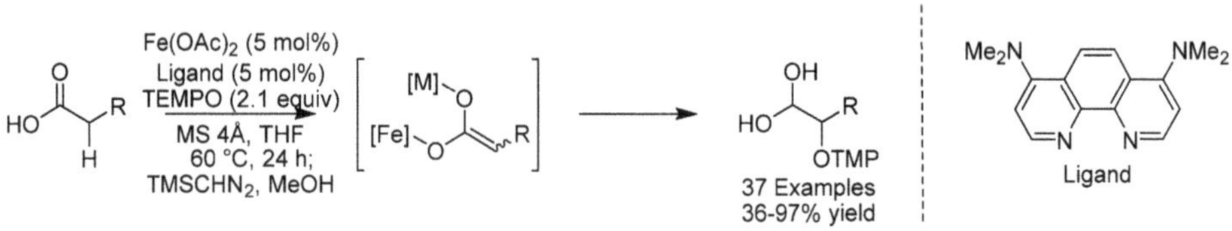

Scheme 32. Ohshima's α-oxidation of carboxylic acids.

3. Conclusions

This review has highlighted recent advances in iron-catalysis with respect to organic synthesis. This field has expanded greatly in recent years and giving a detailed account of each reaction category is an impossible task for a short review of this type. Herein, only key representative examples are given, and interested readers are advised to look into the reviews referenced throughout the text.

As is clear from the examples given in this review, iron complexes are capable of catalyzing a wide range of reactions in organic synthesis. This potential of iron catalysis has not been fully realized, however, and noble transition metal protocols continue to define the state of the art when it comes to transition metal catalysis. The field is still in its infancy, but it is fair to say that iron (and other base metals) may very well challenge this dominance of noble transition metals in the foreseeable future.

Author Contributions: Conceptualization, writing, editing and reviewing—A.G. and J.-E.B. All authors have read and agreed to the published version of the manuscript.

Funding: This research was funded by the Swedish Research Council (grant number 2016-03897), The European Union, and the Knut and Alice Wallenberg Foundation (grant number KAW 2016.0072).

Conflicts of Interest: The authors declare no conflict of interest.

References

1. Frey, P.A.; Reed, G.H. The Ubiquity of Iron. *ACS Chem. Biol.* **2012**, *7*, 1477–1481. [CrossRef] [PubMed]
2. Sessions, A.L.; Doughty, D.M.; Welander, P.V.; Summons, R.E.; Newman, D.K. The Continuing Puzzle of the Great Oxidation Event. *Curr. Biol.* **2009**, *19*, R567–R574. [CrossRef] [PubMed]
3. Enthaler, S.; Junge, K.; Beller, M. Sustainable Metal Catalysis with Iron: From Rust to a Rising Star. *Angew. Chem. Int. Ed.* **2008**, *47*, 3317–3321. [CrossRef]
4. Bolm, C. A new iron age. *Nat. Chem.* **2009**, *1*, 420. [CrossRef]
5. Fürstner, A. Iron Catalysis in Organic Synthesis: A Critical Assessment of What It Takes To Make This Base Metal a Multitasking Champion. *ACS Cent. Sci.* **2016**, *2*, 778–789. [CrossRef]
6. Bauer, I.; Knölker, H.-J. Iron Catalysis in Organic Synthesis. *Chem. Rev.* **2015**, *115*, 3170–3387. [CrossRef]
7. Mond, L.; Quinke, F. Note on a Volatile Compound of Iron with Carbonic Oxide. *J. Chem. Soc.* **1891**, *59*, 604–607. [CrossRef]
8. Berthelot, M.C.R. Sur une combinaison volatile de fer et d'oxyde de carbone, le fer carbonyl, et sur le nickel-carbonyle. *Compt. Rend. Acad. Sci.* **1891**, *112*, 1343–1348.
9. Reppe, W.; Vetter, H. Über die Umsetzung von Acetylen mit Kohlenoxyd und Verbindungen mit reationsfähigen Wasserstoffatomen Synthesen α,β-ungesättigter Carbonsäuren und ihrer Derivate. *Liebigs. Ann. Chem* **1953**, *582*, 1–37. [CrossRef]
10. Kealy, T.J.P.; Pauson, P.L. A New Type of Organo-Iron Compound. *Nature* **1951**, *168*, 1039. [CrossRef]
11. Wilkinson, G.; Rosenblum, M.; Whiting, M.C.; Woodward, R.B. The Structure of Iron Bis-Cyclopentadienyl. *J. Am. Chem. Soc.* **1952**, *74*, 2125–2126. [CrossRef]
12. Fischer, E.O.; Pfab, W. Cyclopentadien-Metallkomplexe, ein neuer Typ metallorganischer Verbindungen. *Z. Neturforsch. B* **1952**, *7*, 676. [CrossRef]
13. Haber, F.; Le Rossignol, R. Über die technische Darstellung von Ammoniak aus den Elementen. *Elektrochem. Angew. Phys. Chem.* **1913**, *19*, 53.
14. Cherkasov, N.; Ibhadon, A.O.; Fitzpatrick, P. A review of the existing and alternative methods for greener nitrogen fixation. *Chem. Eng. Process.* **2015**, *90*, 24–33. [CrossRef]
15. Bornschein, C.; Gustafson, K.; Verho, O.; Beller, M.; Bäckvall, J.-E. Evaluation of Fe and Ru Pincer-Type Complexes as Catalysts for the Racemi¬zation of Secondary Benzylic Alcohols. *Chem. Eur. J.* **2016**, *22*, 11583–11586. [CrossRef]
16. Gustafson, K.P.J.; Guðmundsson, A.; Lewis, K.; Bäckvall, J.-E. Chemoenzymatic Dynamic Kinetic Resolution of Secondary Alochols Using an Air- and Moisture-Stable Iron Racemization Catalyst. *Chem. Eur. J.* **2017**, *23*, 1048–1051. [CrossRef]
17. El-Sepelgy, O.; Alandini, N.; Rueping, M. Merging Iron Catalysis and Biocatalysis-Iron Carbonyl Complexes as Efficient Hydrogen Autotransfer Catalysts in Dynamic Kinetic Resolutions. *Angew. Chem. Int. Ed.* **2016**, *55*, 13602–13605. [CrossRef]
18. Yang, Q.; Zhang, N.; Liu, M.; Zhou, S. New air-stable iron catalyst for efficient dynamic kinetic resolution of secondary benzylic and aliphatic alcohols. *Tetrahederon. Lett.* **2017**, *58*, 2487–2489. [CrossRef]
19. Quintard, A.; Rodriguez, J. Iron Cyclopentadienone Complexes: Discovery, Properties, and Catalytic Reactivity. *Angew. Chem. Int. Ed.* **2014**, *53*, 4044–4055. [CrossRef]
20. Xu, H.; Li, Y.-P.; Cai, Y.; Wang, G.-P.; Zhu, S.-F.; Zhou, Q.-L. Highly Enantioselective Copper- and Iron-Catalyzed Intramolecular Cyclopropanation of Indoles. *J. Am. Chem. Soc.* **2017**, *139*, 7697–7700. [CrossRef]
21. Ma, W.; Zhang, X.; Fan, J.; Liu, Y.; Tang, W.; Xue, D.; Li, C.; Xiao, J.; Wang, C. Iron-Catalyzed Anti-Markovnikov Hydroamination and Hydroamidation of Allylic Alcohols. *J. Am. Chem. Soc.* **2019**, *141*, 13506–13515. [CrossRef] [PubMed]
22. Qian, B.; Chen, S.; Wang, T.; Zhang, X.; Bao, H. Iron-Catalyzed Carboamination of Olefins: Synthesis of Amines and Disubstituted β-Amino Acids. *J. Am. Chem. Soc.* **2017**, *139*, 13076–13082. [CrossRef] [PubMed]
23. Yang, Y.; Song, R.-J.; Ouyang, X.-H.; Wang, C.-Y.; Li, J.-H.; Luo, S. Iron-Catalyzed Intermolecular 1,2-Difunctionalization of Styrenes and Conjugated Alkenes with Silanes and Nucleophiles. *Angew. Chem. Int. Ed.* **2017**, *56*, 7916–7919. [CrossRef] [PubMed]
24. Dimroth, O. Ueber die Mercurirung aromatischer Verbindungen. *Ber. Dtsch. Chem. Ges.* **1902**, *35*, 2853–2873. [CrossRef]

25. Murahashi, S. Synthesis of Phthalimidines from Schiff Bases and Carbon Monoxide. *J. Am. Chem. Soc.* **1955**, *77*, 6403–6404. [CrossRef]

26. Khrushch, A.P.; Tokina, L.A.; Shilov, A.E. *Kinetika i Kataliz*; Pleiades Publishing: Warrensburg, MO, USA, 1966; Volume 7, p. 901.

27. Goldman, A.S.; Goldberg, K.I. Organometallic C-H Bond Activation: An Introduction. In *Activation and Functionalization of C-H Bonds*; Goldberg, K.I., Goldman, A.S., Eds.; ACS Symposium Series; American Chemical Society: Washington, DC, USA, 2004; pp. 4–8.

28. Norinder, J.; Matsumoto, A.; Yoshikai, N.; Nakamura, E. Iron-Catalyzed Direct Arylation through Directed C-H Bond Activation. *J. Am. Chem. Soc.* **2008**, *130*, 5858–5859. [CrossRef]

29. Shang, R.; Ilies, L.; Nakamura, E. Iron-Catalyzed C-H Bond Activation. *Chem. Rev.* **2017**, *117*, 9086–9139. [CrossRef]

30. Jia, F.; Li, Z. Iron-catalyzed/mediated oxidative transformation of C-H bonds. *Org. Chem. Front.* **2014**, *1*, 194–214. [CrossRef]

31. Landwehr, M.; Hochrein, L.; Otey, C.R.; Kasrayan, A.; Bäckvall, J.-E.; Arnold, F.H. Enantioselective α-Hydroxylation of 2-Arylacetic Acid Derivatives and Buspirone Catalyzed by Engineered Cytochrome P450 BM-3. *J. Am. Chem. Soc.* **2006**, *128*, 6058–6059. [CrossRef]

32. Prier, C.K.; Zhang, R.K.; Buller, A.R.; Brinkmann-Chen, S.; Arnold, F.H. Enantioselective, intermolecular benzylic C-H amination catalyzed by an engineered iron-haem enzyme. *Nat. Chem.* **2017**, *9*, 629–634. [CrossRef]

33. Zhou, X.-G.; Yu, X.-Q.; Huang, J.-S.; Che, C.-M. Asymmetric amidation of saturated C-H bonds catalyzed by chiral ruthenium and manganese porphyrins. *Chem. Commun.* **1999**, 2377–2378. [CrossRef]

34. Zhang, R.K.; Chen, K.; Huang, X.; Wohlschlager, L.; Renata, H.; Arnold, F.H. Enzymatic assembly of carbon-carbon bonds via iron-catalyzed sp^3 C-H functionalization. *Nature* **2019**, *565*, 67–72. [CrossRef] [PubMed]

35. Griffin, J.R.; Wendell, C.I.; Garwin, J.A.; White, M.C. Catalytic C(sp3)-H Alkylation via an Iron Carbene Intermediate. *J. Am. Chem. Soc.* **2017**, *139*, 13624–13627. [CrossRef] [PubMed]

36. Mo, J.; Muller, T.; Oliveira, J.C.A.; Ackermann, L. 1,4-Iron Migration for Expedient Allene Annulations through Iron-Catalyzed C-H/N-H/C-O/C-H Functionalizations. *Angew. Chem. Int. Ed.* **2018**, *57*, 7719–7723. [CrossRef] [PubMed]

37. Wang, Y.; Zhu, J.; Durham, A.C.; Lindberg, H.; Wang, Y.-M. α-C-H Functionalization of π-Bonds Using Iron Complexes: Catalytic Hydroxyalkylation of Alkynes and Alkenes. *J. Am. Chem. Soc.* **2019**, *141*, 19594–19599. [CrossRef] [PubMed]

38. Cutler, A.; Ehnholt, D.; Lennon, P.; Nicholas, K.; Marten, D.F.; Madhavarao, M.; Raghu, S.; Rosan, A.; Rosenblum, M. Chemistry of dicarbonyl η5-Cyclopentadienyliron Complexes. General Syntheses of Monosubstituted η2-Olefin Complexes and of 1-Substituted η1-Allyl Complexes. Conformational Effects on the Course of Deprotonation of (η2-Olefin) Cations. *J. Am. Chem. Soc.* **1975**, *97*, 3149. [CrossRef]

39. Guan, H.; Sun, S.; Mao, Y.; Chen, L.; Lu, R.; Huang, J.; Liu, L. Iron(II)-Catalyzed Site-Selective Functionalization of Unactivated C(sp3)-H Bonds Guided by Alkoxyl Radicals. *Angew. Chem. Int. Ed.* **2018**, *57*, 11413–11417. [CrossRef]

40. Tamura, M.; Kochi, J. Vinylation of Grignard reagents. Catalysis by iron. *J. Am. Chem. Soc.* **1971**, *93*, 1487–1489. [CrossRef]

41. Mako, T.L.; Byers, J.A. Recent advances in iron-catalyzed cross coupling reactions and their mechanistic underpinning. *Inorg. Chem. Front.* **2016**, *3*, 766–790. [CrossRef]

42. Piontek, A.; Bisz, E.; Szostak, M. Iron-Catalyzed Cross-Couplings in the Synthesis of Pharmaceuticals: In Pursuit of Sustainability. *Angew. Chem. Int. Ed.* **2018**, *57*, 11116–11128. [CrossRef]

43. Kessler, S.N.; Hundemer, F.; Bäckvall, J.-E. A Synthesis of Substituted α-Allenols via Iron-Catalyzed Cross-Coupling of Propargyl Carboxylates with Grignard Reagents. *ACS Catal.* **2016**, *6*, 7448–7451. [CrossRef] [PubMed]

44. Kessler, S.N.; Bäckvall, J.-E. Iron-catalyzed Cross-Coupling of Propargyl Carboxylates and Grignard Reagents: Synthesis of Substituted Allenes. *Angew. Chem. Int. Ed.* **2016**, *55*, 3734–3738. [CrossRef] [PubMed]

45. Rivera, A.C.P.; Still, R.; Frantz, D.E. Iron-Catalyzed Stereoselective Cross-Coupling Reactions of Stereodefined Enol Carbamates with Grignard Reagents. *Angew. Chem. Int. Ed.* **2016**, *55*, 6689–6693. [CrossRef] [PubMed]

46. Adak, L.; Kawamura, S.; Toma, G.; Takenaka, T.; Isozaki, K.; Takaya, H.; Orita, A.; Li, H.C.; Shing, T.K.M.; Nakamura, M. Synthesis of Aryl C-Glycosides via Iron-Catalyzed Cross Coupling of Halosugars: Stereoselective Anomeric Arylation of Glycosyl Radicals. *J. Am. Chem. Soc.* **2017**, *139*, 10693–10701. [CrossRef] [PubMed]

47. Adams, K.; Ball, A.K.; Birkett, J.; Brown, L.; Chappell, B.; Gill, D.M.; Lo, P.K.T.; Patmore, N.J.; Rice, C.R.; Ryan, J.; et al. An iron-catalyzed C-C bond-forming spirocyclization cascade providing sustainable access to new 3D heterocyclic frameworks. *Nat. Chem.* **2017**, *9*, 396–401. [CrossRef] [PubMed]

48. O'Brien, H.M.; Manzotti, M.; Abrams, R.D.; Elorriaga, D.; Sparkes, H.A.; Davis, S.A.; Bedford, R.B. Iron-Catalyzed substrate-directed Suzuki biaryl cross-coupling. *Nat. Catal.* **2018**, *1*, 429–437. [CrossRef]

49. Kennedy, C.R.; Zhong, H.; Macaulay, R.L.; Chirik, P.J. Regio- and Diastereoselective Iron-Catalyzed [4+4]-Cycloaddition of 1,3-dienes. *J. Am. Chem. Soc.* **2019**, *141*, 8557–8573. [CrossRef]

50. Wang, Q.; Qi, X.; Lu, L.-Q.; Li, T.-R.; Yuan, Z.-G.; Zhang, K.; Li, B.-J.; Lan, Y.; Xiao, W.-J. Iron-Catalyzed Decarboxylative (4+1) Cycloadditions: Exploiting the Reactivity of Ambident Iron-Stabilized Intermediates. *Angew. Chem. Int. Ed.* **2016**, *55*, 2840–2844. [CrossRef]

51. He, R.; Huang, Z.-T.; Zheng, Q.-Y.; Wang, C. Isoquinoline skeleton synthesis via chelation-assisted C-H activation. *Tetrahedron Lett.* **2014**, *55*, 5705–5713. [CrossRef]

52. Jia, T.; Zhao, C.; He, R.; Chen, H.; Wang, C. Iron-Carbonyl-Catalyzed Redox-Neutral [4+2] Annulation of N-H Imines and Internal Alkynes by C-H Bond Activation. *Angew. Chem. Int. Ed.* **2016**, *55*, 5268–5271. [CrossRef]

53. Brenna, D.; Villa, M.; Gieshoff, T.N.; Fischer, F.; Hapke, M.; Jacobi von Wangelin, A. Iron-Catalyzed Cyclotrimerization of Terminal Alkynes by Dual Catalyst Activation in the Absence of Reductants. *Angew. Chem. Int. Ed.* **2017**, *56*, 8451–8454. [CrossRef] [PubMed]

54. Emerson, G.F.; Pettit, R. π-Allyl-Iron Tricarbonyl Cations. *J. Am. Chem. Soc.* **1962**, *84*, 4591. [CrossRef]

55. Guðmundsson, A.; Gustafson, K.P.J.; Mai, B.K.; Yang, B.; Himo, F.; Bäckvall, J.-E. Efficient Formation of 2,3-dihydrofurans via Iron-Catalyzed Cycloisomerization of α-Allenols. *ACS Catal.* **2018**, *8*, 12–16. [CrossRef]

56. Guðmundsson, A.; Gustafson, K.P.J.; Mai, B.K.; Hobiger, V.; Himo, F.; Bäckvall, J.-E. Diastereoselective Synthesis of N-Protected 2,3-Dihydropyrroles via Iron-Catalyzed Cycloisomerization of α-Allenic Sulfonamides. *ACS Catal.* **2019**, *9*, 1733–1737. [CrossRef]

57. El-Sepelgy, O.; Brzozowska, A.; Azofra, L.M.; Jang, Y.K.; Cavallo, L.; Rueping, M. Experimental and Computational Study of an Unexpected Iron-Catalyzed Carboetherification by Cooperative Metal and Ligand Substrate Interaction and Proton Shuttling. *Angew. Chem. Int. Ed.* **2017**, *56*, 14863–14867. [CrossRef] [PubMed]

58. El-Sepelgy, O.; Brzozowska, A.; Sklyaruk, J.; Kyung Jang, Y.; Zubar, V.; Rueping, M. Cooperative Metal-Ligand Catalyzed Intramolecular Hydroamination and Hydroalkoxylation of Allenes Using a Stable Iron Catalyst. *Org. Lett.* **2018**, *20*, 696–699. [CrossRef]

59. Yang, B.; Zhu, C.; Qiu, Y.; Bäckvall, J.-E. Enzyme- and Ruthenium-Catalyzed Enantioselective Transformation of α-Allenic Alcohols into 2,3-Dihydrofurans. *Angew. Chem. Int. Ed.* **2016**, *55*, 5568–5572. [CrossRef]

60. Woodward, R.R.; Hoffmann, R. The Conservation of Orbital Symmetry. *Angew. Chem. Int. Ed. Engl.* **1969**, *8*, 781. [CrossRef]

61. Kramm, F.; Teske, J.; Ullwer, F.; Frey, W.; Plietker, B. Annelated Cyclobutanes by Fe-Catalyzed Cycloisomerization of Enyne Acetates. *Angew. Chem. Int. Ed.* **2018**, *57*, 13335–13338. [CrossRef]

62. Cherkaoui, H.; Soufiaoui, M.; Grée, R. From allylic alcohols to saturated carbonyls using Fe(CO)$_5$ as catalyst: Scope and limitation studies and preparation of two perfume components. *Tetrahedron* **2001**, *57*, 2379–2383. [CrossRef]

63. Xia, T.; Wei, Z.; Spiegelberg, B.; Jiao, H.; Hinze, S.; De Vries, J.G. Isomerization of Allylic Alcohols to Ketones Catalyzed by Well-Defined Iron PNP Pincer Catalysts. *Chem. Eur. J.* **2018**, *24*, 4043–4049. [CrossRef] [PubMed]

64. Escheverria, P.-G.; Fürstner, A. An Iron-Catalyzed Bond-Making/Bond-Breaking Cascade Merges Cycloisomerization and Cross-Coupling Chemistry. *Angew. Chem. Int. Ed.* **2016**, *55*, 11188–11192. [CrossRef] [PubMed]

65. Reed-Berendt, B.G.; Polidano, K.; Morrill, L.C. Recent advances in homogeneous borrowing hydrogen catalysis using earth-abundant first row transition metals. *Org. Biomol.* **2019**, *17*, 1595–1607. [CrossRef] [PubMed]

66. Wei, D.; Darcel, C. Iron Catalysis in Reduction and Hydrometalation Reactions. *Chem. Rev.* **2019**, *119*, 2550–2610. [CrossRef]

67. Okamura, M.; Kondo, M.; Kuga, R.; Kurashige, Y.; Yanai, T.; Hayami, S.; Praneeth, V.K.K.; Yoshida, M.; Yoneda, K.; Kawata, S.; et al. A pentanuclear iron catalyst designed for water oxidation. *Nature* **2016**, *530*, 465–468. [CrossRef]

68. Guðmundsson, A.; Schlipköter, K.E.; Bäckvall, J.-E. Iron(II)-Catalyzed Biomimetic Aerobic Oxidation of Alcohols. *Angew. Chem. Int. Ed.* **2020**, *59*. [CrossRef]

69. Jiang, X.; Zhang, J.; Ma, S. Iron Catalysis for Room-Temperature Aerobic Oxidation of Alcohols to Carboxylic Acids. *J. Am. Chem. Soc.* **2016**, *138*, 8344–8347. [CrossRef]

70. Cussó, O.; Cianfanelli, M.; Ribas, X.; Gebbink, R.J.M.K.; Costas, M. Iron Catalyzed Highly Enantioselective Epoxidation of Cyclic Aliphatic Enones with Aqueous H_2O_2. *J. Am. Chem. Soc.* **2016**, *138*, 2732–2738. [CrossRef]

71. Blasius, C.K.; Vasilenko, V.; Gade, L.H. Ultrafast Iron-Catalyzed Reduction of Functionalized Ketones: Highly Enantioselective Synthesis of Halohydrines, Oxaheterocycles and Aminoalcohols. *Angew. Chem. Int. Ed.* **2018**, *57*, 10231–10235. [CrossRef]

72. Farrar-Tobar, R.A.; Wozniak, B.; Savini, A.; Hinze, S.; Tin, S.; de Vries, J.G. Base-Free Iron Catalyzed Transfer Hydrogenation of Esters Using EtOH as Hydrogen Source. *Angew. Chem. Int. Ed.* **2019**, *58*, 1129–1133. [CrossRef]

73. Lator, A.; Gaillard, S.; Poater, A.; Renaud, J.-L. Iron-Catalyzed Chemoselective Reduction of α,β-Unsaturated Ketones. *Chem. Eur. J.* **2018**, *24*, 5770–5774. [CrossRef] [PubMed]

74. Tanaka, T.; Yazaki, R.; Ohshima, T. Chemoselective Catalytic α-Oxidation of Carboxylic Acids: Iron/Alkali Metal Cooperative Redox Active Catalysis. *J. Am. Chem. Soc.* **2020**, *142*, 4517–4524. [CrossRef] [PubMed]

 *molecules*

Review

Iron Catalysts in Atom Transfer Radical Polymerization

Sajjad Dadashi-Silab and Krzysztof Matyjaszewski *

Department of Chemistry, Carnegie Mellon University, 4400 Fifth Avenue, Pittsburgh, PA 15213, USA;
sdadashi@andrew.cmu.edu
* Correspondence: km3b@andrew.cmu.edu; Tel.: +1-(412)-268-3209

Academic Editor: Hans-Joachim Knölker
Received: 11 March 2020; Accepted: 1 April 2020; Published: 3 April 2020

Abstract: Catalysts are essential for mediating a controlled polymerization in atom transfer radical polymerization (ATRP). Copper-based catalysts are widely explored in ATRP and are highly efficient, leading to well-controlled polymerization of a variety of functional monomers. In addition to copper, iron-based complexes offer new opportunities in ATRP catalysis to develop environmentally friendly, less toxic, inexpensive, and abundant catalytic systems. Despite the high efficiency of iron catalysts in controlling polymerization of various monomers including methacrylates and styrene, ATRP of acrylate-based monomers by iron catalysts still remains a challenge. In this paper, we review the fundamentals and recent advances of iron-catalyzed ATRP focusing on development of ligands, catalyst design, and techniques used for iron catalysis in ATRP.

Keywords: iron catalyst; ATRP; controlled radical polymerization; external stimuli

---

1. Introduction

Reversible deactivation radical polymerization (RDRP) techniques have provided access to advanced polymers with precise control over molecular weight, dispersity, composition, and structure. Typical approaches to control the growth of polymer chains in radical polymerizations include reversible deactivation of propagating radicals or using degenerative transfer processes to exchange radicals with dormant species in the presence of chain transfer agents. For example, atom transfer radical polymerization (ATRP) employs, primarily, Cu-based catalysts to control the growth of polymer chains via a reversible redox process that involves transfer of halogen atoms to activate dormant species generating initiating radicals and also to deactivate propagating chains [1–4]. In regard to degenerative transfer processes, reversible addition fragmentation chain transfer (RAFT) [5–8] and iodine-mediated [9,10] polymerizations have been significantly explored in controlled radical polymerizations.

ATRP catalysis has advanced based on developing new catalytic systems with the aim of progressively increasing the activity, efficiency, and selectivity of catalysts through designing ligands, using external stimuli to control the catalytic process, and also decreasing the amount of catalysts needed for achieving a controlled polymerization [11,12]. For instance, the L/Cu^I activator for ATRP can be generated in situ via reduction of air stable L/Cu^{II}-X using external stimuli. Regeneration of the activator allows use of low concentration of the catalyst and overcoming the consumption of the activator as a result of radical termination reactions. In addition, an important aspect of using external stimuli is to exert control over the catalytic system to enable spatial and temporal control over the growth of polymer chains [13–15].

ATRP catalysis involves generation of radicals via activation of halogen chain ends by L/Cu^I activator and reversible deactivation of propagating radicals by a halogen atom transfer from L/Cu^{II}-X deactivator (X = Br or Cl). While copper-based complexes have been widely explored and used for polymerization of a wide range of vinyl monomers with high efficiency, ATRP catalysis comprises other transition metal-based catalysts [16,17] such as iron [18–21], ruthenium [22], osmium [23], and iridium [24] complexes as well as organic-based photoredox catalysts [25–28].

Iron complexes are particularly interesting due to the abundance of iron and its involvement in important biological processes that make iron-based catalytic systems less toxic, inexpensive, and biocompatible. Iron complexes mainly favor one-electron redox chemistry between +2 and +3 oxidation states, suitable for ATRP catalysis. In iron-catalyzed ATRP, L/Fe^{II} species activate dormant halogen chain ends, whereas L/Fe^{III}-X deactivate propagating radicals via a reversible halogen atom transfer process. Interestingly, iron complexes possessing anionic properties (for example those obtained in the presence of halide anions used as the ligand) have often performed well in ATRP reactions, whereas in copper-catalyzed ATRP, the efficient activator L/Cu^{I} complexes are cationically charged. In addition to atom transfer reactions, iron catalysts (L/Fe^{II}) can also react with the propagating radicals to form organometallic species (P_n-L/Fe^{III}). While the organometallic species can be involved in an organometallic mediated radical polymerization (OMRP) mechanism [29,30], termination reactions may also be promoted through catalytic chain transfer (CCT) [31] or catalytic radical termination (CRT) processes involving P_n-L/Fe^{III} species and propagating radicals to yield terminated chains [32]. In copper-catalyzed ATRP, CRT was reported only for the most active catalysts. Nevertheless, due to the low concentration of L/Cu^{I} present in these highly active systems, the overall contribution of OMRP and CRT processes in the polymerization is not important [33]. However, the contribution of organometallic species and CRT in iron-catalyzed systems may be more significant, leading to termination, especially in the polymerization of acrylate monomers.

In this paper, we review the principles and recent advances in iron-catalyzed ATRP. Iron catalytic systems are presented based on the structure and properties of the ligands and their effect in polymerization catalysis. Different ATRP initiating systems based on developing activator regeneration techniques and challenges and opportunities offered by iron complexes in ATRP are discussed.

2. Ligands and Iron Complexes in Iron-Catalyzed ATRP

Ligands play an important role in controlling the efficiency and performance of ATRP catalysts, affecting polymerization control. A variety of different families of ligands have been developed and explored in ATRP with a special focus on understanding the catalysts' behavior in polymerization and underlying mechanisms.

2.1. Halide Salts

Halide salts with bulky, non-coordinating cations have been used as simple and robust ligands in promoting iron-catalyzed ATRP with high efficiency. Halide salts containing organic onium counter cations, ionic liquids, or inorganic based salts can be used [34–41]. Active iron complexes in the presence of halide salts possess overall anionic charge. Disproportionation (or dismutation) of $Fe^{II}Br_2$ forms a catalytically active anionic complex for ATRP and a cationic species [42], a much less active (or inactive) form of the catalyst, according to Equation (1):

$$3\,Fe^{II}Br_2 \rightarrow Fe^{II} + 2\,[FeBr_3]^- \tag{1}$$

in the presence of halide salts, various iron complexes may be generated that may or may not be the active catalyst for ATRP, depending on the nature of the reaction medium and/or the amount of the ligand used. For example, iron complexes with cationic charges are less effective as ATRP catalysts, whereas anionic complexes efficiently catalyze polymerization due to the presence of high electron density around the metal center attained by complexation of anionic species (Scheme 1) [43]. Tri-coordinated monoanionic $[Fe^{II}Br_3/L]^-$ (L = solvent or monomer) species was proposed to be the active species for catalyzing ATRP [39]. In the presence of excess bromide salts, dianionic $[Fe^{II}Br_4]^{2-}$ complexes may also be generated, which contain four coordinated bromide anions rendering this species less effective to undergo a halogen abstraction during the activation of dormant chains. Therefore, the rate of polymerization decreased in the presence of excess amounts of halide salt ligands [34,35].

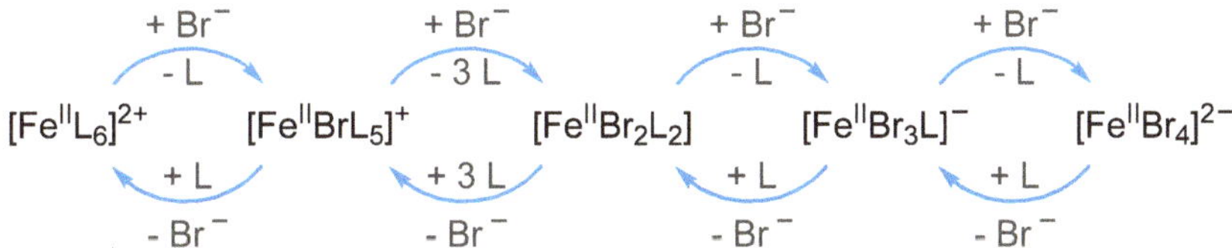

Scheme 1. Iron species in the presence of bromide anions as ligand and solvent (L) generating a number of mononuclear cationic and anionic species in atom transfer radical polymerization (ATRP). Reproduced with slight modification and permission from Ref [43].

2.2. Nitrogen-Based Ligands

Alkyl amines with mono or multidentate coordinating sites or bipyridine derivatives were used for conducting iron ATRP [19,44–47]. A series of Fe^{II} complexes with *N,N,N*-trialkyl-1,4,9-triazacyclononane (TACN) ligands were investigated in ATRP [48–51]. The complexes formed either a mononuclear iron species in the presence of TACN ligands substituted with bulky groups, or trinuclear ionic species with a cationic dinuclear and an anionic mononuclear complex in the presence of less bulky substitution groups to form coordinatively saturated compounds. The mononuclear species performed well in catalyzing ATRP. Interestingly, an iron complex with cyclopentyl-substituted TACN ligand exchanged between mono and dinuclear species, therefore providing efficient catalysis and yielding well-controlled polymers, and also allowing easy purification and reusing the catalyst because of the formation of ionic structures [50].

Iminopyridine [52,53] and α-diimine [54–56] ligands were also developed for iron-based polymerization catalysts. Gibson and coworkers investigated the electronic and steric properties of the iron complexes with diimine ligands that affected the underlying mechanism of polymerization and therefore control over the growth of polymer chains. The electronic properties of the catalyst tuned via different *N*-substitution groups influenced the catalysts' performance and promoted either ATRP or catalytic chain transfer (CCT) processes [31]. The CCT process was proposed to result from *N*-aryl substituted catalysts and involved formation of organometallic species followed by a hydrogen elimination to give low molecular weight, terminated polymers. *N*-Alkyl substituted complexes favored the ATRP mechanism and therefore yielded well-controlled polymers.

Bis(imino)pyridine ligands were immobilized on a polymer chain to synthesize iron catalysts for controlled polymerization [57]. The bis(imino)pyridine-containing polymer ligand was obtained using an amphiphilic copolymer comprising poly(ethylene glycol) (PEG), dodecyl, and urea/aniline pendant groups that reacted with 2,6-pyridinedicarboxaldehyde to yield self-folded polymers. Iron was immobilized on the polymer ligand and used in controlling polymerization of functional methacrylate monomers. Importantly, the amphiphilic nature of the self-folded polymers with immobilized iron catalyst allowed for easy purification of the products by rinsing with water, as the presence of hydrophilic PEG chains transferred the catalyst to aqueous phase and yielded pure polymethacrylates in the organic phase.

Shaver and co-workers developed amine-bis(phenolate) iron (III) complexes for polymerization of styrene and methacrylate monomers [58,59]. These iron complexes possess dianionic bisphenolate moieties with fixed and strong multidentate metal-ligand bonds that are less prone to undergo dissociation of the metal center during catalysis. Complexes with electron-withdrawing Cl substitution (especially in the para position of the aromatic groups) resulted in well controlled polymerizations, whereas catalysts with alkyl substituted aromatic groups gave uncontrolled results. Polymerizations were performed under reverse ATRP conditions using azobisisobutyronitrile (AIBN) as a source of radical initiator and equimolar ratios of the iron complexes. Fast and well-controlled polymerization of styrene and methacrylate monomers was obtained using the amine-bis(phenolate) iron complexes. Both Br- and Cl-based catalysts afforded well-controlled results. Control over chain growth was proposed to mainly involve the ATRP pathway as well as minor contribution of the OMRP mechanism. In the absence of ATRP alkyl halide initiators, use of an amine-bis(phenolate) iron (II) complex enabled moderate control over polymerization of methacrylate or styrene monomers via OMRP mechanism [60–62]. A phenolate-bis(pyridyl)amine ligand with monoanionic phenolate group was also used for iron-catalyzed ATRP but showed less efficient control over polymerization compared to the dianionic bisphenolate ligands [63].

Proteins and enzymes such as horseradish peroxidase, catalase, and hemoglobin that contain iron protoporphyrin centers were used as naturally occurring, bio-inspired iron catalysts for ATRP [64–68]. Hemin was also used as a catalyst mimicking the enzymatic catalytic systems. Further modification of the hemin structure was performed by hydrogenation of the vinyl bonds to prevent incorporation of the catalyst in the polymer chains by addition of radicals to the double bonds. Moreover, the catalyst was functionalized by attaching PEG, imidazole, or thioether arms that enabled solubility of the catalyst in aqueous media and therefore performing well-controlled polymerizations in water (Scheme 2) [69–71].

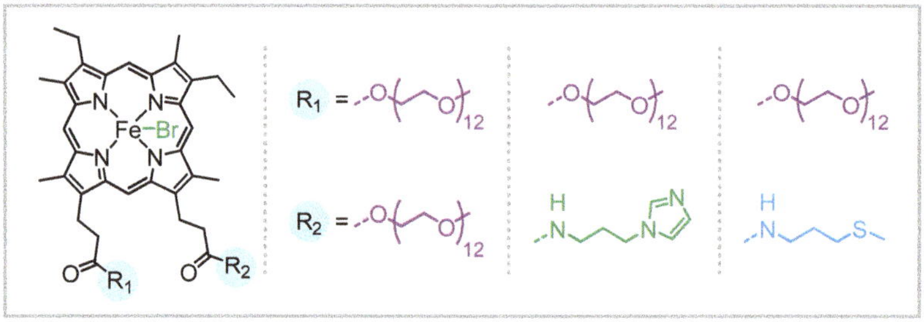

Scheme 2. Iron porphyrin catalysts functionalized with poly(ethylene glycol) (PEG) and imidazole or thioether groups used in iron-catalyzed ATRP.

2.3. Phosphorus-Based Ligands

Ligands containing phosphorus such as various alkyl and aryl compounds with mono- or bi-dentate phosphine or phosphite are among the most widely used ligands in iron-catalyzed ATRP [18,19,72–76]. The activity of triphenylphosphine ligands in iron-catalyzed ATRP depends on the electronic properties of the ligand, increasing in the presence of electron-donating functionalities [77]. The effect of substitution of electron donating methoxy groups on triphenylphosphine was attributed to an increase in the activity of the iron catalyst in the presence of ligands possessing more electron-donating properties, which is also reported for copper-catalyzed ATRP (Scheme 3) [78,79]. Moreover, the activity of triphenylphosphine compounds as a Lewis base increases by introducing electron donating functional groups leading to stronger reducing agents to generate L/Fe^{II} catalyst [80]. Therefore, the enhanced polymerization can be related to the electron-donating ability of the phosphine-based ligands in ATRP. In addition, similar behavior was observed in photoinduced iron-catalyzed ATRP in the presence of triphenylphosphine ligands where better control over polymerization was achieved in the presence of ligands with more electron donating properties, indicating increased activity of the iron catalyst applied in the ppm levels under blue light irradiation [81].

The effect of substitution on the phosphine-based ligands in catalyzing ATRP was further explored [77,82]. In the presence of alkyl substituted phosphines, polymerization of methyl methacrylate (MMA) was uncontrolled with high dispersity and broad or bimodal molecular weight distributions. Polymerization of MMA with triphenylphosphine was well-controlled but reached low monomer conversions. Introduction of electron donating groups in the para position of triphenylphosphine increased the activity of the resulting iron catalyst showing well-controlled polymerizations with dispersities < 1.2 and yielding high monomer conversions. The iron catalysts were obtained by reacting the phosphine ligands with $FeBr_2$ for 12 h yielding diphosphine iron(II) complexes, followed by addition of monomer and initiator to start the polymerization. As expected, the iron catalyst with chloro-substituted triphenylphosphine having electron withdrawing properties was inefficient in initiating polymerization of MMA. A PEG-substituted triphenylphosphine ligand was effective in controlling polymerization of functional monomers such as hydroxyethyl methacrylate (HEMA). The iron complex with PEG-substituted triphenylphosphine tolerated functional groups due to the bulkiness of the PEG chains, which protected the catalyst center from poisoning by the functional groups. These results further signify the importance of electronic and steric properties of functional groups in designing ligands for achieving a successful and well-controlled polymerization.

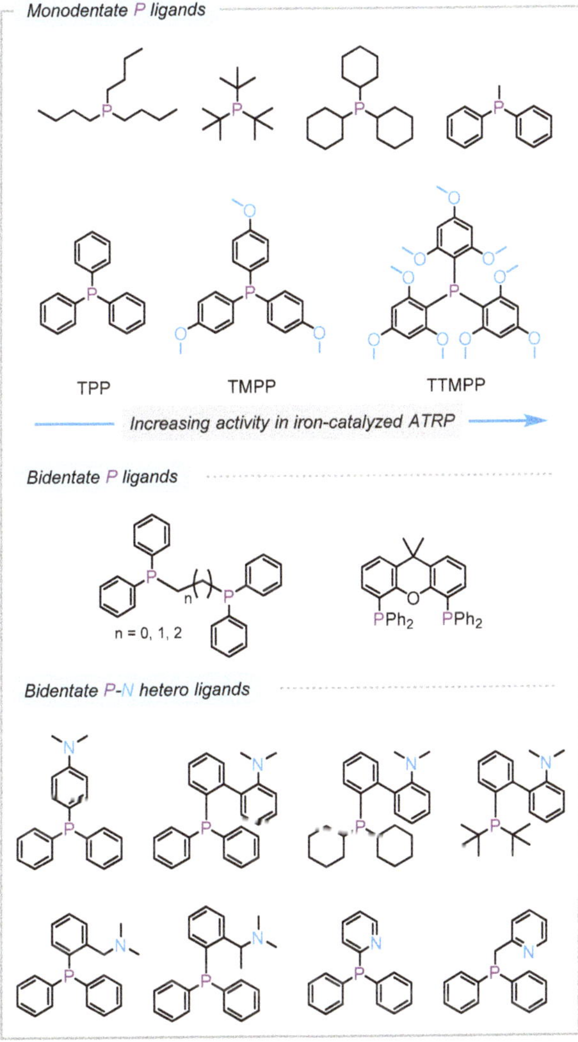

Scheme 3. Examples of mono and bidentate phosphorus-containing ligands used in iron-catalyzed ATRP. Substitution of triphenylphosphine ligands with electron-donating methoxy groups increases the activity of the iron catalyst in ATRP.

In addition, bidentate ligands containing P-P homo [73,74] and P-N [83,84] or P-carbonyl [85] hetero chelating sites were reported to perform well in iron-catalyzed ATRP. The presence of a second coordinating site (P or N), increased the catalytic activity and efficiency of the iron complex. Scheme 3 shows examples of phosphorus-containing ligands developed for controlled polymerization in iron-catalyzed ATRP.

2.4. Miscellaneous Ligands

Iron-catalyzed ATRP can be also successfully performed in the absence of additional ligands with polar solvents that can act as a ligand for stabilizing the iron catalyst [86]. For example, iron-catalyzed ATRP was controlled in the presence of solvents such as acetonitrile, *N,N*-dimethylformamide (DMF), *N*-methyl-2-pyrrolidone (NMP) without the use of other ligands. These solvents resulted in disproportionation of the iron complexes to form inactive cationic species and catalytically active anionic (Equation (1) and Scheme 1) [86,87].

N-heterocyclic carbenes (NHC) were used as ligands in iron-catalyzed ATRP [88]. The redox potential of the iron complexes with NHC ligands were more negative than those obtained using halide salts, indicating higher activity of the catalysts with NHC ligands. As a result, control over polymerization was better with low concentrations of Fe/NHC catalysts. Interestingly, polymerizations using Br-based systems resulted in improving polymerization control compared to the Cl-based catalysts. The effect of Br vs. Cl initiating systems on control over polymerization may be related to the activity of dormant chain ends regarding the activation process, as well as halidophilicity of the iron catalyst, which influences the deactivation process and hence control over polymerization. More analysis of the Br- and Cl-based initiating systems is required to understand the effect of halide nature in the activation and deactivation processes and therefore polymerization control with iron catalysts.

3. Iron-Catalyzed ATRP Initiating Systems

A common feature of new ATRP initiating systems is the in situ regeneration of the activator catalyst used at ppm amount by applying various reduction processes. Regeneration of the activator species via reduction of the oxidized form of the catalyst (i.e., the deactivator) allows for use of low amounts of the catalyst starting with its more stable oxidized form, and also overcome the problems associated with the consumption of the activator as a result of termination of radicals. These techniques mainly include use of reducing agents as in activators regenerated by electron transfer (ARGET), conventional radical initiators as in initiators for continuous activator regeneration (ICAR), zerovalent metals as in supplemental activator and reducing agent (SARA) systems, or external stimuli such as photo and electrochemical approaches (Scheme 4). These systems arc applicable in both copper- and iron-based ATRP reactions [38].

ICAR ATRP uses conventional thermal initiators to form radicals that can react with L/Fe^{III}-X to generate the ATRP activator, L/Fe^{II} complexes, and initiate the ATRP process. Generation of radicals from decomposition of the initiator species ensures continuous activator regeneration throughout the reaction and hence results in a steady rate of polymerization and well-controlled polymers [88–91].

Reducing agents are used in ARGET ATRP to generate the activator catalyst via electron transfer processes by reducing more stable L/Fe^{III}-X species [92]. Well-controlled ATRP was obtained using phosphorus ligands in the absence of conventional reducing agents. The generation of L/Fe^{II} activator was proposed to occur in the presence of phosphorus ligands that acted as a reducing agent facilitating transfer of bromine radical from L/Fe^{III}-Br to the monomer and therefore generation of both the activator and the ATRP initiator in situ [77,93,94]. Decamethylferrocene (FeCp*$_2$) was used as a co-catalyst to improve efficiency of polymerizations catalyzed by iron catalysts [95]. FeCp*$_2$ was proposed to act as a reducing agent for FeBr$_3$ in the presence of tetrabutylammonium bromide (TBABr) generating FeBr$_2$ activator and the ferrocenium salt with a bromide anion. Indeed, the lower redox potential of decamethylferrocene enabled reduction of Fe^{III} to improve the kinetics of the polymerization whereas use of ferrocene having a higher redox potential than the main iron catalyst showed no effect in the polymerization. Deactivation of growing radicals was improved by contribution of the ferrocenium salt with the bromide anion.

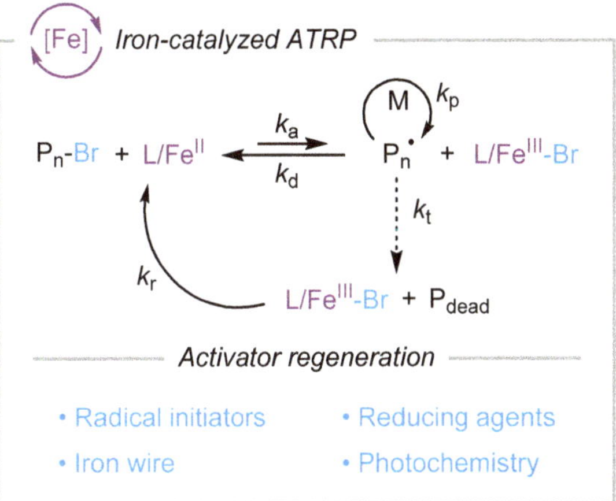

Scheme 4. Iron-catalyzed ATRP initiating systems with activator regeneration techniques developed for generation of L/Fe^{II} activator species via reduction (k_r, rate constant of reduction) of L/Fe^{III}-Br.

Zerovalent metal species such as iron wire or plates can be used to promote ATRP [38,96]. Recently, surface-initiated ATRP was investigated under SARA ATRP conditions using iron catalysts and an iron plate that acted as a source of activator (re)generation [97]. This surface functionalization technique was performed under ambient conditions without the need for inert atmosphere, as oxygen was consumed in the presence of iron plate, thereby simplifying the functionalization process. Importantly, the polymerization was conducted in the presence of mammalian cell cultures that showed high viability under polymerization conditions, indicating the potential of iron catalyzed ATRP to be applied under biologically relevant conditions (Scheme 5).

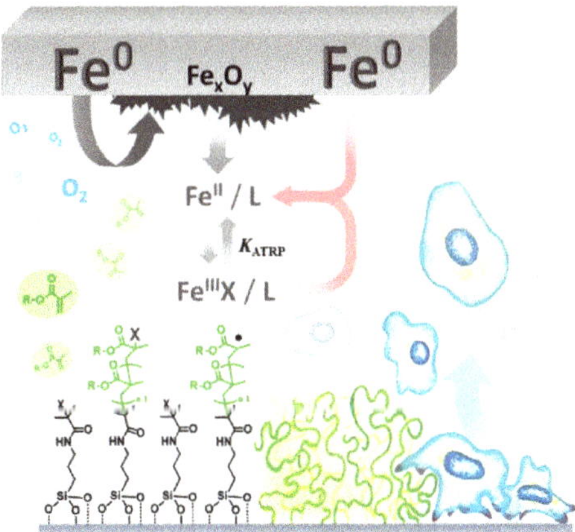

Scheme 5. Iron-catalyzed surface-initiated ATRP using an iron plate to generate L/Fe^{II} activators and remove oxygen allowing facile surface functionalization showing high cytocompatibility. Reproduced with permission from Ref [97].

Photochemistry has resulted in great advancements in the field of controlled polymerizations and enabled unique opportunities for synthesis of functional polymeric materials and asserting spatiotemporal control over polymerization [98–102]. Photoinduced iron-catalyzed ATRP was developed for controlled polymerization of methacrylate monomers under light irradiation, which promoted generation of the L/Fe^{II} catalyst [103,104]. Photoreduction of FeBr$_3$ proceeded through a homolytic cleavage of the Fe-Br bond upon photoexcitation to generate FeBr$_2$ and a Br$^\bullet$ radical. The Br$^\bullet$ radical added to methacrylate monomers and generated new initiating chains. A simplified iron-catalyzed ATRP was reported by using only monomer and FeBr$_3$ catalyst under UV light irradiation without using any conventional alkyl halide ATRP initiators or ligands (Scheme 6) [105]. Indeed, polymerizations relied on the formation of initiators in situ via photoreduction of FeBr$_3$ generating Br$^\bullet$ radicals that reacted with methacrylate monomers. Moreover, photoinduced iron-catalyzed ATRP was further improved upon by using ppm amounts of iron catalyst while yielding well-controlled polymers under blue light irradiation [81].

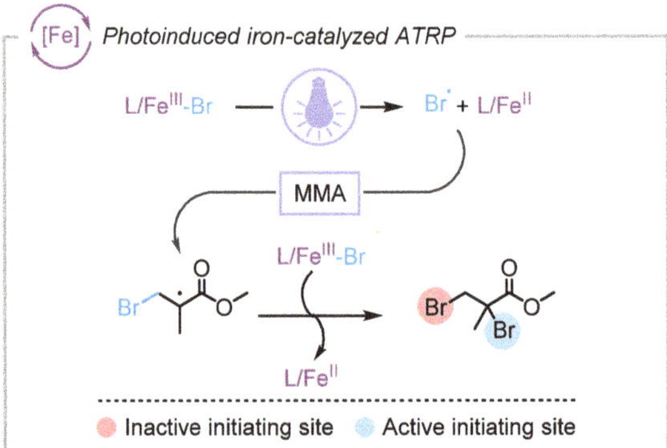

Scheme 6. Photoinduced iron-catalyzed ATRP undergoing a homolytic cleavage of the Fe-Br bond under visible light irradiation to generate the activator L/Fe^{II} and a Br$^\bullet$ radical, which can add to the monomer (MMA) and initiate new chains [103,105]. Polymerization can be performed without the need for use of conventional alkyl halide initiators using just the monomer and the catalyst.

Iron-catalyzed ATRP was conducted in the presence of residual oxygen to control polymerizations without performing conventional deoxygenation procedures [81]. For example, polymerization of MMA was well-controlled using 10 mol% FeBr$_3$ (with respect to initiator) to give high monomer conversions (>90%) and well-controlled polymers with low dispersities (<1.20) under blue light irradiation. Importantly, block copolymers were synthesized with well-controlled properties using photoinduced iron-catalyzed ATRP in the presence of residual oxygen, indicating the potential of this system to simplify the polymerization procedure or apply in areas where no deoxygenation is desired. Although the exact mechanism of consumption of oxygen was not determined in detail, it is possible that the initiator radicals and/or the iron catalyst would contribute in removing oxygen from the solution to allow a well-controlled polymerization to proceed, as previously shown for photoinduced copper-catalyzed ATRP systems [106].

Recently, iron-catalyzed ATRP was applied for the polymerization of semi-fluorinated methacrylate monomers controlled under blue light irradiation [41]. The use of iron as a catalyst was advantageous compared to copper-based catalysts, as the use of nitrogen-ligands in these systems promoted undesired transesterification reactions between the fluorinated monomer and protic solvent leading to a loss of control over the polymerization. However, polymerization of fluorinated monomers using iron catalyst was well-controlled in the presence of a variety of fluorinated and non-fluorinated solvents without undergoing side reactions or requiring special solvent systems to control the polymerization. A variety of semi-fluorinated methacrylate monomers were polymerized in a controlled manner using FeBr$_3$/TBABr as the catalyst under blue light irradiation resulting in high monomer conversions and polymers showing low dispersities (<1.20). Importantly, the use of blue light to trigger polymerization through generation of the $FeBr_2$ activator catalyst enabled gaining temporal control over the growth of polymer chains. The polymerization was initiated under blue light irradiation that generated the activator catalyst. Removal of the light stopped continuous (re)generation of the activator catalyst, continuously consumed as a result of radicals' termination, leading the polymerization to stop in the dark periods. Therefore, by decreasing the concentration of initially added catalyst, better temporal control was achieved as a result of low concentration of the activator, which required less time to be consumed during the light-off periods (Figure 1) [107].

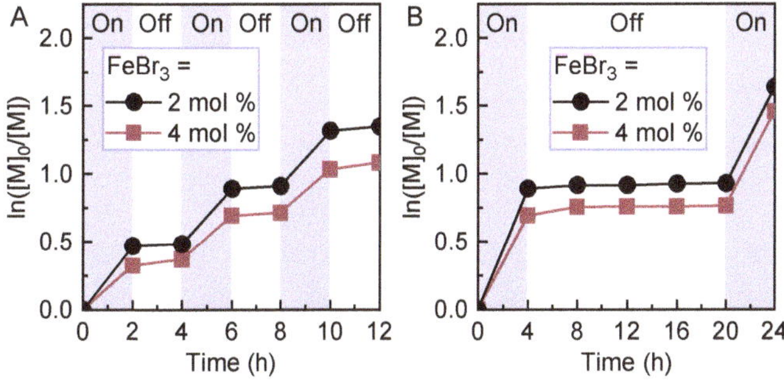

Figure 1. Temporal control in photoinduced iron-catalyzed ATRP of trifluoroethyl methacrylate (TFEMA). Better temporal control was achieved by decreasing the concentration of the iron catalyst from 4 to 2 mol%. Reaction conditions: [TFEMA]/[EBPA]/[FeBr$_3$]/[TBABr] = 50/1/x/2x (x = 0.02 or 0.04) in 50 vol% solvent (trifluoroethanol/anisole = 9/1), irradiated under blue LEDs (465 nm, 12 mW/cm^2). Reproduced with permission from Ref [41].

Photoinduced iron-catalyzed ATRP was attempted in aqueous media using $FeBr_3$ and water-soluble ligands (TBABr, tris[2-(2-methoxyethoxy)ethyl]amine, and triphenylphosphine-3,3′,3″-trisulfonic acid trisodium) under UV light irradiation [108]. Polymerizations in water were controlled only in the presence of high concentrations of the iron catalyst as much as 6–10 equiv. with respect to initiator (or 20000–33000 ppm with respect to monomer) giving polymers with dispersities < 1.40.

Iron complexes have been used as photocatalysts to enable ATRP under photoredox catalysis conditions [17,109]. Under visible light irradiation, the catalysts were promoted to the excited state in which the iron catalysts were effective in activating the alkyl halide initiators and generate initiating/propagating radicals. Although these iron complexes were strongly reducing in the photoexcited state (<-2.1 V), they showed very short excited state life-time of <10 ns, and therefore resulted in moderate control over polymerization.

In photoinduced ATRP, direct photolysis of the Fe-Br bond to generate the activator results in the formation of Br$^\bullet$ radicals, which are also capable of reacting with monomers and initiate polymerization of new chains. As a result, targeting high molecular weight polymers may be not accessible under these conditions, as generation of new initiating species may lead to formation of new low molecular weight chains and high dispersities. Therefore, developing photocatalytic processes where (re)generation of the activator is realized via electron or energy transfer events would be advantageous to both expand the applicability of photochemically mediated ATRP to a variety of iron complexes and also achieve well-controlled synthesis of polymers with high molecular weights.

Furthermore, electrochemically-mediated ATRP [14] has not yet been fully investigated in iron-based catalytic systems [110]. Electrochemistry would enable direct use of electrons to generate activator species without contaminating the polymer mixture with side products of the chemical reduction processes, and also better characterize the catalytic processes.

4. Monomer Scope in Iron-Catalyzed ATRP

Iron-catalyzed ATRP has been successfully applied in well-controlled polymerization of various methacrylate and styrene monomers. However, polymerization of monomers that contain polar functional groups can be challenging under iron-catalyzed conditions as the polar functional groups may interact with catalyst and therefore affect its catalytic efficiency.

ATRP of methacrylic acid (MAA) was controlled using heme catalysts [111]. Direct polymerization of MAA by ATRP was challenging due to termination of chain ends via lactonization. Therefore, direct ATRP of MAA required development of special conditions such as low pH to control the polymerization using a copper catalyst [112]. However, the heme catalyst was effective in controlling the polymerization of MAA to yield well-controlled polymers in aqueous solutions. Importantly, the heme-based catalysts performed well in the presence of acidic groups and in aqueous media, suggesting the potential of these compounds as robust iron based ATRP catalysts.

ATRP of other families of vinyl monomers such as acrylate and (meth)acrylamides, using iron catalysts has been challenging and less efficient. Iron-catalyzed ATRP of acrylates often leads to low monomer conversions, low initiation efficiency and polymers with high dispersities [34,55,59]. Acrylate-based secondary radicals are more susceptible to undergo formation of organometallic species with the L/Fe^{II} activators than the tertiary methacrylate-based or styrenic radicals. As a result, high degree of the formation of organometallic species and subsequent catalytic radical termination or catalytic chain transfer reactions, result in termination of polymers chains. Furthermore, the lower activity of dormant acrylate chain end compared to the methacrylate chain end may also contribute for the lower efficiency of control observed in the polymerization of acrylate monomers using iron catalysts. Considering the high propagation rate constants of acrylate monomers, the deactivation should also be fast enough to provide well-controlled polymerization. The use of iron catalysts that typically have lower activity than copper complexes may not be efficient for promoting fast initiation and activation of the chain ends in polymerization of acrylates.

ATRP of methyl acrylate (MA) using iron catalysts in the presence of halide salts gave low monomer conversions (~30%) and polymers with relatively high dispersities (1.3–1.9) [34]. Similarly, polymerization of MA using diimine iron complexes was slow and resulted in high dispersity vales (~1.5–1.6). Polymerization of butyl acrylate was well-controlled using cyclopentyl-functionalized TACN iron complexes yielding quantitative monomer conversions, and low dispersity values (1.2) [50].

An iodine-based initiating system was developed for the polymerization of acrylate monomers using iron catalysis. Dicarbonylcyclopentadienyliodoiron(II) in conjunction with a metal alkoxide such as $Al(Oi\text{-}Pr)_3$ was used in the presence of an alkyl iodide initiator to control polymerization of acrylate monomers [113]. The nature of the alkyl halide initiator greatly affected control over the polymerization as use of less active Br or Cl chain ends resulted in a loss of control with polymers having high dispersities, whereas I-based initiating system afforded well-controlled polymers with low dispersity (<1.2). The control was attributed to the synergistic effect of the iron catalyst and the metal alkoxide additive, which enhanced activation of chain ends, as well as using the more active iodine-based initiating system. However, for alkyl iodides, a degenerative transfer is also possible and can provide additional pathway for controlling polymerization [114].

Polymerization of vinyl acetate was attempted using iron(II) acetate catalyst and carbon tetrachloride (CCl_4) as the initiator [115]. The polymerization mechanism was redox-initiated radical telomerization and not ATRP, resulting in polymers with molecular weights that did not increase with conversion but followed quite precisely $[VOAc]_0/[CCl_4]_0$ ratios due to a C_{tr}~1 for CCl_4. Under OMRP conditions, polymerization of vinyl acetate using iron(II) acetate catalyst showed better controlled results [116].

5. Conclusions and Outlook

Iron complexes form an important class of ATRP catalysts that provide efficient and well-controlled polymerization of various functional monomers. There is a great potential to develop biocompatible, less toxic, and inexpensive iron-based catalytic systems for ATRP. Iron catalysts have been mainly applied for well-controlled ATRP of methacrylate and styrene monomers. Despite many successful studies on designing and use of iron catalysts in ATRP, there are still challenges that need to be addressed to take a full advantage of iron catalysis in ATRP. For instance, the use of iron catalysts has been rarely successful for controlling polymerization of acrylates, due to the possibility of formation of organometallic species and subsequent termination events. In addition to developing new iron-based catalytic systems for ATRP and broadening their utility in polymerization of new functional monomers, iron catalysts need to be explored and studied further to better understand their behavior in catalyzing ATRP reactions. For example, establishing structure-reactivity relationships in iron catalysts regarding the activation and deactivation processes would enable a better understanding of iron-catalyzed ATRP and therefore developing more efficient catalytic systems. Establishing computational approaches and high throughput experimentation would enable understanding, designing, and examining efficient iron catalysts for ATRP. Considering the biological relevance of iron compounds, it would be advantageous to develop iron-based catalytic systems, especially those derived or inspired from biological resources, for synthesis of polymer-bioconjugates or other functional polymeric materials by iron-catalyzed ATRP.

Author Contributions: S.D.-S. and K.M. discussed and constructed the outline of the manuscript and wrote the final draft. All authors have read and agreed to the published version of the manuscript.

Funding: This research received funding from the NSF.

Acknowledgments: Financial support from the NSF (CHE 1707490) is acknowledged.

Conflicts of Interest: The authors declare no conflict of interest.

References

1. Wang, J.-S.; Matyjaszewski, K. Controlled/"living" radical polymerization. Atom transfer radical polymerization in the presence of transition-metal complexes. *J. Am. Chem. Soc.* **1995**, *117*, 5614–5615. [CrossRef]
2. Matyjaszewski, K.; Xia, J. Atom transfer radical polymerization. *Chem. Rev.* **2001**, *101*, 2921–2990. [CrossRef] [PubMed]
3. Matyjaszewski, K. Atom Transfer Radical Polymerization (ATRP): Current status and future perspectives. *Macromolecules* **2012**, *45*, 4015–4039. [CrossRef]
4. Matyjaszewski, K.; Tsarevsky, N.V. Macromolecular engineering by atom transfer radical polymerization. *J. Am. Chem. Soc.* **2014**, *136*, 6513–6533. [CrossRef]
5. Perrier, S. 50th Anniversary perspective: RAFT polymerization—A user guide. *Macromolecules* **2017**, *50*, 7433–7447. [CrossRef]
6. Xu, J.; Jung, K.; Atme, A.; Shanmugam, S.; Boyer, C. A robust and versatile photoinduced living polymerization of conjugated and unconjugated monomers and its oxygen tolerance. *J. Am. Chem. Soc.* **2014**, *136*, 5508–5519. [CrossRef]
7. Moad, G.; Rizzardo, E.; Thang, S.H. Living radical polymerization by the RAFT process—A third update. *Aust. J. Chem.* **2012**, *65*, 985–1076. [CrossRef]
8. Hill, M.R.; Carmean, R.N.; Sumerlin, B.S. Expanding the scope of RAFT polymerization: Recent advances and new horizons. *Macromolecules* **2015**, *48*, 5459–5469. [CrossRef]
9. David, G.; Boyer, C.; Tonnar, J.; Ameduri, B.; Lacroix-Desmazes, P.; Boutevin, B. Use of iodocompounds in radical polymerization. *Chem. Rev.* **2006**, *106*, 3936–3962. [CrossRef]
10. Goto, A.; Ohtsuki, A.; Ohfuji, H.; Tanishima, M.; Kaji, H. Reversible generation of a carbon-centered radical from alkyl iodide using organic salts and their application as organic catalysts in living radical polymerization. *J. Am. Chem. Soc.* **2013**, *135*, 11131–11139. [CrossRef]
11. Ribelli, T.G.; Lorandi, F.; Fantin, M.; Matyjaszewski, K. Atom transfer radical polymerization: Billion times more active catalysts and new initiation systems. *Macromol. Rapid Commun.* **2019**, *40*, 1800616. [CrossRef] [PubMed]
12. Lorandi, F.; Matyjaszewski, K. Why do we need more active ATRP catalysts? *Isr. J. Chem.* **2020**, *60*, 108–123. [CrossRef]

13. Pan, X.; Fantin, M.; Yuan, F.; Matyjaszewski, K. Externally controlled atom transfer radical polymerization. *Chem. Soc. Rev.* **2018**, *47*, 5457–5490. [CrossRef] [PubMed]

14. Magenau, A.J.D.; Strandwitz, N.C.; Gennaro, A.; Matyjaszewski, K. Electrochemically mediated atom transfer radical polymerization. *Science* **2011**, *332*, 81–84. [CrossRef] [PubMed]

15. Dadashi-Silab, S.; Lorandi, F.; Fantin, M.; Matyjaszewski, K. Redox-switchable atom transfer radical polymerization. *Chem. Commun.* **2019**, *55*, 612–615. [CrossRef]

16. di Lena, F.; Matyjaszewski, K. Transition metal catalysts for controlled radical polymerization. *Prog. Polym. Sci.* **2010**, *35*, 959–1021. [CrossRef]

17. Telitel, S.; Dumur, F.; Campolo, D.; Poly, J.; Gigmes, D.; Pierre Fouassier, J.; Lalevée, J. Iron complexes as potential photocatalysts for controlled radical photopolymerizations: A tool for modifications and patterning of surfaces. *J. Polym. Sci. Part A Polym. Chem.* **2016**, *54*, 702–713. [CrossRef]

18. Ando, T.; Kamigaito, M.; Sawamoto, M. Iron(II) chloride complex for living radical polymerization of methyl methacrylate. *Macromolecules* **1997**, *30*, 4507–4510. [CrossRef]

19. Matyjaszewski, K.; Wei, M.; Xia, J.; McDermott, N.E. Controlled/"living" radical polymerization of styrene and methyl methacrylate catalyzed by iron complexes. *Macromolecules* **1997**, *30*, 8161–8164. [CrossRef]

20. Poli, R.; Allan, L.E.N.; Shaver, M.P. Iron-mediated reversible deactivation controlled radical polymerization. *Prog. Polym. Sci.* **2014**, *39*, 1827–1845. [CrossRef]

21. Xue, Z.; He, D.; Xie, X. Iron-catalyzed atom transfer radical polymerization. *Polym. Chem.* **2015**, *6*, 1660–1687. [CrossRef]

22. Kato, M.; Kamigaito, M.; Sawamoto, M.; Higashimura, T. Polymerization of Methyl methacrylate with the carbon Tetrachloride/Dichlorotris- (triphenylphosphine)ruthenium(II)/Methylaluminum Bis(2,6-di-tert-butylphenoxide) initiating system: Possibility of living radical polymerization. *Macromolecules* **1995**, *28*, 1721–1723. [CrossRef]

23. Braunecker, W.A.; Itami, Y.; Matyjaszewski, K. Osmium-mediated radical polymerization. *Macromolecules* **2005**, *38*, 9402–9404. [CrossRef]

24. Fors, B.P.; Hawker, C.J. Control of a living radical polymerization of methacrylates by light. *Angew. Chem. Int. Ed.* **2012**, *51*, 8850–8853. [CrossRef] [PubMed]

25. Discekici, E.H.; Anastasaki, A.; Read de Alaniz, J.; Hawker, C.J. Evolution and future directions of metal-free atom transfer radical polymerization. *Macromolecules* **2018**, *51*, 7421–7434. [CrossRef]

26. Treat, N.J.; Sprafke, H.; Kramer, J.W.; Clark, P.G.; Barton, B.E.; Read de Alaniz, J.; Fors, B.P.; Hawker, C.J. Metal-free atom transfer radical polymerization. *J. Am. Chem. Soc.* **2014**, *136*, 16096–16101. [CrossRef]

27. Pan, X.; Fang, C.; Fantin, M.; Malhotra, N.; So, W.Y.; Peteanu, L.A.; Isse, A.A.; Gennaro, A.; Liu, P.; Matyjaszewski, K. Mechanism of photoinduced metal-free atom transfer radical polymerization: Experimental and computational studies. *J. Am. Chem. Soc.* **2016**, *138*, 2411–2425. [CrossRef]

28. Theriot, J.C.; Lim, C.-H.; Yang, H.; Ryan, M.D.; Musgrave, C.B.; Miyake, G.M. Organocatalyzed atom transfer radical polymerization driven by visible light. *Science* **2016**, *352*, 1082–1086. [CrossRef]

29. Allan, L.E.N.; Perry, M.R.; Shaver, M.P. Organometallic mediated radical polymerization. *Prog. Polym. Sci.* **2012**, *37*, 127–156. [CrossRef]

30. Poli, R. New phenomena in Organometallic-Mediated Radical Polymerization (OMRP) and perspectives for control of less active monomers. *Chem. Eur. J.* **2015**, *21*, 6988–7001. [CrossRef]

31. Shaver, M.P.; Allan, L.E.N.; Rzepa, H.S.; Gibson, V.C. Correlation of metal spin state with catalytic reactivity: Polymerizations mediated by α-Diimine–Iron complexes. *Angew. Chem. Int. Ed.* **2006**, *45*, 1241–1244. [CrossRef]

32. Lake, B.R.M.; Shaver, M.P. The Interplay of ATRP, OMRP and CCT in Iron-Mediated Controlled Radical Polymerization. In *Controlled Radical Polymerization: Mechanisms*; American Chemical Society: Washington, DC, USA, 2015; Volume 1187, pp. 311–326.

33. Fantin, M.; Lorandi, F.; Ribelli, T.G.; Szczepaniak, G.; Enciso, A.E.; Fliedel, C.; Thevenin, L.; Isse, A.A.; Poli, R.; Matyjaszewski, K. Impact of organometallic intermediates on copper-catalyzed atom transfer radical polymerization. *Macromolecules* **2019**, *52*, 4079–4090. [CrossRef]

34. Teodorescu, M.; Gaynor, S.G.; Matyjaszewski, K. Halide anions as ligands in iron-mediated atom transfer radical polymerization. *Macromolecules* **2000**, *33*, 2335–2339. [CrossRef]

35. Wang, Y.; Matyjaszewski, K. ATRP of MMA catalyzed by FeIIBr2 in the presence of triflate anions. *Macromolecules* **2011**, *44*, 1226–1228. [CrossRef]

36. Sarbu, T.; Matyjaszewski, K. ATRP of Methyl methacrylate in the presence of ionic liquids with ferrous and cuprous anions. *Macromol. Chem. Phys.* **2001**, *202*, 3379–3391. [CrossRef]
37. Ishio, M.; Katsube, M.; Ouchi, M.; Sawamoto, M.; Inoue, Y. Active, versatile, and removable iron catalysts with phosphazenium salts for living radical polymerization of methacrylates. *Macromolecules* **2009**, *42*, 188–193. [CrossRef]
38. Wang, Y.; Zhang, Y.; Parker, B.; Matyjaszewski, K. ATRP of MMA with ppm levels of iron catalyst. *Macromolecules* **2011**, *44*, 4022–4025. [CrossRef]
39. Wang, J.; Han, J.; Xie, X.; Xue, Z.; Fliedel, C.; Poli, R. FeBr2-Catalyzed bulk ATRP promoted by simple inorganic salts. *Macromolecules* **2019**, *52*, 5366–5376. [CrossRef]
40. Wang, J.; Xie, X.; Xue, Z.; Fliedel, C.; Poli, R. Ligand- and solvent-free ATRP of MMA with FeBr3 and inorganic salts. *Polym. Chem.* **2020**, *11*, 1375–1385. [CrossRef]
41. Dadashi-Silab, S.; Matyjaszewski, K. Iron-Catalyzed atom transfer radical polymerization of semifluorinated methacrylates. *ACS Macro Lett.* **2019**, *8*, 1110–1114. [CrossRef]
42. Ding, K.; Zannat, F.; Morris, J.C.; Brennessel, W.W.; Holland, P.L. Coordination of N-methylpyrrolidone to iron(II). *J. Organomet. Chem.* **2009**, *694*, 4204–4208. [CrossRef]
43. Schroeder, H.; Buback, J.; Demeshko, S.; Matyjaszewski, K.; Meyer, F.; Buback, M. Speciation analysis in iron-mediated ATRP studied via FT-Near-IR and mössbauer spectroscopy. *Macromolecules* **2015**, *48*, 1981–1990. [CrossRef]
44. Matyjaszewski, K.; Coca, S.; Gaynor, S.G.; Wei, M.; Woodworth, B.E. Zerovalent metals in controlled/"living" radical polymerization. *Macromolecules* **1997**, *30*, 7348–7350. [CrossRef]
45. Zhang, H.; Marin, V.; Fijten, M.W.M.; Schubert, U.S. High-throughput experimentation in atom transfer radical polymerization: A general approach toward a directed design and understanding of optimal catalytic systems. *J. Polym. Sci. Part A Polym. Chem.* **2004**, *42*, 1876–1885. [CrossRef]
46. Bergenudd, H.; Jonsson, M.; Malmström, E. Investigation of iron complexes in ATRP: Indications of different iron species in normal and reverse ATRP. *J. Mol. Catal. A Chem.* **2011**, *346*, 20–28. [CrossRef]
47. Aoshima, H.; Satoh, K.; Umemura, T.; Kamigaito, M. A simple combination of higher-oxidation-state FeX3 and phosphine or amine ligand for living radical polymerization of styrene, methacrylate, and acrylate. *Polym. Chem.* **2013**, *4*, 3554–3562. [CrossRef]
48. Niibayashi, S.; Hayakawa, H.; Jin, R.-H.; Nagashima, H. Reusable and environmentally friendly ionic trinuclear iron complex catalyst for atom transfer radical polymerization. *Chem. Commun.* **2007**. [CrossRef]
49. Kawamura, M.; Sunada, Y.; Kai, H.; Koike, N.; Hamada, A.; Hayakawa, H.; Jin, R.-H.; Nagashima, H. New Iron(II) complexes for atom-transfer radical polymerization: The ligand design for triazacyclononane results in high reactivity and catalyst performance. *Adv. Synth. Catal.* **2009**, *351*, 2086–2090. [CrossRef]
50. Nakanishi, S.-I.; Kawamura, M.; Kai, H.; Jin, R.-H.; Sunada, Y.; Nagashima, H. Well-defined Iron complexes as efficient catalysts for "green" atom-transfer radical polymerization of styrene, Methyl Methacrylate, and Butyl Acrylate with low catalyst loadings and catalyst recycling. *Chem. Eur. J.* **2014**, *20*, 5802–5814. [CrossRef]
51. Nakanishi, S.-I.; Kawamura, M.; Sunada, Y.; Nagashima, H. Atom transfer radical polymerization by solvent-stabilized (Me3TACN)FeX2: A practical access to reusable iron(ii) catalysts. *Polym. Chem.* **2016**, *7*, 1037–1048. [CrossRef]
52. Gibson, V.C.; O'Reilly, R.K.; Wass, D.F.; White, A.J.P.; Williams, D.J. Iron complexes bearing iminopyridine and aminopyridine ligands as catalysts for atom transfer radical polymerisation. *Dalton Trans.* **2003**. [CrossRef]
53. Göbelt, B.; Matyjaszewski, K. Diimino- and diaminopyridine complexes of CuBr and FeBr2 as catalysts in atom transfer radical polymerization (ATRP). *Macromol. Chem. Phys.* **2000**, *201*, 1619–1624. [CrossRef]
54. Gibson, V.C.; O'Reilly, R.K.; Reed, W.; Wass, D.F.; White, A.J.P.; Williams, D.J. Four-coordinate iron complexes bearing α-diimine ligands: Efficient catalysts for Atom Transfer Radical Polymerisation (ATRP). *Chem. Commun.* **2002**. [CrossRef]
55. O'Reilly, R.K.; Shaver, M.P.; Gibson, V.C.; White, A.J.P. α-Diimine, diamine, and diphosphine iron catalysts for the controlled radical polymerization of styrene and acrylate monomers. *Macromolecules* **2007**, *40*, 7441–7452. [CrossRef]
56. Gibson, V.C.; O'Reilly, R.K.; Wass, D.F.; White, A.J.P.; Williams, D.J. Polymerization of Methyl methacrylate using four-coordinate (α-Diimine)iron catalysts: Atom transfer radical polymerization vs catalytic chain transfer. *Macromolecules* **2003**, *36*, 2591–2593. [CrossRef]

57. Azuma, Y.; Terashima, T.; Sawamoto, M. Self-folding polymer iron catalysts for living radical polymerization. *ACS Macro Lett.* **2017**, *6*, 830–835. [CrossRef]
58. Allan, L.E.N.; MacDonald, J.P.; Reckling, A.M.; Kozak, C.M.; Shaver, M.P. Controlled radical polymerization mediated by Amine–Bis(phenolate) Iron(III) complexes. *Macromol. Rapid Commun.* **2012**, *33*, 414–418. [CrossRef]
59. Allan, L.E.N.; MacDonald, J.P.; Nichol, G.S.; Shaver, M.P. Single component iron catalysts for atom transfer and organometallic mediated radical polymerizations: Mechanistic studies and reaction scope. *Macromolecules* **2014**, *47*, 1249–1257. [CrossRef]
60. Coward, D.L.; Lake, B.R.M.; Shaver, M.P. Understanding organometallic-mediated radical polymerization with an Iron(II) Amine–Bis(phenolate). *Organometallics* **2017**, *36*, 3322–3328. [CrossRef]
61. Schroeder, H.; Lake, B.R.M.; Demeshko, S.; Shaver, M.P.; Buback, M. A synthetic and multispectroscopic speciation analysis of controlled radical polymerization mediated by Amine–Bis(phenolate)iron complexes. *Macromolecules* **2015**, *48*, 4329–4338. [CrossRef]
62. Poli, R.; Shaver, M.P. Atom Transfer Radical Polymerization (ATRP) and Organometallic Mediated Radical Polymerization (OMRP) of styrene mediated by diaminobis(phenolato)iron(II) complexes: A DFT study. *Inorg. Chem.* **2014**, *53*, 7580–7590. [CrossRef]
63. Nishiura, C.; Williams, V.; Matyjaszewski, K. Iron and copper based catalysts containing anionic phenolate ligands for atom transfer radical polymerization. *Macromol. Res.* **2017**, *25*, 504–512. [CrossRef]
64. Sigg, S.J.; Seidi, F.; Renggli, K.; Silva, T.B.; Kali, G.; Bruns, N. Horseradish peroxidase as a catalyst for atom transfer radical polymerization. *Macromol. Rapid Commun.* **2011**, *32*, 1710–1715. [CrossRef]
65. Pollard, J.; Bruns, N. Biocatalytic ATRP. In *Reversible Deactivation Radical Polymerization: Mechanisms and Synthetic Methodologies*; American Chemical Society: Washington, DC, USA, 2018; Volume 1284, pp. 379–393.
66. Rodriguez, K.J.; Gajewska, B.; Pollard, J.; Pellizzoni, M.M.; Fodor, C.; Bruns, N. Repurposing biocatalysts to control radical polymerizations. *ACS Macro Lett.* **2018**, *7*, 1111–1119. [CrossRef]
67. Ng, Y.-H.; di Lena, F.; Chai, C.L.L. PolyPEGA with predetermined molecular weights from enzyme-mediated radical polymerization in water. *Chem. Commun.* **2011**, *47*, 6464–6466. [CrossRef]
68. Silva, T.B.; Spulber, M.; Kocik, M.K.; Seidi, F.; Charan, H.; Rother, M.; Sigg, S.J.; Renggli, K.; Kali, G.; Bruns, N. Hemoglobin and red blood cells catalyze atom transfer radical polymerization. *Biomacromolecules* **2013**, *14*, 2703–2712. [CrossRef]
69. Simakova, A.; Mackenzie, M.; Averick, S.E.; Park, S.; Matyjaszewski, K. Bioinspired iron-based catalyst for atom transfer radical polymerization. *Angew. Chem. Int. Ed.* **2013**, *52*, 12148–12151. [CrossRef]
70. Fu, L.; Simakova, A.; Park, S.; Wang, Y.; Fantin, M.; Matyjaszewski, K. Axially ligated mesohemins as bio-mimicking catalysts for atom transfer radical polymerization. *Molecules* **2019**, *24*, 3969. [CrossRef]
71. Smolne, S.; Buback, M.; Demeshko, S.; Matyjaszewski, K.; Meyer, F.; Schroeder, H.; Simakova, A. Kinetics of Fe–Mesohemin–(MPEG500)2-Mediated RDRP in aqueous solution. *Macromolecules* **2016**, *49*, 8088–8097. [CrossRef]
72. Uchiike, C.; Terashima, T.; Ouchi, M.; Ando, T.; Kamigaito, M.; Sawamoto, M. Evolution of iron catalysts for effective living radical polymerization: Design of phosphine/halogen ligands in FeX2(PR3)2. *Macromolecules* **2007**, *40*, 8658–8662. [CrossRef]
73. Xue, Z.; Linh, N.T.B.; Noh, S.K.; Lyoo, W.S. Phosphorus-containing ligands for Iron(III)-catalyzed atom transfer radical polymerization. *Angew. Chem. Int. Ed.* **2008**, *47*, 6426–6429. [CrossRef]
74. Xue, Z.; Oh, H.S.; Noh, S.K.; Lyoo, W.S. Phosphorus ligands for Iron(III)-Mediated atom transfer radical polymerization of Methyl methacrylate. *Macromol. Rapid Commun.* **2008**, *29*, 1887–1894. [CrossRef]
75. Xue, Z.; Noh, S.K.; Lyoo, W.S. 2-[(Diphenylphosphino)methyl]pyridine as ligand for iron-based atom transfer radical polymerization. *J. Polym. Sci. Part A Polym. Chem.* **2008**, *46*, 2922–2935. [CrossRef]
76. Xue, Z.; He, D.; Noh, S.K.; Lyoo, W.S. Iron(III)-Mediated atom transfer radical polymerization in the absence of any additives. *Macromolecules* **2009**, *42*, 2949–2957. [CrossRef]
77. Wang, Y.; Kwak, Y.; Matyjaszewski, K. Enhanced activity of ATRP Fe catalysts with phosphines containing electron donating groups. *Macromolecules* **2012**, *45*, 5911–5915. [CrossRef]
78. Schröder, K.; Mathers, R.T.; Buback, J.; Konkolewicz, D.; Magenau, A.J.D.; Matyjaszewski, K. Substituted Tris(2-pyridylmethyl)amine ligands for highly active ATRP catalysts. *ACS Macro Lett.* **2012**, *1*, 1037–1040. [CrossRef]

79. Ribelli, T.G.; Fantin, M.; Daran, J.-C.; Augustine, K.F.; Poli, R.; Matyjaszewski, K. Synthesis and characterization of the most active copper ATRP catalyst based on tris[(4-dimethylaminopyridyl)methyl]amine. *J. Am. Chem. Soc.* **2018**, *140*, 1525–1534. [CrossRef]

80. Schroeder, H.; Matyjaszewski, K.; Buback, M. Kinetics of Fe-Mediated ATRP with triarylphosphines. *Macromolecules* **2015**, *48*, 4431–4437. [CrossRef]

81. Dadashi-Silab, S.; Pan, X.; Matyjaszewski, K. Photoinduced Iron-Catalyzed atom transfer radical polymerization with ppm levels of iron catalyst under blue light irradiation. *Macromolecules* **2017**, *50*, 7967–7977. [CrossRef]

82. Nishizawa, K.; Ouchi, M.; Sawamoto, M. Phosphine–ligand decoration toward active and robust iron catalysts in LRP. *Macromolecules* **2013**, *46*, 3342–3349. [CrossRef]

83. Uchiike, C.; Ouchi, M.; Ando, T.; Kamigaito, M.; Sawamoto, M. Evolution of iron catalysts for effective living radical polymerization: P–N chelate ligand for enhancement of catalytic performances. *J. Polym. Sci. Part A Polym. Chem.* **2008**, *46*, 6819–6827. [CrossRef]

84. Khan, M.Y.; Zhou, J.; Chen, X.; Khan, A.; Mudassir, H.; Xue, Z.; Lee, S.W.; Noh, S.K. Exploration of highly active bidentate ligands for iron (III)-catalyzed ATRP. *Polymer* **2016**, *90*, 309–316. [CrossRef]

85. Ishio, M.; Terashima, T.; Ouchi, M.; Sawamoto, M. Carbonyl–phosphine heteroligation for pentamethylcyclopentadienyl (Cp*)–Iron complexes: Highly active and versatile catalysts for living radical polymerization. *Macromolecules* **2010**, *43*, 920–926. [CrossRef]

86. Wang, Y.; Matyjaszewski, K. ATRP of MMA in polar solvents catalyzed by FeBr2 without additional ligand. *Macromolecules* **2010**, *43*, 4003–4005. [CrossRef]

87. Eckenhoff, W.T.; Biernesser, A.B.; Pintauer, T. Structural characterization and investigation of iron(III) complexes with nitrogen and phosphorus based ligands in atom transfer radical addition (ATRA). *Inorg. Chim. Acta* **2012**, *382*, 84–95. [CrossRef]

88. Okada, S.; Park, S.; Matyjaszewski, K. Initiators for continuous activator regeneration atom transfer radical polymerization of methyl methacrylate and styrene with n-heterocyclic carbene as ligands for fe-based catalysts. *ACS Macro Lett.* **2014**, *3*, 944–947. [CrossRef]

89. Zhang, L.; Miao, J.; Cheng, Z.; Zhu, X. Iron-mediated ICAR ATRP of styrene and Methyl Methacrylate in the absence of thermal radical initiator. *Macromol. Rapid Commun.* **2010**, *31*, 275–280. [CrossRef]

90. Zhu, G.; Zhang, L.; Zhang, Z.; Zhu, J.; Tu, Y.; Cheng, Z.; Zhu, X. Iron-mediated ICAR ATRP of Methyl methacrylate. *Macromolecules* **2011**, *44*, 3233–3239. [CrossRef]

91. Mukumoto, K.; Wang, Y.; Matyjaszewski, K. Iron-based ICAR ATRP of styrene with ppm amounts of FeIIIBr3 and 1,1′-Azobis(cyclohexanecarbonitrile). *ACS Macro Lett.* **2012**, *1*, 599–602. [CrossRef]

92. Luo, R.; Sen, A. Electron-transfer-induced iron-based atom transfer radical polymerization of styrene derivatives and copolymerization of styrene and methyl methacrylate. *Macromolecules* **2008**, *41*, 4514–4518. [CrossRef]

93. Khan, M.Y.; Chen, X.; Lee, S.W.; Noh, S.K. Development of new atom transfer radical polymerization system by iron (III)-metal salts without using any external initiator and reducing agent. *Macromol. Rapid Commun.* **2013**, *34*, 1225–1230. [CrossRef] [PubMed]

94. He, D.; Xue, Z.; Khan, M.Y.; Noh, S.K.; Lyoo, W.S. Phosphorus ligands for iron(III)-mediated ATRP of styrene via generation of activators by monomer addition. *J. Polym. Sci. Part A Polym. Chem.* **2010**, *48*, 144–151. [CrossRef]

95. Fujimura, K.; Ouchi, M.; Sawamoto, M. Ferrocene cocatalysis for iron-catalyzed living radical polymerization: Active, robust, and sustainable system under concerted catalysis by two iron complexes. *Macromolecules* **2015**, *48*, 4294–4300. [CrossRef]

96. Qin, J.; Cheng, Z.; Zhang, L.; Zhang, Z.; Zhu, J.; Zhu, X. A highly efficient iron-mediated AGET ATRP of Methyl methacrylate using Fe(0) powder as the reducing agent. *Macromol. Chem. Phys.* **2011**, *212*, 999–1006. [CrossRef]

97. Layadi, A.; Kessel, B.; Yan, W.; Romio, M.; Spencer, N.D.; Zenobi-Wong, M.; Matyjaszewski, K.; Benetti, E.M. Oxygen tolerant and cytocompatible Iron(0)-mediated ATRP enables the controlled growth of polymer brushes from mammalian cell cultures. *J. Am. Chem. Soc.* **2020**, *142*, 3158–3164. [CrossRef]

98. Dadashi-Silab, S.; Doran, S.; Yagci, Y. Photoinduced electron transfer reactions for macromolecular syntheses. *Chem. Rev.* **2016**, *116*, 10212–10275. [CrossRef]

99. Pan, X.; Tasdelen, M.A.; Laun, J.; Junkers, T.; Yagci, Y.; Matyjaszewski, K. Photomediated controlled radical polymerization. *Prog. Polym. Sci.* **2016**, *62*, 73–125. [CrossRef]

100. Chen, M.; Zhong, M.; Johnson, J.A. Light-controlled radical polymerization: Mechanisms, methods, and applications. *Chem. Rev.* **2016**, *116*, 10167–10211. [CrossRef]

101. Corrigan, N.; Shanmugam, S.; Xu, J.; Boyer, C. Photocatalysis in organic and polymer synthesis. *Chem. Soc. Rev.* **2016**, *45*, 6165–6212. [CrossRef]

102. Dadashi-Silab, S.; Atilla Tasdelen, M.; Yagci, Y. Photoinitiated atom transfer radical polymerization: Current status and future perspectives. *J. Polym. Sci. Part A Polym. Chem.* **2014**, *52*, 2878–2888. [CrossRef]

103. Pan, X.; Malhotra, N.; Zhang, J.; Matyjaszewski, K. Photoinduced Fe-based atom transfer radical polymerization in the absence of additional ligands, reducing agents, and radical initiators. *Macromolecules* **2015**, *48*, 6948–6954. [CrossRef]

104. Zhou, Y.-N.; Guo, J.-K.; Li, J.-J.; Luo, Z.-H. Photoinduced Iron(III)-mediated atom transfer radical polymerization with in situ generated initiator: Mechanism and kinetics studies. *Ind. Eng. Chem. Res.* **2016**, *55*, 10235–10242. [CrossRef]

105. Pan, X.; Malhotra, N.; Dadashi-Silab, S.; Matyjaszewski, K. A simplified Fe-based PhotoATRP using only monomers and solvent. *Macromol. Rapid Commun.* **2017**, *38*, 1600651. [CrossRef] [PubMed]

106. Rolland, M.; Whitfield, R.; Messmer, D.; Parkatzidis, K.; Truong, N.P.; Anastasaki, A. Effect of polymerization components on Oxygen-Tolerant Photo-ATRP. *ACS Macro Lett.* **2019**, *8*, 1546–1551. [CrossRef]

107. Dadashi-Silab, S.; Matyjaszewski, K. Temporal control in atom transfer radical polymerization using zerovalent metals. *Macromolecules* **2018**, *51*, 4250–4258. [CrossRef]

108. Bian, C.; Zhou, Y.-N.; Guo, J.-K.; Luo, Z.-H. Photoinduced Fe-mediated atom transfer radical polymerization in aqueous media. *Polym. Chem.* **2017**, *8*, 7360–7368. [CrossRef]

109. Bansal, A.; Kumar, P.; Sharma, C.D.; Ray, S.S.; Jain, S.L. Light-induced controlled free radical polymerization of methacrylates using iron-based photocatalyst in visible light. *J. Polym. Sci. Part A Polym. Chem.* **2015**, *53*, 2739–2746. [CrossRef]

110. Wang, J.; Tian, M.; Li, S.; Wang, R.; Du, F.; Xue, Z. Ligand-free iron-based electrochemically mediated atom transfer radical polymerization of methyl methacrylate. *Polym. Chem.* **2018**, *9*, 4386–4394. [CrossRef]

111. Fu, L.; Simakova, A.; Fantin, M.; Wang, Y.; Matyjaszewski, K. Direct ATRP of methacrylic acid with iron-porphyrin based catalysts. *ACS Macro Lett.* **2018**, *7*, 26–30. [CrossRef]

112. Fantin, M.; Isse, A.A.; Venzo, A.; Gennaro, A.; Matyjaszewski, K. Atom transfer radical polymerization of methacrylic acid: A won challenge. *J. Am. Chem. Soc.* **2016**, *138*, 7216–7219. [CrossRef]

113. Onishi, I.; Baek, K.-Y.; Kotani, Y.; Kamigaito, M.; Sawamoto, M. Iron-catalyzed living radical polymerization of acrylates: Iodide-based initiating systems and block and random copolymerizations. *J. Polym. Sci. Part A Polym. Chem.* **2002**, *40*, 2033–2043. [CrossRef]

114. Matyjaszewski, K.; Gaynor, S.; Wang, J.-S. Controlled radical polymerizations: The use of Alkyl iodides in degenerative transfer. *Macromolecules* **1995**, *28*, 2093–2095. [CrossRef]

115. Xia, J.; Paik, H.-J.; Matyjaszewski, K. Polymerization of Vinyl acetate promoted by iron complexes. *Macromolecules* **1999**, *32*, 8310–8314. [CrossRef]

116. Xue, Z.; Poli, R. Organometallic mediated radical polymerization of vinyl acetate with Fe(acac)2. *J. Polym. Sci. Part A Polym. Chem.* **2013**, *51*, 3494–3504. [CrossRef]

 *molecules*

Review

C-H Functionalization via Iron-Catalyzed Carbene-Transfer Reactions

Claire Empel, Sripati Jana and Rene M. Koenigs *

Institute of Organic Chemistry, RWTH Aachen University, Landoltweg 1, D-52074 Aachen, Germany;
claire.empel@rwth-aachen.de (C.F.); janasripati@gmail.com (S.J.)
* Correspondence: rene.koenigs@rwth-aachen.de

Academic Editor: Hans-Joachim Knölker
Received: 4 January 2020; Accepted: 5 February 2020; Published: 17 February 2020

Abstract: The direct C-H functionalization reaction is one of the most efficient strategies by which to introduce new functional groups into small organic molecules. Over time, iron complexes have emerged as versatile catalysts for carbine-transfer reactions with diazoalkanes under mild and sustainable reaction conditions. In this review, we discuss the advances that have been made using iron catalysts to perform C-H functionalization reactions with diazoalkanes. We give an overview of early examples employing stoichiometric iron carbene complexes and continue with recent advances in the C-H functionalization of $C(sp^2)$-H and $C(sp^3)$-H bonds, concluding with the latest developments in enzymatic C-H functionalization reactions using iron-heme-containing enzymes.

Keywords: iron; carbene; diazoalkane; C-H functionalization

---

1. Introduction

C-H bonds belong to the most common motifs in organic molecules and their direct functionalization is one of the main challenges in synthesis methodology, which impacts on the step economy and sustainability of chemical processes. In recent decades, this research area has flourished and different approaches have been realized to enable C-H functionalization reactions via different strategies: (a) by the directing group-assisted C-H activation, (b) via the innate reactivity of organic molecules, or (c) via direct C-H functionalization (Scheme 1) [1–4]. While the introduction and subsequent removal of directing groups is a prerequisite for directed C-H activation, direct C-H functionalization allows the introduction of new functional groups into organic molecules without the need for directing groups. It thus represents the most efficient and step-economic strategy by which to conduct C-H functionalization reaction. The challenge of this strategy lies within the selective activation of only one type of C-H bond; elegant methods have been developed in recent years, mainly relying on the use of expensive and mostly toxic precious metal catalysts based on rhodium, iridium, palladium, and others to direct the C-H functionalization reaction as either the catalyst or the substrate [5–7].

Scheme 1. Strategies for the functionalization of C-H bonds. (**a**) directing-group-assisted C-H activation, (**b**) innate reactivity of organic molecules, (**c**) direct C-H functionalization of C-H bonds.

High costs of catalysts, limited resources, and toxicity of precious metals are the main drivers in the exploration of new synthesis methods based on third-row transition metals. Among these, iron plays a pivotal role. It is the second most abundant metal in the Earth's crust following aluminum, and catalysts based on iron are currently emerging as new and important catalysts able to improve the environmental impact of chemical processes [8–13]. In nature, iron-containing enzymes play an important role, e.g., in the hemoglobin or myoglobin enzymes that are important for oxygen transport in mammals. They can also be found in cytochrome P450 enzymes or iron-sulfur clusters, amongst others, and are pivotal for the detoxification, e.g., by C-H oxidation reactions, of xenobiotics [14]. Against this background, the development of C-H functionalization reactions with iron catalysts has received significant attention over the years with the goal of reducing the economic and environmental footprint of organic synthesis methodologies, but also that of accessing new reactivity that is unique to iron. In this review, we discuss the advances made in this research area with a focus on C-H functionalization reactions with carbenes.

2. Iron in Carbene-Transfer Reactions

Carbene-transfer reactions are one of the key strategies used today to conduct highly efficient and selective C-H functionalization reactions. The earliest examples date back to reports by Meerwein and Doering from 1942 and 1959, in which they described metal-free photochemical C-H functionalization reactions with diazoalkanes using high-energy UV light via free carbene intermediates (Scheme 2) [15,16]. However, the high reactivity of the free carbene intermediate led to unselective reactions and, as a consequence, a lack of applications in organic synthesis. In the subsequent decades, these shortcomings led to the development of metal-catalyzed carbine-transfer reactions using noble metals such as Rh(II), Ru(II), Ir(III), Au(I), Pd(II), or Cu(I) [17–21].

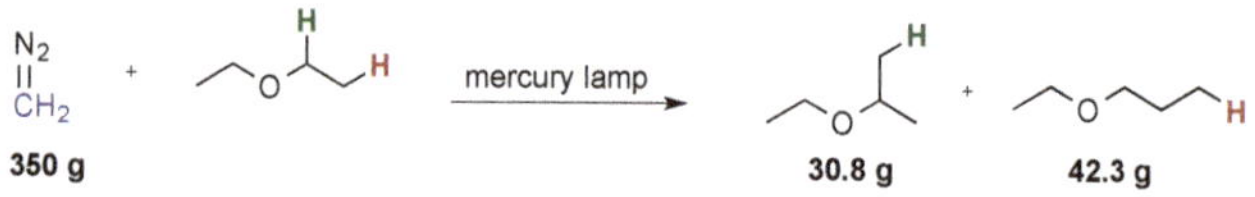

Scheme 2. Photochemical C-H functionalization of diethyl ether with diazomethane.

More recently, carbene-transfer reactions with iron complexes have gained significant attention in organic synthesis methodology for their potential to overcome the limitations of precious metal complexes. In 1992, Hossain et al. described the first iron-catalyzed carbene-transfer reaction using the cationic CpFe(CO)$_2$(THF)BF$_4$ complex in cyclopropanation reactions of styrene **1** and ethyl diazoacetate **2** (Scheme 3a) [22]. Following this strategy, different groups have since reported on their efforts to conduct iron-catalyzed carbene-transfer reactions [11–13].

In the following years, several groups reported their efforts in the field of iron-catalyzed carbine-transfer reactions. Woo et al. reported an important milestone in this research area when they uncovered the cyclopropanation of styrenes with ethyl diazoacetate **2** using an iron porphyrin complex (**4a**) in 1995 [23]. In this report and in further reports by the Che (**4c**) and Aviv (**5**) groups, the fine-tuning of iron porphyrin complexes by the axial ligand and/or electronics of the porphyrin ring system was demonstrated (Scheme 3b) [24–26]. Today, iron porphyrin complexes and derivatives thereof are important cheap and readily available catalysts to enable carbine-transfer reactions under mild conditions.

Khade and Zhang studied the formation of iron carbene complexes via quantum chemical studies using ethyl phenyldiazoacetate (**8**) as a model compound. They were able to identify a reaction intermediate (**9a**) in which ethyl phenyldiazoacetate (**8**) coordinated to the iron porphyrin complex **7** via the carbon atom. In a next step, nitrogen was expelled via transition state **9b** and the iron carbene complex **9c** was formed. In these studies, the authors were also able to show that *N*-methyl imidazole reduced both the reaction energy ΔG and the activation energy ΔG_{TS} (Scheme 4) [27].

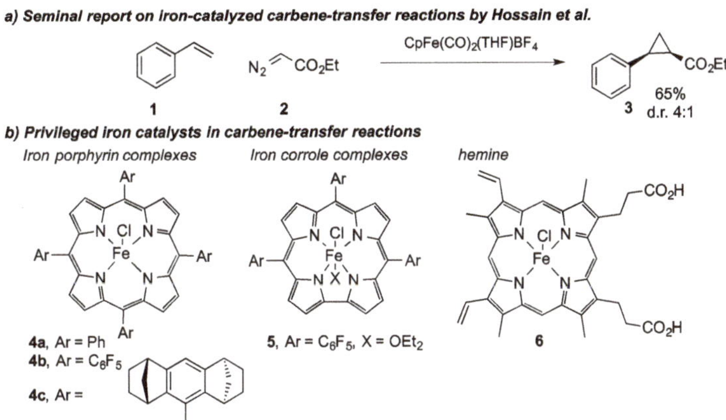

Scheme 3. (**a**) Seminal iron-catalyzed cyclopropanation by Hossain et al. [22]. (**b**) Privileged iron catalysts in carbene-transfer reactions.

Scheme 4. Mechanism of the formation of iron carbene complexes.

3. Stoichiometric C-H Functionalization

Initial applications of iron carbene complexes in C-H functionalization reactions involved the use of preformed iron carbene complexes **10** that were used in intramolecular C-H insertion reactions. The nature of the preformation of the iron carbene complex renders these applications stoichiometric in iron. Following this methodology, C-H insertion occurs selectively to form cyclopentane ring systems (**13,14**). The corresponding bi- and tricyclic target structures can be obtained with high *trans*-selectivity (Scheme 5) [28]. The same authors investigated the selectivity of this process in further studies and explained the observed reaction outcome via a concerted C-H insertion reaction that proceeds via a chair-like cyclic transition state (**12**, Scheme 5) [29,30].

Scheme 5. Intramolecular C-H insertion reactions for the synthesis of cyclopentane derivatives (**13,14**) by Helquist and coworkers [29,30].

4. C-H Functionalization Reactions of Aromatic C-H Bonds

The direct C-H functionalization of benzene or alkyl benzenes (**15**) via carbene or metal-carbene intermediates can give access to three different products, which arise from (a) insertion into a $C(sp^2)$-H bond (**17**), (b) insertion into a side-chain $C(sp^3)$-H bond (**18**), or (c) a cycloaddition reaction with the aromatic system to give a norcaradiene **19** that can undergo a Buchner reaction to give a cycloheptatriene **20** (Scheme 6). The challenge lies in the chemoselective differentiation of these three reaction pathways. Moreover, a C-H insertion into a $C(sp^2)$-H bond may give rise to different regioisomers.

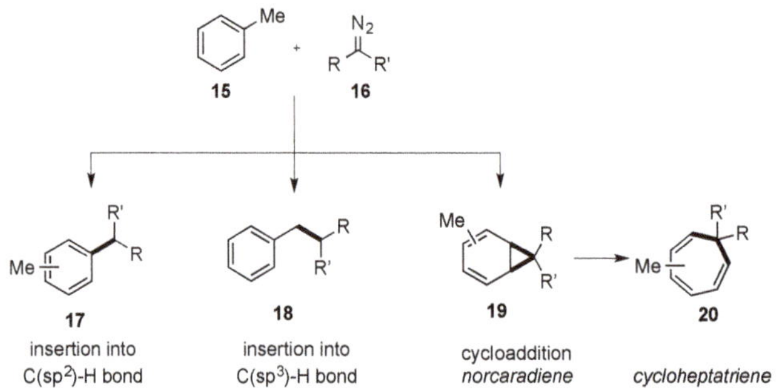

Scheme 6. Possible products in the iron-catalyzed functionalization of toluene **15** by carbene insertion.

In this context, Luis, Perez, and coworkers reported on the use of iron(II)-complex (**23**) bearing a pytacn ligand in the reaction of ethyl diazoacetate **2** with aromatic hydrocarbons (**21**). Under these reaction conditions, ethyl diazoacetate (**2**) underwent a chemoselective C-H functionalization of the aromatic $C(sp^2)$-H bond of benzene and alkyl benzene derivatives, and only trace amounts of the norcaradiene/cycloheptatriene reaction pathway were observed [31]. The substrate scope of benzene derivatives was further studied by the same authors and, in all cases, exclusive $C(sp^2)$-H bond insertion occurred in moderate to good yields as a mixture of regioisomers. Electron-withdrawing groups (e.g., Cl) had a detrimental effect on the C-H functionalization reaction, and the corresponding products were obtained with reduced yields (Scheme 7) [32].

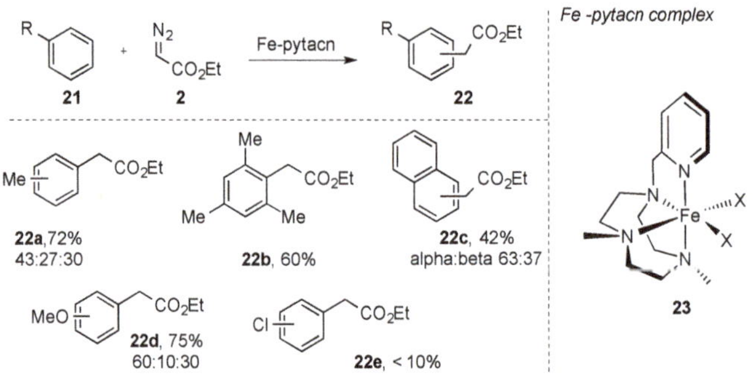

Scheme 7. (**Left**): functionalization of arenes **21** with ethyl diazoacetate **2**; (**Right**): iron-pytacn complex.

The reaction mechanism of this reaction was studied in detail by the authors. In a first step, the iron complex **23** undergoes a counterion exchange with $NaBAr_F$ to give the dicationic complex **23a**. The latter undergoes formation of the initial coordination of ethyl diazoacetate (**23b**), which leads to the formation of an iron carbene complex **23c** upon extrusion of nitrogen. Next, the arene **21** undergoes nucleophilic addition to **23c** under formation of a Wheland-type intermediate **23d**, which undergoes a 1,2-H shift to release product **22** and the active catalyst **23a** (Scheme 8) [32]. The proposed mechanism was in line with a previous report by Doyle, Padwa, and coworkers [33].

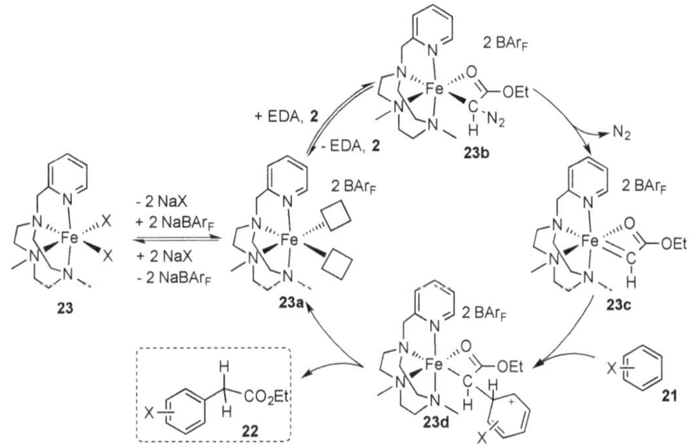

Scheme 8. Proposed reaction pathway for the C-H functionalization reaction of arenes (**21**) using ethyl diazoacetate **2**.

The direct C-H functionalization reaction of aromatic compounds (used as solvents) was also studied by Woo and coworkers using dimethyl diazomalonate **25** and an iron(III) porphyrin complex (**4a**) as a catalyst at elevated reaction temperatures. However, only moderate chemoselectivity of $C(sp^2)$-H bond vs. $C(sp^3)$-H bond insertion could be achieved in this reaction (**26**:**27**, 2:1 to 10:1). Different substituted aromatic systems such as toluene, mesitylene, 4-chloro-toluene, or anisole were shown to be compatible with the present reaction conditions. In contrast to the previously described reaction, the product from the $C(sp^3)$-H bond insertion reaction was obtained as the major product (Scheme 9) [34].

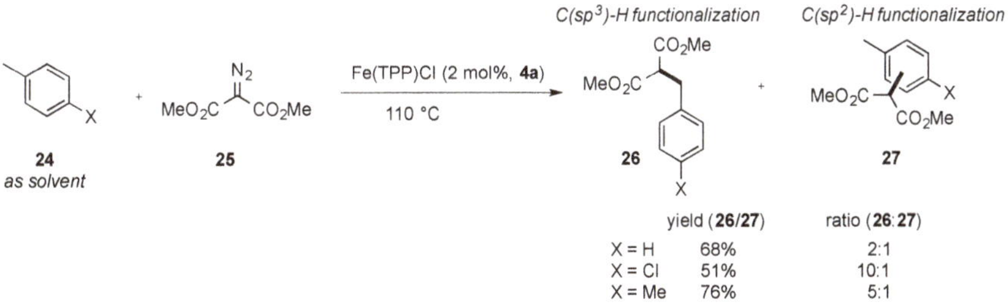

Scheme 9. Chemoselectivity between $C(sp^2)$-H bond and $C(sp^3)$-H bond insertion reaction investigated by Woo and coworkers [34].

In 2015, Zhou and coworkers studied the reaction of donor–acceptor diazo compounds **29** with electron-rich aromatics **28**. Using a simple, non-porphyrin iron complex, the authors were able to demonstrate the $C(sp^2)$-H functionalization reaction of *N,N*-dialkyl aniline derivatives (**28**) with high yield and selectivity. In all cases, selective C-H functionalization in the 4 position occurred, which could be rationalized by the high nucleophilicity of *N,N*-dialkyl aniline derivatives in the *para* position. From the viewpoint of the catalyst, a very simple iron catalyst was used that could be obtained in situ from FeCl$_3$, 1,10-phenanthroline, and NaBAr$_F$ (Scheme 10) [35]. More recently, Deng and coworkers reported on a similar transformation using a *bis*(imino)pyridine iron complex, which could be used in the direct C-H functionalization of *N,N*-dimethylaniline **28** to give **30a** with a 71% yield [36].

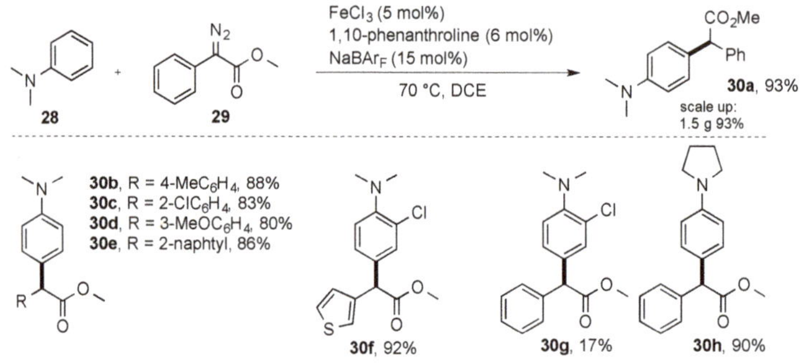

FeCl$_3$ (5 mol%)
1,10-phenanthroline (6 mol%)
NaBAr$_F$ (15 mol%)

70 °C, DCE

30a, 93%
scale up:
1.5 g 93%

30b, R = 4-MeC$_6$H$_4$, 88%
30c, R = 2-ClC$_6$H$_4$, 83%
30d, R = 3-MeOC$_6$H$_4$, 80%
30e, R = 2-naphtyl, 86%

30f, 92%   **30g**, 17%   **30h**, 90%

Scheme 10. Iron-catalyzed arylation of donor-acceptor diazo compound **29** with *N,N*-dialkyl anilines **28** by the Zhou group [36].

5. C-H Functionalization Reactions of Heteroaromatic C-H Bonds

N-heterocycles like indoles or pyrroles are privileged motifs in drugs and natural products, and a broad variety of strategies for their construction or their functionalization has been described in recent decades. A particularly intriguing transformation is the direct C-H bond functionalization of heterocycles, which would streamline current synthesis strategies of drugs or natural products. Furthermore, it would also allow the introduction of molecular diversity into late-stage functionalization reactions. This approach, in combination with the benefits of iron catalysts, would enable important applications of C-H functionalization strategies in medicinal chemistry, agrochemistry, or total synthesis.

In 2006, Woo et al. reported on the iron-catalyzed functionalization of *N*-heterocycles with ethyl diazoacetate **2** as a carbene precursor. While no reaction was observed with indole **31**, the closely related pyrrole **32** underwent C-H functionalization without concomitant N-H functionalization to give **33** in a 37% yield (Scheme 11) [37].

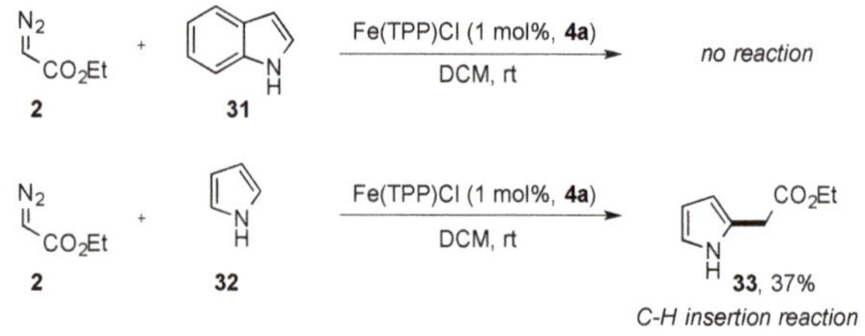

Fe(TPP)Cl (1 mol%, **4a**)
DCM, rt

no reaction

Fe(TPP)Cl (1 mol%, **4a**)
DCM, rt

33, 37%
C-H insertion reaction

Scheme 11. Iron-catalyzed reaction of ethyl diazoacetate **2** with indole **31** and pyrrole **32**.

Thereafter, Zhou and-coworkers reported on an enantioselective C-H functionalization of indole heterocycles. For this purpose, they studied the reaction of silyl-protected indole heterocycles (**34**) with α-aryl-α-diazoesters (**29**) in the presence of Fe(ClO$_4$)$_2$ and a chiral spirobisoxazoline ligand **36**. This strategy allowed the C-H functionalization of indole heterocycles under mild reaction conditions with moderate enantioselectivity. Importantly, different substituents at the indole heterocycles were tolerated, such as chlorine and methyl (Scheme 12) [38].

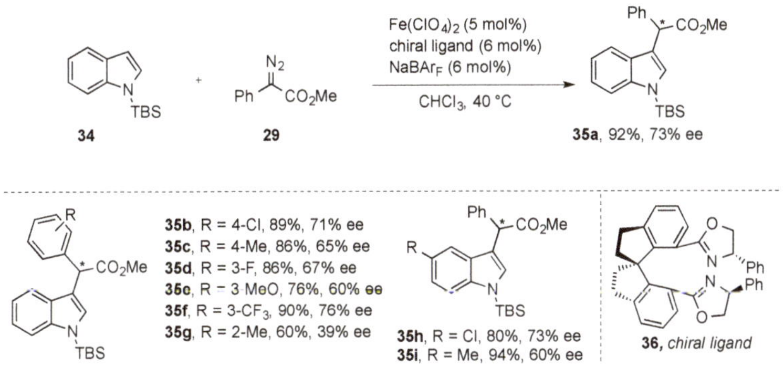

Scheme 12. Enantioselective iron-catalyzed C-H functionalization of TBS-indole (**34**) with α-aryl-α-diazoesters (**29**), described by Zhou et al. [38].

As described by Woo and coworkers, the direct functionalization of indole heterocycles in the C3 position does not readily occur when using ethyl diazoacetate [37]. The direct functionalization of indole with acceptor-only diazoalkanes would open up avenues towards the efficient synthesis of tryptamines and related alkaloids. In recent years, different groups have focused their efforts on identifying iron catalysts and/or reaction conditions that would allow this important transformation. In 2019, Koenigs and Weissenborn et al. reported the iron-catalyzed C-H functionalization reaction of indoles **37** in the C3 position using the same catalyst as the Woo group, but using diazoacetonitrile **38**. Using this methodology, protected and unprotected indole and indazole heterocycles underwent selective functionalization in the C3 position. Importantly, no reaction was observed when blocking the C3 position (Scheme 13). This approach has now enabled a two-step approach towards the synthesis of important tryptamine derivatives [39].

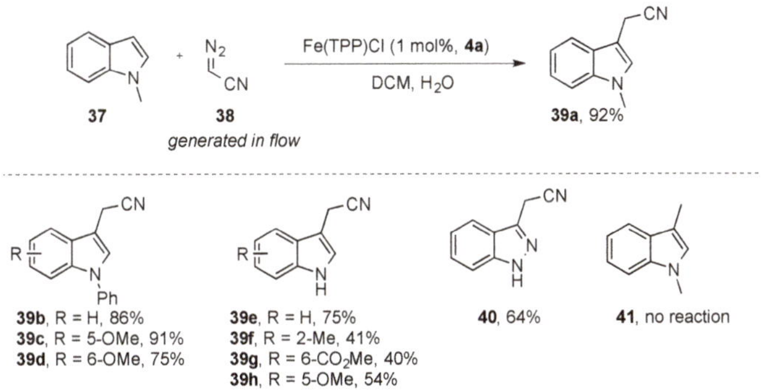

Scheme 13. Reaction of diazoacetonitrile **38** with indole heterocycles using Fe(TPP)Cl, reported by Koenigs, Weissenborn, and coworkers [39].

To date, various proposals for the C-H functionalization of indole heterocycles have been reported in the literature. One important mechanistic proposal suggests the initial formation of an iron carbene complex **43** that undergoes nucleophilic addition of an indole followed by a [1,2]-proton transfer reaction (Scheme 14a). Zhou and coworkers reported a KIE (= kinetic isotope effect) of 5.06 in the reaction of *N*-methyl indole and deuterated *N*-methyl indole, which led them to conclude a proton-transfer reaction to be the rate-determining step (Scheme 14b) [38]. Koenigs, Weissenborn, and coworkers observed a similar trend in the reaction of *N*-methyl indole and deuterated *N*-methyl indole with diazoacetonitrile using both a synthetic and an enzymatic iron catalyst. However, investigations using TEMPO **46** as a radical scavenger revealed a complete inhibition of the C-H functionalization reaction. Based on these experimental data, the authors assumed a radical reaction, yet it was not yet clear in which particular reaction step radical intermediates were involved (Scheme 14c) [39].

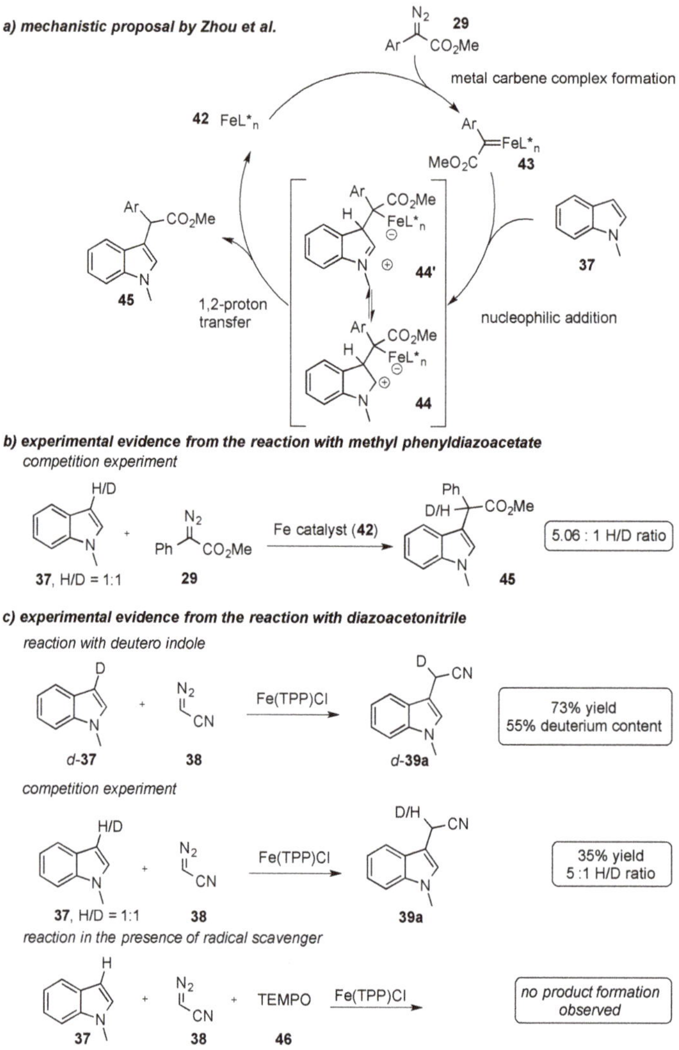

Scheme 14. Hypothesized mechanism for the C-H functionalization of indole with iron carbene complexes and experimental evidence. (**a**) mechanistic proposal by Zhou et al., [38], (**b**) experimental evidence from the reaction with methyl phenyldiazoacetate [38], (**c**) evidence from the reaction with diazoacetonitrile [39].

The reaction mechanism of the C-H functionalization reaction of indole heterocycles with iron carbene intermediates still remains unclear. It would be valuable to investigate this reaction via DFT calculations to gain knowledge of the exact spin state of the participating iron catalyst and to identify the exact reaction mechanism.

6. C-H Functionalization Reactions of Aliphatic C-H Bonds

The direct C-H functionalization reaction of aliphatic C-H bonds is an important strategy for the introduction of new functional groups onto a hydrocarbon skeleton; the difficulties lie in the differentiation of chemically very similar C-H bonds. Over the years, metal-catalyzed insertion reactions of carbene fragments have emerged as a promising strategy for this purpose, and a variety of precious-metal-catalyzed $C(sp^3)$-H bond functionalization reactions have been described in the literature, ranging from site-selective C-H functionalization to late-stage functionalization of complex molecules [5,40,41].

The application of non-toxic and highly active iron catalysts has blossomed in the last few years, and today, iron-catalyzed carbine-transfer reactions have emerged as an important, environmentally benign strategy by which to conduct $C(sp^3)$-H bond functionalization reactions. In this context, Woo and coworkers reported their studies in C-H functionalization reactions of cyclohexane **47** and tetrahydrofuran **49** as substrates. Using a donor–acceptor diazoalkanes (**29**) as a carbene precursor and a Fe(TPP)Cl (**4a**) catalyst, the authors demonstrated proof-of-concept studies in this research area. Cyclohexane **47** underwent smooth C-H functionalization to give product **48** with a 66% yield. No C-H functionalization was observed when ethyl diazoacetate **2** or dimethyl diazomalonate **25** were used as carbene precursors. Similarly, tetrahydrofuran **49** underwent selective C-H functionalization reaction in the β position (**50**). A byproduct arising from ring-opening of the THF ring was observed as a minor reaction product **51** with ~20% yield. In a competition experiment with mesitylene (**52**) as a substrate, the authors obtained the product from $C(sp^3)$-H bond and $C(sp^2)$-H bond in a 1.5:1 ratio (Scheme 15) [34].

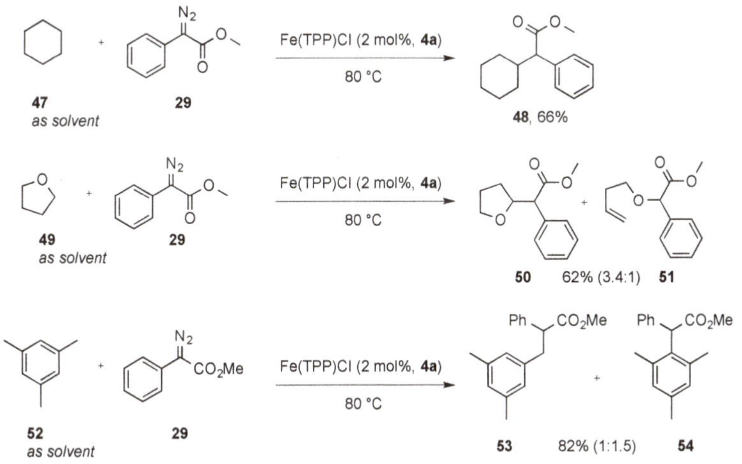

Scheme 15. Reaction of α-aryl-α-diazoesters with cyclohexane (**47**), THF (**49**), and mesitylene (**52**).

The C-H functionalization of cyclohexane represents one of the simplest examples, as all C-H bonds within cyclohexane are identical. Contrarily, linear or branched hydrocarbons are more challenging, as different C-H bonds are present that need to be differentiated by the catalyst. Zhu et al. studied the reaction of these substrates with donor–acceptor diazoalkanes using an iron catalyst. When *n*-hexane **56** was used as a substrate, selective C-H functionalization of secondary C-H groups occurred, but without any meaningful selectivity of the C2 vs. C3 positon. When studying 2-metylpentane **59**, the C-H functionalization occurred preferentially at the tertiary C-H group. No C-H functionalization of primary C-H bonds was observed, and the reactivity order for the C-H functionalization decreased from tertiary > secondary > allylic/benzylic >> primary C-H bonds (Scheme 16). It is of note that this catalyst can be applied in the C-H functionalization of cyclohexane with turnover numbers of up to 690 on gram scale. In this reaction, the authors observed a KIE of 2.0 for k_H/k_D [42].

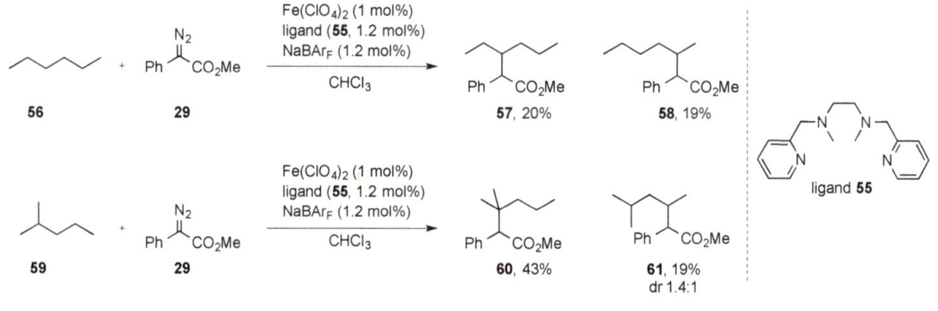

Scheme 16. Studies on the selectivity of iron-catalyzed $C(sp^3)$-H insertion reactions.

While intermolecular $C(sp^3)$-H functionalization reactions suffer from missing selectivity and reactivity, the development of intramolecular processes would enable important steps to be taken in the understanding of iron-catalyzed $C(sp^3)$-H functionalization. In this context, the White group explored an intramolecular cyclization via C-H functionalization using an iron phthalocyanin complex as the carbene-transfer catalyst; importantly, weakly coordinating counterions needed to be used to increase the electrophilicity of the iron complex. Under these conditions, the *bis*-acceptor diazoalkanes (**62**) underwent smooth intramolecular C-H functionalization reaction to yield product **63** in moderate to good yields with broad functional group tolerance (Scheme 17). It is important to note that the C-H functionalization reaction selectivity occurred in the allylic, benzylic position or in the α position to a heteroatom [43].

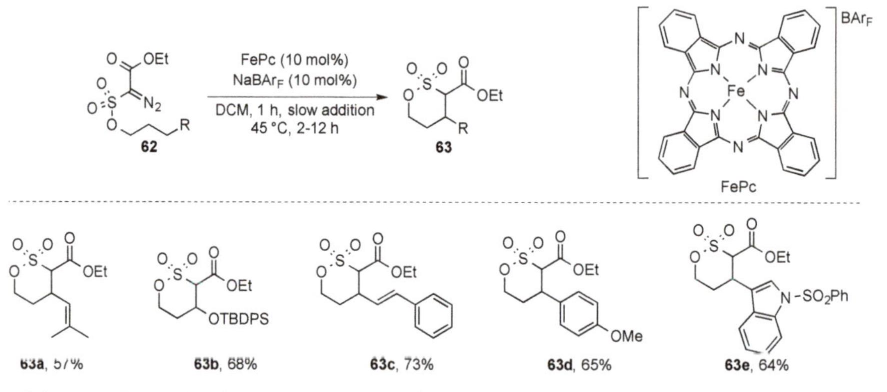

Scheme 17. Intramolecular cyclisation via C-H functionalization for the synthesis of sulfonate esters.

Recently, Costas and coworkers reported on the intramolecular functionalization of $C(sp^3)$-H bonds using the electrophilic [Fe(^{F}pda)-(THF)]$_2$ complex (**67**) as catalyst and a lithium salt with a weakly coordinating counterion as a co-catalyst. Importantly, the lithium salt is required to activate the diazoester under mild conditions, while the iron complex is needed for the actual carbene-transfer reaction. Following this strategy, intramolecular C-H functionalization reactions of $C(sp^3)$-H could be realized to selectively obtain the five membered ring products (**65**). Following this strategy, bi- and spirocyclic systems were synthesized (Scheme 18). Under these conditions, the five membered ring was is favored, e.g., over benzylic C-H functionalization (**65f**). The formation of α,β-unsaturated esters (**66**) occurred via a β-hydride elimination reaction [44].

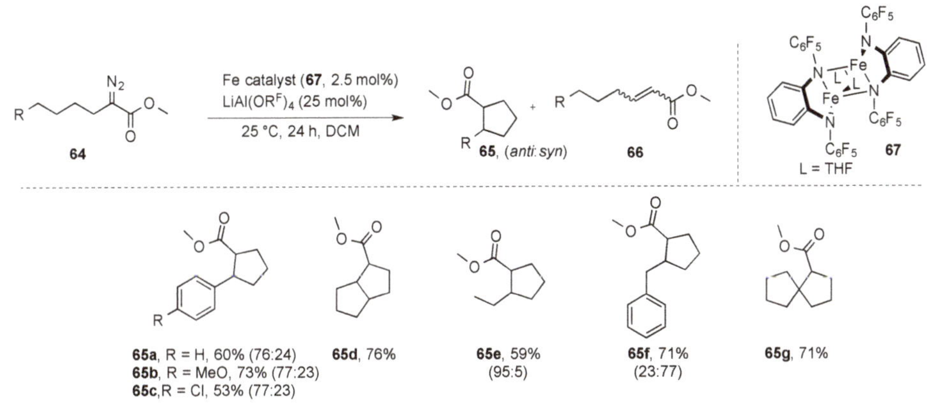

Scheme 18. Intramolecular alkylation reaction for the synthesis of five membered rings.

7. Biocatalytic C-H Functionalization Reactions of Aromatic C-H Bonds

Over the years, the development of non-natural activity of enzymes has developed as an important strategy by which to conduct highly efficient carbene-transfer reactions [45,46]. In this section, we have focused on the recent developments in this research area, with a focus on applications in C-H functionalization reactions.

In the report of Koenigs, Weissenborn, and coworkers, the reaction of *N*-methyl indole (**37**) with diazoacetonitrile (**38**) was also studied with the bacterial dye-decolorizing peroxidase YfeX from *E. coli*. A library of YfeX variants and the wild-type protein were studied and, under the best conditions, a turnover number of 37 was achieved with the wild-type enzyme. This TON (= turnover number) could be improved to 80 by using the I230A variant of YfeX. This variant was also studied in the reaction of *N*-methyl indole with ethyl diazoacetate **2**, which led to an increased TON of 236 (Scheme 17) [39]. At the same time, the Fasan group reported on the application of an engineered myoglobin enzyme in the C3-functionalization reaction of unprotected indoles using ethyl diazoacetate **2** as a carbene precursor. The Mb(H64V,V68A) variant gave the highest conversion and a TON of 82 in a whole-cell setup, and **68a** was obtained as the only product of C-H functionalization. In their report, Fasan and coworkers also investigated the functional group tolerance of this reaction, and different substitution patterns on the indole heterocycle were tolerated. Notably, in the case of *N*-protected indoles, the Mb(L29F,H64V) variant was required to achieve a similar activity of the enzyme (Scheme 19) [47].

Fasan and coworkers studied the reaction mechanism of the enzymatic C-H functionalization reaction. For this purpose, the reaction of 3-deutero-indole *d*-**31** and ethyl diazoacetate **2** was investigated. The C-H functionalization product **68a** was obtained as the protonated product exclusively, and no deuterium label was found in the reaction product. Moreover, in a competition experiment, no difference in reaction kinetics of the deuterated and non-deuterated starting material was observed. The absence of a kinetic isotope effect led the authors to the conclusion that no C-H insertion mechanism occurred. Based on this evidence, Fasan and coworkers concluded a nucleophilic attack by the indole substrate **31** to generate a zwitterionic intermediate **69/69′**, which underwent a solvent-assisted proton-transfer reaction to give the desired C-H functionalization product (Scheme 20) [47]. At this point, it is noteworthy that Koenigs and Weissenborn et al. were able to observe a retention of the deuterium label in the reaction of deuterated *N*-methyl indole *d*-**47** with diazoacetonitrile [39].

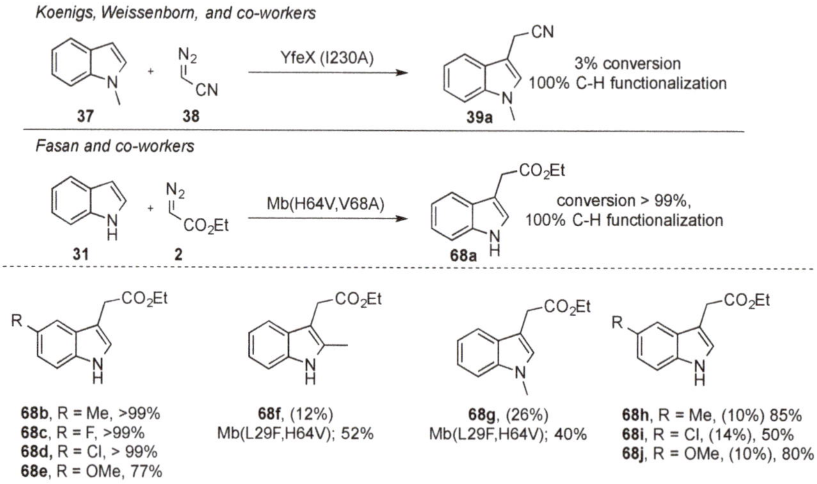

Scheme 19. Enzyme-catalyzed C3-alkylation reaction of indoles.

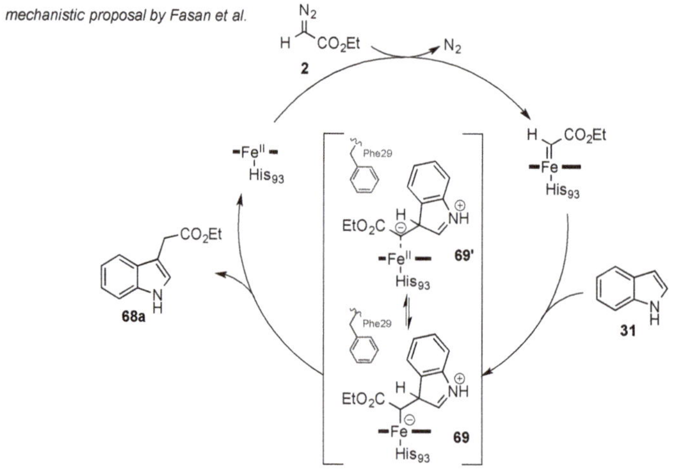

Scheme 20. Mechanistic investigations and proposed catalytic cycle by the Fasan group.

Shortly after the reports by Koenigs, Weissenborn. and Fasan, the enzymatic C-H functionalization of indole and other heterocycles with an engineered cytochrome P411 enzyme was studied by Arnold and coworkers. They reported on the directed evolution of a "carbene transferase" enzyme that enables the highly efficient, chemoselective, and biocatalytic C-H functionalization of indole and pyrrole heterocycles. Using a UV-Vis spectrophotometry-based high-throughput screening, the authors were able study several thousand mutants to engineer enzymes to conduct chemoselective C-H functionalization reactions. Cytochrome P411 variants were tested in reactions with *N*-Me pyrrole (**71**), and two different variants allowed the regioselective C-H functionalization in the C3 position or the C2 position as well as enantioselective C-H functionalization of indole heterocycles with diazopropionate (Scheme 21) [48].

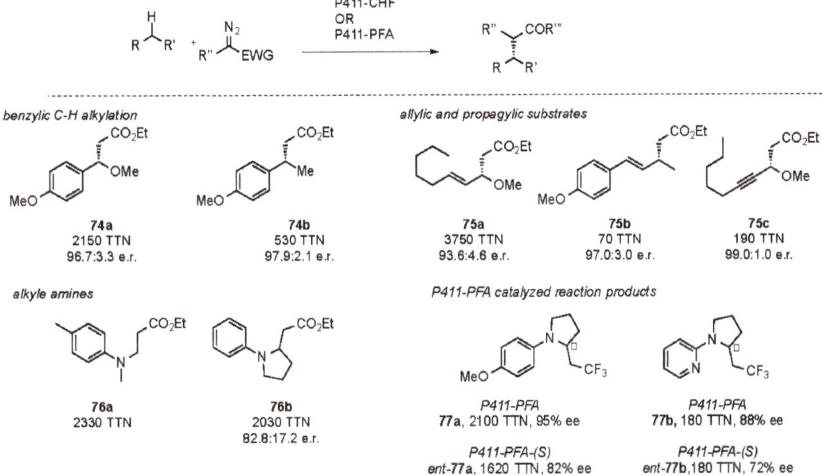

Scheme 21. Indole C3-alkylation with engineered P411-HF enzymes; regioselective alkylation of 1-methylpyrrole **53**.

In further reports, the Arnold group studied engineered cytochrome P411 enzymes in benzylic C-H functionalization reactions of alkyl arenes with ethyl diazoacetate (**2**). A cytochrome P411 variant with an axial serine ligand served as the starting point for the directed evolution, which led to the P411-CHF variant that gave a TTN (= total turnover number) of 2020 and an enantiomeric ratio of 96.7:3.3 (**74a**, Scheme 22). In further studies, the C-H functionalization of allylic and propargylic substrates as well as alkyl amines was studied (**75, 76**, Scheme 22) [49]. Only recently, Arnold et al. were able to further extend the substrate scope of carbene precursors to trifluoro diazoethane, which was studied in the α-C(sp^3)-H functionalization of N,N-dialkyl aniline derivatives. By directed evolution of the wild-type cytochrome P411 enzyme the C-H functionalization reaction with trifluoro diazoethane was realized. Further studies focused on the access of the opposite stereoisomer. Starting from a P411 variant that provided the opposite stereochemistry, direct evolution led to a variant providing the inverse stereochemistry. This example showcases the power of biocatalytic transformations and the fact that both enantiomers of a desired product can be obtained by means of directed evolution (**77**, Scheme 22) [50].

Scheme 22. C-H functionalization using engineered P411-CHF and P411-PFA enzymes.

8. Conclusions and Perspectives

Over the years, the application of iron complexes has gained significant attention for the conduction of C-H functionalization reactions with diazoalkanes under mild and sustainable reaction conditions. A variety of iron complexes have been described to date to perform C-H functionalization reactions with high efficiency, which have leveraged iron-catalyzed carbine-transfer reactions as an important tool for organic chemists to functionalize unreactive C-H bonds. In light of the recent developments in enzyme-catalyzed carbene-transfer reactions and directed evolution, iron-heme enzymes are now developing as important tools with which to conduct highly chemo-, regio-, and stereoselective C-H functionalization reactions. Building upon these advances, future developments of iron-catalyzed C-H functionalization might include, for example, the site-selective activation of unactivated $C(sp^3)$-H bonds, the broadening of the substrate scope to include non-activated aromatic systems, or development of an understanding of the reaction mechanisms and the spin state of iron within the catalytic cycle.

Author Contributions: R.M.K., C.E. and S.J. wrote the manuscript and schemes. All authors have read and agreed to the published version of the manuscript.

Funding: This article was funded by Deutsche Forschungsgemeinschaft.

Conflicts of Interest: The authors declare no conflict of interest.

References

1. Norinder, J.; Matsumoto, A.; Yishikai, N.; Nakamura, E. Iron-Catalyzed Direct Arylation trough Directed C-H Bond Activation. *J. Am. Chem. Soc.* **2008**, *130*, 5858–5859. [CrossRef] [PubMed]
2. Wencel-Delord, J.; Dröger, T.; Liu, F.; Glorius, F. Towads mild metal-catalyzed C-H activation. *Chem. Soc. Rev.* **2011**, *40*, 4740–4761. [CrossRef] [PubMed]
3. Zhang, M.; Zhang, Y.; Jie, X.; Zhao, H.; Lim, G.; Su, W. Recent advantages in directed C-H functionalizations using monodentate nitrogen-based directing groups. *Org. Chem. Front.* **2014**, *1*, 843–895. [CrossRef]
4. Jia, C.; Kitamura, T.; Fujiwara, Y. Catalytic Functionalization of Arenes and Alkanes via C-H Bond Activation. *Acc. Chem. Res.* **2001**, *34*, 633–639. [CrossRef]
5. Guptill, D.M.; Davies, H.M.L. 2,2,2-Trichloroethyl Aryldiazoacetates as Robust Reagents for the Enantioselective C-H Functionalization of Methyl Ethers. *J. Am. Chem. Soc.* **2014**, *136*, 17718–17721. [CrossRef]
6. Pan, S.; Shibata, T. Recent Advances in Iridum-Catalyzed Alkylation of C-H and N-H Bonds. *Acs Catal.* **2013**, *3*, 704–712. [CrossRef]
7. Ammann, S.E.; Rice, G.T.; White, M.C. Terminal Olefins to Chromans, Isochromans and Pyrans via Allylic C-H Oxidation. *J. Am. Chem. Soc.* **2014**, *136*, 10834–10837. [CrossRef]
8. Burbidge, E.; Burbidge, G.R.; Fowler, A.W.; Hoyle, F. Synthesis of the Elements in stars. *Rev. Mod. Phys.* **1957**, *29*, 547–560. [CrossRef]
9. Gandeepan, P.; Müller, T.; Zell, D.; Cera, G.; Warratz, S.; Ackermann, L. 3d Tansition Metals for C-H Activation. *Chem. Rev.* **2019**, *119*, 2192–2452. [CrossRef]
10. Shang, R.; Illies, L.; Nakamura, E. Iron-Catalyzed C-H Bond Activation. *Chem. Rev.* **2017**, *117*, 9086–9139. [CrossRef]
11. Zhu, S.-F.; Zhou, Q.-L. Iron-catalyzed transformations of diazo compounds. *Natl. Sci. Rev.* **2014**, *1*, 580–603. [CrossRef]
12. Bauer, I.; Knölker, H.-J. Iron Catalysis in Organic Synthesis. *Chem. Rev.* **2015**, *115*, 3170–3387. [CrossRef] [PubMed]
13. Plietker, B. *Iron Catalysis in Organic Chemistry: Reactions and Applications*, 2nd ed.; Wiley-VCH: Weinheim, Germany, 2008.
14. Felg, A.L.; Lippard, S.J. Reactions of Non-Heme Iron(II) Centers with Dioxygen in Biology and Chemistry. *Chem. Rev.* **1994**, *94*, 759–805.
15. Meerwein, H.; Rathjen, H.; Werner, H. Die Methylierung von RH-Verbindungen mittels Diazomethan unter Mitwirkung des Lichtes. *Ber. Dtsch. Chem. Ges.* **1942**, *24*, 136–137. [CrossRef]

16. von E. Doering, W.; Knox, L.; Jones, M. Notes. Reaction of Methylene with Diethyl Ether and Tetrahydrofuran. *J. Org. Chem.* **1959**, *24*, 136–137. [CrossRef]

17. Ford, A.; Miel, M.; Ring, A.; Slattery, C.N.; Maguire, A.R.; McKervey, M.A. Modern Organic Synthesis with α-Doazocarbonyl Compounds. *Chem. Rev.* **2015**, *115*, 9981–10080. [CrossRef]

18. Doyle, M.P.; Duffy, R.; Ratnikov, M.; Zhou, L. Catalytic Carbene Insertion into C-H Bonds. *Chem. Rev.* **2010**, *110*, 704–724. [CrossRef]

19. Davies, H.M.L.; Manning, J.R. Catalytic C-H functionalization by metal carbenoid and nitrenoid insertion. *Nature* **2008**, *451*, 417–424. [CrossRef]

20. Davies, H.M.L.; Morton, D. Guiding principles for site selective and stereoselective intramolecular C-H functionalization by donor/acceptor rhodium carbenes. *Chem. Soc. Rev.* **2011**, *40*, 1857–1869. [CrossRef]

21. Empel, C.; Koenigs, R.M. Sustainable Carbene Transfer Reactions with Iron and Light. *Synlett* **2019**, *30*, 1929–1934. [CrossRef]

22. Seitz, W.J.; Saha, A.K.; Hossain, M.M. Iron Lewis acid catalyzed cyclopropanation reaction of ethyl diazoacetate and olefins. *Organometallics* **1993**, *12*, 2604–2608. [CrossRef]

23. Wolf, J.R.; Hamaker, C.G.; Djukic, J.-P.; Kodadek, T.; Woo, L.K. Shape and stereoselective cyclopropanation of alkenes catalyzed by iron porphyrins. *J. Am. Chem. Soc.* **1995**, *117*, 9194–9199. [CrossRef]

24. Lai, T.-S.; Chan, F.-Y.; So, P.-K.; Ma, D.-L.; Wong, K.Y.; Che, C.-M. Alkene cyclopropanation catalyzed by Halterman iron porphyrin: Participation of organic based axial ligands. *Dalton Trans.* **2006**, *40*, 4845–4851. [CrossRef] [PubMed]

25. Aviv, I.; Gross, Z. Corrole-based applications. *Chem. Commun.* **2007**, *20*, 1987–1999. [CrossRef] [PubMed]

26. Morandi, B.; Carreira, E.M. Iron-Catalyzed Cyclopropanation with Trifluoroethylamine Hydrochloride and Olefines in Aqueous Media: In Situ Generation of Trifluoromethyl Diazomethane. *Angew. Chem. Int. Ed.* **2010**, *49*, 938–941. [CrossRef]

27. Khade, R.L.; Zhang, Y. C-H Insertions by Iron Porphyrin Carbene: Basis Mechanism and Origin of Substrate Selectivity. *Chem. Eur. J.* **2017**, *23*, 17654–17658. [CrossRef]

28. Ishii, S.; Helquist, P. Intramolecular C-H Insertion Reaction of Iron Carbene Complexes as a General Method for Synthesis of Bicyclo[n.3.0]alkanones. *Synlett* **1997**, *4*, 508–510. [CrossRef]

29. Ishii, S.; Zhao, S.; Nehta, G.; Knors, C.J.; Helquist, P. Intramolecular C-H insertion Reactions of (η^5-Cyclopentadienyl)dicabonyliron Carbene Complexes: Scope of the Reaction and Application to the Synthesos of (±)-Sterpurene and (±)-Pentalene. *J. Org. Chem.* **2001**, *66*, 3449–3458. [CrossRef]

30. Ishii, S.; Zhao, S.; Helquist, P. Stereochemical Probes of Intramolecular C-H Insertion Reaction of Iron-Carbene Complexes. *J. Am. Chem. Soc.* **2000**, *122*, 5897–5898. [CrossRef]

31. Postils, V.; Rodriguez, M.; Sabenya, G.; Conde, A.; Mar Diaz-Requejo, M.; Pérez, P.J.; Costas, M.; Solà, M.; Luis, J.M. Mechanism of the Selective Fe-Catalyzed Arene Carbon-Hydrogen Bond Functionalization. *Acs Catal.* **2018**, *8*, 4313–4322. [CrossRef]

32. Conde, A.; Sabenya, G.; Rodriguez, M.; Postil, V.; Luis, J.M.; Mar Diaz-Requejo, M.; Costas, M.; Pérez, P.J. Iron and Manganese Catalysts for the Selective Functionalization of Arene C(sp^2)-H Bonds by Carbene Insertion. *Angew. Chem Int. Ed.* **2016**, *55*, 6530–6534. [CrossRef] [PubMed]

33. Padwa, A.; Austin, D.J.; Price, A.T.; Semones, M.A.; Doyle, M.P.; Protopopova, M.N.; Winchester, W.P.; Tran, A. Ligand effects on dirhodium(II) carbene reactivities. Highly effective switch between competitive carbenoid transformations. *J. Am. Chem. Soc.* **1993**, *115*, 8669–8680. [CrossRef]

34. Mbuvi, H.M.; Woo, L.K. Catalytic C-H Insertions Using Iron(III) Porphyrin Complexes. *Organometallics* **2008**, *27*, 637–645. [CrossRef]

35. Yang, J.-M.; Cai, Y.; Zhu, S.-F.; Zhou, Q.-L. Iron-catalyzed arylation of α-aryl-α -diazoesters. *Org. Biomol. Chem.* **2016**, *14*, 5516–5519. [CrossRef]

36. Wang, B.; Howard, I.G.; Pope, J.W.; Conte, E.D.; Deng, Y. Bis(imino)pyridine iron complexes for catalytic carbene transfer reactions. *Chem. Sci.* **2019**, *10*, 7958–7963. [CrossRef]

37. Baumann, L.K.; Mbuvi, H.M.; Du, G.; Woo, L.K. Iron Porphyrin Catalyzed N-H Insertion Reactions with Ethyl Diazoacetate. *Organometallics* **2007**, *26*, 3995–4002. [CrossRef]

38. Cai, Y.; Zhu, S.-F.; Wang, G.-P.; Zhou, Q.-L. Iron-Catalyzed C-H Functionalization of Indoles. *Adv. Synth. Catal.* **2011**, *353*, 2939–2944. [CrossRef]

39. Hock, K.J.; Knorrscheidt, A.; Hommelsheim, R.; Ho, J.; Weissenborn, M.J.; Koenigs, R.M. Tryptamine Synthesis by Iron Porphyrin Catalyzed C-H Functionalization of Indoles with Diazoacetonitrile. *Angew. Chem. Int. Ed.* **2019**, *58*, 3630–3634. [CrossRef]

40. Wang, B.; Qui, D.; Zhang, Y.; Wang, J. Recent advantages in C(sp^3)-H bond functionalization via metal-carbene insertion. *Beilstein J. Org. Chem.* **2016**, *12*, 796–804. [CrossRef]

41. He, J.; Harmann, L.G.; Davies, H.M.L.; Beckwith, R.E.J. Late-stage C-H functionalization of complex alkaloids and drug molecules via intramolecular rhodium-carbenoid insertion. *Nat. Commun.* **2015**, *6*, 5943–5952. [CrossRef]

42. Chen, Q.-Q.; Yang, J.-M.; Xu, H.; Zhu, S.-F. Iron-Catalyzed Carbenoid Insertion into C(sp^3)-H Bonds. *Synlett* **2017**, *28*, 1327–1330.

43. Griffin, J.R.; Wendell, C.I.; Garwin, J.A.; White, M.C. Catalytic C(sp^3)-H Alkylation via an Iron Carbene Intermediate. *J. Am. Chem. Soc.* **2017**, *139*, 13624–13627. [CrossRef]

44. Hernán-Gómez, A.; Rodriges, M.; Parella, T.; Costas, M. Electrophilic Iron Catalyst Paired with a Lithium Cation Enables Selective Functionalization of Non-Activated Aliphatic C-H Bonds via Metallocarbene Intermediates. *Angew. Chem. Int. Ed.* **2019**, *58*, 13904–13911. [CrossRef] [PubMed]

45. Brandenberg, O.F.; Fasan, R.; Arnold, F.H. Exploiting and engineering hemoproteins for abiological carbene and nitrene transfer reactions. *Curr. Opin. Biotechnol.* **2017**, *47*, 102–111. [CrossRef] [PubMed]

46. Weissenborn, M.J.; Koenigs, R.M. Iron-porphyrin catalyzed carbene transfer reactions—An evolution from biomimetic catalysis towards chemistry-inspired non-natural reactivities of enzymes. *ChemCatChem* **2019**. [CrossRef]

47. Vargas, D.V.; Tinoco, A.; Tyagi, V.; Fasan, R. Myoglobin-Catalyzed C-H Functionalization of Unprotected Indoles. *Angew. Chem. Int. Ed.* **2018**, *57*, 9911–9915. [CrossRef]

48. Brandenberg, O.F.; Chen, K.; Arnold, F.H. Directed Evolution of a Cytochrome P450 Carbene Transferase for Selective Functionalization of Cyclic Compounds. *J. Am. Chem. Soc.* **2019**, *141*, 8989–8995. [CrossRef]

49. Zhang, R.K.; Chen, K.; Huang, X.; Wohlschlager, L.; Renata, H.; Arnold, F.H. Enzymatic assembly of carbon-carbon bonds via iron-catalysed *sp^3* C-H functionalization. *Nature* **2019**, *565*, 67–72. [CrossRef]

50. Zhang, J.; Huang, X.; Zhang, R.K.; Arnold, F.H. Enantiodivergent α-Amino C-H Fluoroalkylation Catalysed by Engineered Cytochrome P450s. *J. Am. Chem. Soc.* **2019**, *141*, 9798–9802. [CrossRef]

 *molecules*

Communication

Cyclopentadienone Iron Tricarbonyl Complexes-Catalyzed Hydrogen Transfer in Water

Daouda Ndiaye [1,2], Sébastien Coufourier [1], Mbaye Diagne Mbaye [2], Sylvain Gaillard [1] and Jean-Luc Renaud [1,*]

[1] Normandie Univ., LCMT, ENSICAEN, UNICAEN, CNRS, 6 boulevard du Maréchal Juin, 14050 Caen, France; daouda.ndiaye79@gmail.com (D.N.); coufourier.sebastien@gmail.com (S.C.); sylvain.gaillard@ensicaen.fr (S.G.)

[2] Université Assane Seck de Ziguinchor, BP 523, Ziguinchor, Senegal; mmbaye@univ-zig.sn

* Correspondence: jean-luc.renaud@ensicaen.fr; Tel.: +33-231-452842

Academic Editor: Hans-Joachim Knölker
Received: 30 December 2019; Accepted: 17 January 2020; Published: 20 January 2020

Abstract: The development of efficient and low-cost catalytic systems is important for the replacement of robust noble metal complexes. The synthesis and application of a stable, phosphine-free, water-soluble cyclopentadienone iron tricarbonyl complex in the reduction of polarized double bonds in pure water is reported. In the presence of cationic bifunctional iron complexes, a variety of alcohols and amines were prepared in good yields under mild reaction conditions.

Keywords: iron complexes; hydrogen transfer; reductive amination; alcohols; amines

1. Introduction

The reduction of polarized C=X bonds is an important process, both in industry and in academia, for the synthesis of fine chemicals, perfumes, agrochemicals, and pharmaceuticals [1–8]. To avoid the use of stoichiometric amounts of borohydrides or aluminium hydrides, metal-catalyzed pathways to amines and alcohols have been introduced [9–14]. These procedures consist of hydrogenation, hydrosilylation, and transfer hydrogenation (with *iso*-propanol or formic acid) and involve mainly platinum complexes [9,10], but recent contributions highlighted the rise of Earth-abundant metals for such reductions [11–14]. Hydrogenation is the most atom economical approach, but requires hydrogen gas handling and consequently implies some safety issues. Hydrogen transfer (TH) is an alternative pathway and a more practical tool. Alcohols and formic acid (or formates) are among the most advantageous hydride donors.

Water is a non-toxic, non-flammable, non-explosive and also an economically relevant solvent [15,16]. Water-soluble organometallic complexes have attracted some interest because of the environmentally acceptable process, the simple product separation and, in some reactions, the possibility to control the selectivity by adjusting the pH [15,16]. Despite these advantages, the use of water in catalysis, and more specifically, in reduction, still constitutes a challenge and is underexplored compared to organic solvent [17,18]. Hydrogenation of ketones and imines [19], and reductive amination [20] in water have been reported with few iron complexes. As an example; our group has disclosed the first water-soluble and well-defined cyclopentadienone iron complex able to catalyze the reduction of aldehydes, ketones, and 2-substituted dihydroisoquinolines in pure water at 85–100 °C under hydrogen pressure (Figure 1, for the first synthesis of a water-soluble cyclopentadienone iron complex, see [19]). Little is known on the transfer hydrogenation with Earth-abundant complexes, while formates are used by enzymes for enantioselective reduction. To the best of our knowledge, excepted the hydrogen transfer reduction of heterocyclic compounds with formic acid catalyzed by a cobalt-phosphines complex [20], no reduction of polarized C=X bonds (aldehydes, ketones, and imines) with formic acid derivatives has been yet reported. Toward this objective, we thought of developing new water-soluble non-phosphine ligand iron complexes for the reduction of polarized bonds in the presence of formates or formic acid in pure water.

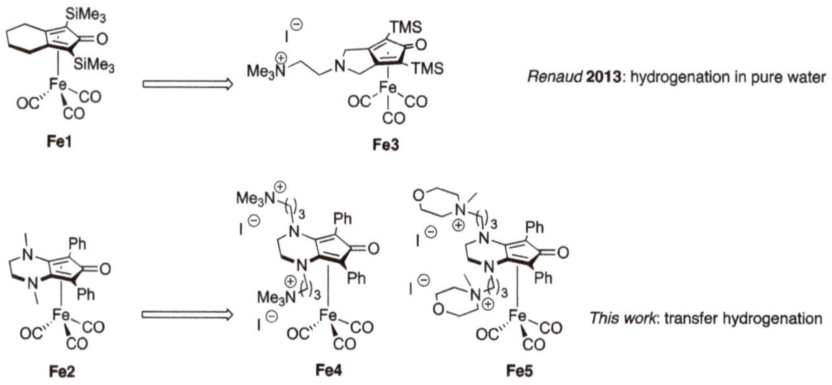

Figure 1. Previous water-soluble cyclopentadienone iron complex and new complexes for application reduction.

In 2015, we introduced in catalysis the tricarbonyl iron complex **Fe2** bearing a diaminocyclopentadienone ligand [21]. Compared to other cyclopentadienone iron carbonyl complexes, this phosphine-free iron complex has, to the best of our knowledge, the highest catalytic activities to date in reductive amination [21], in chemoselective reduction of α,β-unsaturated ketones [22], in the hydrogenation of carbon dioxide [23], in alkylation of ketones [24–27], amines [25,28], oxindoles [29], indoles [25,30] and alcohols [31,32]. In our ongoing interest in reduction and alkylation, we thought that a water-soluble analog of **Fe2** would be more active than our previous water-soluble cyclopentadienone iron complex **Fe3** (Figure 1). In this work, we report on the synthesis and application of two water-soluble cyclopentadienone iron complexes in the reduction of aldehydes and in reductive amination in pure water.

2. Results and Discussion

2.1. Synthesis of Complexes

To develop water-soluble iron complexes, we selected a diaminocyclopentadienone ligand bearing ammonium functionalities [33]. The tetraamines **2** and **4** were prepared from diethyloxalate and *N,N*-dimethylpropylenediamine and *N*-aminopropylenemorpholine via an amidation followed by a reduction in good overall yield (93 and 95%, respectively). The corresponding aminocyclopentadienone ligands **1** and **2** were then prepared by reacting the amines **2** and **4** with the cyclopentatrienone in refluxing methanol for 16 h and were isolated in moderate yield (62% and 88%, respectively, Scheme 1). The complexes **Fe6** and **Fe7** were synthesized in 48% and 76% yield by simple heating of the corresponding amino ligand with [Fe$_2$(CO)$_9$] in refluxing toluene (Scheme 1). Finally, the water-soluble bifunctional iron complexes **Fe4** and **Fe5** bearing ionic frameworks were obtained in almost quantitative yields after a subsequent alkylation of the pendant amines with iodomethane (Scheme 1) [33]. These complexes were fully characterized by ^{1}H-, ^{13}C-NMR, and IR spectroscopies (see Supplementary Materials). These analyses showed that complexes **Fe2** and **Fe4-Fe7** have similar features. The back donation from the metal center to the CO ligands is more significant than in the Knölker's complex **Fe1** [34,35]. Thus, the CO stretching frequencies were at 2032, 1961, and 1919 cm^{-1} and at 2015 and 1967 cm^{-1} in the neutral complexes **Fe6** and **Fe7**, respectively, at 2038 and 1955 cm^{-1} and at 2034 and 1957 cm^{-1} in the ionic complexes **Fe4** and **Fe5**, respectively. These frequencies are comparable to those of the analog **Fe2** (2027, 1962 and 1947 cm^{-1}) and lower than those of the Knölker's complex **Fe1** (2061, 2053, and 1987 cm^{-1}) or its water-soluble analog **Fe3** (2066, 2016, and 1996 cm^{-1}) [19].

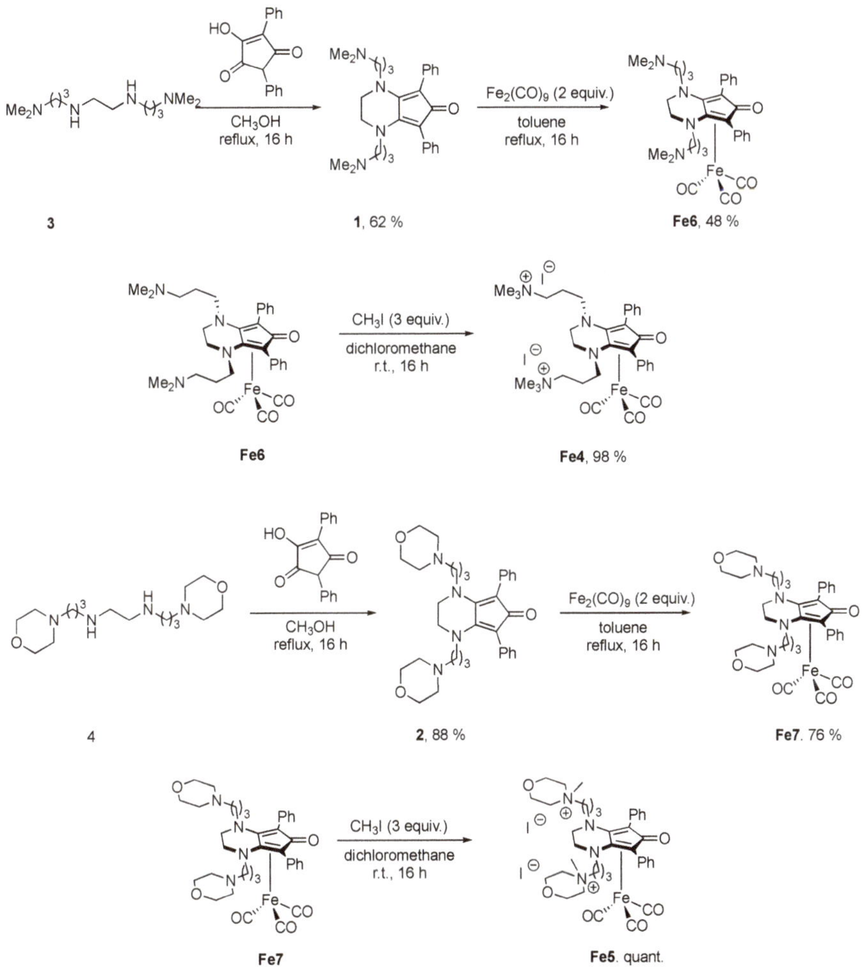

Scheme 1. Synthesis of the iron complexes **Fe4**, **Fe5**, **Fe6**, and **Fe7**.

2.2. Iron-Catalyzed Reduction of Carbonyl Compounds

With these complexes in hands, we evaluated their catalytic activities in the reduction of 4-methoxybenzaldehyde as a benchmark reaction. Various methods of activation can be used with the cyclopentadienone iron carbonyl complexes [34–44]. We applied in this work the activation with Me$_3$NO as oxidant [34–42]. Our first attempt with formic acid, in the presence of 2 mol % of **Fe4** and 2.5 mol % of Me$_3$NO at 100 °C for 24 h in 2 mL of pure water (concentration of 0.5 M), was unsuccessful as no reduction was noticed (entry 1, Table 1). In sharp contrast, in the same reaction conditions, complete conversions were obtained with different formate salts (entries 2–5, Table 1). Without a hydride donor or iron complex, no reduction occurred (entries 6–7, Table 1). Decreasing the reaction time (entries 8 and 11, Table 1), the catalyst loading (entry 13) and the amount of formate (entry 14) led to a drop in the conversion. No variation of the conversion was noticed by lowering the temperature to 80 °C (entries 3 and 9, Table 1), while, at 60 °C, the conversion was only 75% (entry

12). To our surprise, the first generation water-soluble cyclopentadienone iron complex **Fe3** did not catalyze the reduction of 4-methoxybenzaldehyde in these conditions, while **Fe5** appeared as active as **Fe4** (entries 10–15, Table 1). Finally, the best conditions for the reduction of 4-methoxybenzaldehyde into the corresponding alcohol **4a** were: 1 mmol of aldehyde, in the presence of five equivalents of sodium formate in 2 mL of water, 2 mol % of **Fe4** or **Fe5** and 2.5 mol % of Me$_3$NO at 80 °C for 24 h.

Table 1. Optimization of the reaction conditions for the aldehyde reduction [a].

Entry	HCO$_2$X	Fe	Temperature (°C)	Time (h)	Conv. (%) [b]
1	HCO$_2$H	Fe4	100	24	0
2	HCO$_2$H/Et$_3$N (1/1)	Fe4	100	24	100
3	HCO$_2$Na	Fe4	100	24	100
4	HCO$_2$K	Fe4	100	24	100
5	HCO$_2$Cs	Fe4	100	24	100
6	-	Fe4	100	24	0
7	HCO$_2$Na	-	100	24	0
8	HCO$_2$Na	Fe4	100	16	83
9	HCO$_2$Na	Fe4	80	24	100 (99%) [c]
10	HCO$_2$Na	Fe3	80	24	0
11	HCO$_2$Na	Fe4	80	16	81
12	HCO$_2$Na	Fe4	60	24	75
13 [d]	HCO$_2$Na	Fe4	80	24	53
14 [e]	HCO$_2$Na	Fe4	80	24	86
15	HCO$_2$Na	Fe5	80	24	100 (98%) [c]

[a] General conditions: HCO$_2$X (5 mmol, 5 equiv.), 4-methoxybenzaldehyde (1 mmol), pre-catalyst (2 mol %), Me$_3$NO (2.5 mol %), water (2 mL). [b] Conversion was determined by ^{1}H-NMR spectroscopy analysis. [c] Isolated yield in the bracket. [d] **Fe** 0(1 mol %), Me$_3$NO (1.25 mol %) were used. [e] HCO$_2$Na (3 mmol, three equiv.) were used.

Having established the optimized conditions, we delineated the scope of the carbonyl derivatives (Table 2). Both electron-donating (methoxy, methyl, acetal substituents) and electron-withdrawing (nitro, nitrile, and ester substituents) groups were tolerated in this reduction. The corresponding alcohols **4a–m** were isolated in excellent yields in all examples (91–99%, Table 2). No reduction of halogen-carbon bonds in the substituted phenyl group (compounds **4i–j**) was observed. Other reducible functions, such as ester or nitrile, were preserved in these conditions. Heteroaromatic derivatives, such as pyridine or thiophene carboxaldehyde, furfural, did not impede the catalytic activity and provided the alcohols **4n–r** in 75–98%yield (Table 2). Finally, to extend the scope, aliphatic aldehydes were also engaged in this reduction and the corresponding alcohols **4s–v** were isolated in 92–99% yield (Table 2). It is worth to mention that (i) ethanol was used as a co-solvent with some substrates to facilitate the solubility and consequently enhanced the reactivity; and (ii) no reaction occurred in a mixture of water and ethanol without sodium formate.

Table 2. Iron-catalyzed reduction of aldehydes with sodium formate [a].

$$\text{(Het)Aryl, Alkyl}-\text{CHO} \quad \xrightarrow[\text{H}_2\text{O, 80 °C, 24 h}]{\substack{\text{HCO}_2\text{Na (5 equiv.)} \\ \textbf{Fe4} \text{ (2 mol\%)} \\ \text{Me}_3\text{NO (2.5 mol\%)}}} \quad \text{(Het)Aryl, Alkyl}-\text{CH}_2\text{OH}$$

4a, 99 %   **4b**, 98 %   **4c**, 91 %   **4d**, 97 %[b]   **4e**, 95 %[b]

4f, 99 %[b]   **4g**, 96 %[b]   **4h**, 99 %[b]   **4i**, X = Br, 75 %   **4j**, X = Cl, 92 %   **4k**, 96 %

4l, 98 %   **4m**, 96 %   **4n**, 82 %   **4o**, 75 %   **4p**, 91 %

4q, Y = O, 98 %   **4r**, Y = S, 91 %   **4s**, 99 %[b]   **4t**, 99 %[b]   **4u**, 98 %[b]   **4v**, 92 %[b]

[a] General conditions: aldehyde (1 mmol), HCO$_2$Na (5 mmol, 5 equiv.), pre-catalyst **Fe4** (2 mol %), Me$_3$NO (2.5 mol %), water (2 mL). [b] H$_2$O/EtOH 1/1.

2.3. Iron-Catalyzed Reductive Amination

Having established a simple protocol for the reduction of aldehydes in water, we thought to extend this work to the synthesis of amines. Amines are usually prepared via the reduction of C=N bonds either in catalytic conditions under hydrogen pressure or in stoichiometric conditions in the presence of aluminum/boron hydride [45]. However, imines are not always easily prepared and cannot be stable. Reductive amination of aldehydes constitutes a direct route to amines, without requiring any purification of the imine intermediate. Many efforts have been devoted to the development of reductive amination [46–48]. For example, in iron chemistry, Bhanage described that a combination of iron sulfate and ethylenediaminetetraacetic acid (EDTA) catalyzed a reductive amination under hydrogen pressure (400 psi) in water at elevated temperatures (150 °C) [20]. Beller reported a reductive amination with anilines catalyzed by Fe$_2$(CO)$_9$ under high hydrogen pressure and elevate temperature [49]. We have reported that cyclopentadienone iron tricarbonyl complexes [21,34,35] or cyclopentadienyl iron(II) tricarbonyl complex [50] were able to catalyzed the reductive alkylation of various amines and carbonyl derivatives under 5 bar of hydrogen, at 40–70 °C and even room temperature. To avoid the use of a large amount of hydride (and the concomitant formation of wastes) or the handling of gas, hydrogen transfer with formate derivatives appears as a simple and versatile procedure. The reductive alkylation of *N*-methylbenzylamine with citronellal was chosen as a model reaction for the optimization of the reaction conditions. Three formate salts were tested, and the cation appeared to be crucial for the catalytic activity (entries 1–4, Table 3). Indeed, the ammonium favored both the condensation and the

suppliers and used as received. Deuterated solvents for NMR spectroscopy were purchased from Sigma Aldrich (Saint-Quentin Fallavier, France) and used as received. NMR spectra were recorded on a 500 MHz Bruker spectrometer. Proton (^{1}H) NMR information is given in the following format: multiplicity (s, singlet; d, doublet; t, triplet; q, quartet; m, multiplet), coupling constant(s) (J) in Hertz (Hz), number of protons, type. The prefix *app* is occasionally applied when the true signal multiplicity was unresolved, and *br* indicates the signal in question broadened. Carbon (^{13}C) NMR spectra are reported in ppm (δ) relative to CDCl$_3$ unless noted otherwise. Infrared spectra were recorded over a PerkinElmer Spectrum 100 FT-IR Spectrometer using neat conditions. HRMS analyses were performed by using the Laboratoire de Chimie Moléculaire et Thioorganique analytical Facilities.

3.1. General Procedure for the Reduction of Aldehydes

In a dried flamed Schlenk tube under argon, the corresponding aldehyde (1 equiv.) and sodium formate (5 equiv.) were mixed in water (0.5 M solution). The iron complex **Fe4** (2 mol %) and Me$_3$NO (2.5 mol %) were then added. The mixture was stirred and heated at 80 °C for 24 h. After cooling down to room temperature, the resulting solution was quenched with a saturated aqueous solution of sodium bicarbonate and extracted three times with ethyl acetate. The organic phase was dried over MgSO$_4$, filtrated, and concentrated under vacuum to afford the crude product. Purification by flash chromatography on silica gel furnished the alcohol.

3.2. General Procedure for the Reductive Amination of Aldehydes

In a dried flamed Schlenk tube under argon, the aldehyde (1 equiv.), the amine (2 equiv.) and ammonium formate (6.5 equiv.) were mixed in water (0.5 M solution). The iron complex **Fe5** (2 mol %) and Me$_3$NO (2.5 mol %) were then added. The mixture was stirred and heated at 90 °C for 24–48 h. After cooling down to room temperature, the resulting solution was quenched with a saturated aqueous solution of sodium bicarbonate and extracted three times with ethyl acetate. The organic phase was dried over MgSO$_4$, filtrated, and concentrated under vacuum to afford the crude product. Purification by flash chromatography on silica gel furnished the amine.

4. Conclusions

In conclusion, we have described the application of water-soluble cyclopentadienone iron tricarbonyl complexes in the reduction of aldehydes and in reductive amination under hydride transfer conditions in pure water. Recyclability of iron complex **Fe5** was also demonstrated in a model reductive amination. This system tolerated a variety of functional groups such as halides, ethers, heteroaromatic derivatives without impeding the chemical yields. These water-soluble iron complexes allow an efficient, green, and practical procedure for the synthesis of amines and alcohols.

Supplementary Materials: The following are available online. Table S1: Optimization of the reaction conditions for aldehyde reduction by hydride transfer. Table S2: Optimization of the reaction conditions for reductive amination by hydride transfer. Figure S1–S64: The ^{1}H, ^{13}C, and ^{19}F-NMR spectra of compounds.

Author Contributions: Conceptualization—J.-L.R. and S.C.; methodology—J.-L.R.; validation—J.-L.R., S.G. and M.D.M.; formal analysis—S.C. and D.N.; investigation—S.C. and D.N.; resources—J.-L.R.; data curation—S.C. and D.N.; writing—original draft preparation—J.-L.R.; writing—review and editing—J.-L.R., S.G. and M.D.M.; visualization—J.-L.R., S.G. and M.D.M.; supervision—J.-L.R.; project administratio—J.-L.R.; funding acquisition—J.-L.R. and M.D.M. All authors have read and agreed to the published version of the manuscript.

Funding: This research received no external funding.

Acknowledgments: We gratefully acknowledge financial support from the "Ministère de la Recherche et des Nouvelles Technologies", Normandie Université, CNRS, "Région Normandie", and the LABEX SynOrg (ANR-11-LABX-0029). Ademe Agency is acknowledged for a grant to S. C. and Coopération Française-Sénégal for a grant to D. N.

Conflicts of Interest: The authors declare no conflict of interest.

References

1. Klingler, F.D. Asymmetric Hydrogenation of Prochiral Amino Ketones to Amino Alcohols for Pharmaceutical Use. *Accounts Chem. Res.* **2007**, *40*, 1367–1376. [CrossRef]
2. Hems, W.P.; Groarke, M.; Zanotti-Gerosa, A.; Grasa, G.A. [(Bisphosphine) Ru(II) Diamine] Complexes in Asymmetric Hydrogenation: Expanding the Scope of the Diamine Ligand. *Accounts Chem. Res.* **2007**, *40*, 1340–1347. [CrossRef] [PubMed]
3. Johnson, N.B.; Lennon, I.C.; Moran, P.H.; Ramsden, J.A. Industrial-Scale Synthesis and Applications of Asymmetric Hydrogenation Catalysts. *Accounts Chem. Res.* **2007**, *40*, 1291–1299. [CrossRef] [PubMed]
4. Saudan, L.A. Hydrogenation Processes in the Synthesis of Perfumery Ingredients. *Accounts Chem. Res.* **2007**, *40*, 1309–1319. [CrossRef] [PubMed]
5. Shimizu, H.; Nagasaki, I.; Matsumura, K.; Sayo, N.; Saito, T. Developments in Asymmetric Hydrogenation from an Industrial Perspective. *Accounts Chem. Res.* **2007**, *40*, 1385–1393. [CrossRef] [PubMed]
6. Blaser, H.-U.; Studer, M. Cinchona-Modified Platinum Catalysts: From Ligand Acceleration to Technical Processes. *Accounts Chem. Res.* **2007**, *40*, 1348–1356. [CrossRef]
7. Blaser, H.-U.; Pugin, B.; Spindler, F.; Thommen, M. From a Chiral Switch to a Ligand Portfolio for Asymmetric Catalysis. *Accounts Chem. Res.* **2007**, *40*, 1240–1250. [CrossRef]
8. Shultz, C.S.; Krska, S.W. Unlocking the Potential of Asymmetric Hydrogenation at Merck. *Accounts Chem. Res.* **2007**, *40*, 1320–1326. [CrossRef]
9. Wang, N.; Astruc, D. The Golden Age of Transfer Hydrogenation. *Chem. Rev.* **2015**, *115*, 6621–6686. [CrossRef]
10. Wang, C.; Wu, X.; Xiao, J. Broader, Greener, and More Efficient: Recent Advances in Asymmetric Transfer Hydrogenation. *Chem. Asian J.* **2008**, *3*, 1750–1770. [CrossRef]
11. Wei, D.; Darcel, C. Iron Catalysis in Reduction and Hydrometalation Reactions. *Chem. Rev.* **2018**, *119*, 2550–2610. [CrossRef] [PubMed]
12. Liu, W.; Sahoo, B.; Junge, K.; Beller, M. Cobalt Complexes as an Emerging Class of Catalysts for Homogeneous Hydrogenations. *Accounts Chem. Res.* **2018**, *51*, 1858–1869. [CrossRef] [PubMed]
13. Ai, W.; Zhong, R.; Liu, X.; Liu, Q. Hydride Transfer Reactions Catalyzed by Cobalt Complexes. *Chem. Rev.* **2018**, *119*, 2876–2953. [CrossRef] [PubMed]
14. Filonenko, G.A.; Van Putten, R.; Hensen, E.J.M.; Pidko, E.A. Catalytic (de)hydrogenation promoted by non-precious metals—Co, Fe and Mn: Recent advances in an emerging field. *Chem. Soc. Rev.* **2018**, *47*, 1459–1483. [CrossRef] [PubMed]
15. Lindström, U.M. Stereoselective Organic Reactions in Water. *Chem. Rev.* **2002**, *102*, 2751–2772. [CrossRef]
16. Cornils, B.; Hermann, W.A. *Aqueous-Phase Organometallic Catalysis*; Wiley-VCH: Weinheim, Germany, 2002.
17. Robertson, A.; Matsumoto, T.; Ogo, S. The development of aqueous transfer hydrogenation catalysts. *Dalton Trans.* **2011**, *40*, 10304–10310. [CrossRef]
18. Wu, X.; Xiao, J. Aqueous Phase Asymmetric transfer hydrogenation of ketones-a greener approach to chiral alcohols. *Chem. Commun.* **2007**, *24*, 2449–2466. [CrossRef]
19. Mérel, D.S.; Elie, M.; Lohier, J.-F.; Gaillard, S.; Renaud, J.-L. Bifunctional Iron Complexes: Efficient Catalysts for C=O and C=N Reduction in Water. *ChemCatChem* **2013**, *5*, 2939–2945. [CrossRef]
20. Bohr, M.D.; Bhanushali, M.J.; Nandurkar, N.S.; Bhanage, B.M. Direct Reductive Amination of Carbonyl Compounds with Primary/Secondary Amines Using Recyclable Water-Soluble FeII/EDTA Complex as Catalyst. *Tetrahedron Lett.* **2008**, *49*, 965–969. [CrossRef]
21. Thai, T.-T.; Mérel, D.S.; Poater, A.; Gaillard, S.; Renaud, J.-L. Highly active phosphine-free bifunctional iron complex for hydrogenation of bicarbonate and reductive amination. *Chem. Eur. J.* **2015**, *21*, 7066–7070. [CrossRef]
22. Lator, A.; Gaillard, S.; Poater, A.; Renaud, J.-L. Iron-Catalyzed Chemoselective Reduction of α,β-Unsaturated Ketones. *Chem. Eur. J.* **2018**, *24*, 5770–5774. [CrossRef] [PubMed]
23. Coufourier, S.; Gaillard, S.; Clet, G.; Serre, C.; Daturi, M.; Renaud, J.-L. A MOF-assisted phosphine free bifunctional iron complex for the hydrogenation of carbon dioxide, sodium bicarbonate and carbonate to formate. *Chem. Commun.* **2019**, *55*, 4977–4980. [CrossRef] [PubMed]
24. Seck, C.; Mbaye, M.D.; Coufourier, S.; Lator, A.; Lohier, J.-F.; Poater, A.; Ward, T.R.; Gaillard, S.; Renaud, J. Alkylation of Ketones Catalyzed by Bifunctional Iron Complexes: From Mechanistic Understanding to Application. *ChemCatChem* **2017**, *9*, 4410–4416. [CrossRef]

25. Polidano, K.; Allen, B.D.W.; Williams, J.M.J.; Morrill, L.C. Iron-Catalyzed Methylation Using the Borrowing Hydrogen Approach. *ACS Catal.* **2018**, *8*, 6440–6445. [CrossRef]

26. Bettoni, L.; Seck, C.; Mbaye, M.D.; Gaillard, S.; Renaud, J.-L. Iron-Catalyzed Tandem Three-Component Alkylation: Access to α-Methylated Substituted Ketones. *Org. Lett.* **2019**, *21*, 3057–3061. [CrossRef]

27. Latham, D.E.; Polidano, K.; Williams, J.M.J.; Morrill, L.C. One-Pot Conversion of Allylic Alcohols to α-Methyl Ketones via Iron-Catalyzed Isomerization-Methylation. *Org. Lett.* **2019**, *21*, 7914–7918. [CrossRef]

28. Lator, A.; Gaillard, S.; Poater, A.; Renaud, J.-L. Well-Defined Phosphine-Free Iron-Catalyzed N-Ethylation and N-Methylation of Amines with Ethanol and Methanol. *Org. Lett.* **2018**, *20*, 5985–5990. [CrossRef]

29. Dambatta, M.B.; Polidano, K.; Northey, A.D.; Williams, J.M.J.; Morrill, L.C. Iron-Catalyzed Borrowing Hydrogen C-Alkylation of Oxindoles with Alcohols. *ChemSusChem* **2019**, *12*, 2345–2349. [CrossRef]

30. Seck, C.; Mbaye, M.D.; Gaillard, S.; Renaud, J.-L.; Seck-Diouf, C. Bifunctional Iron Complexes Catalyzed Alkylation of Indoles. *Adv. Synth. Catal.* **2018**, *360*, 4640–4645. [CrossRef]

31. Bettoni, L.; Gaillard, S.; Renaud, J.-L. Iron-Catalyzed β-Alkylation of Alcohols. *Org. Lett.* **2019**, *21*, 8404–8408. [CrossRef]

32. Polidano, K.; Williams, J.M.J.; Morrill, L.C. Iron-Catalyzed Borrowing Hydrogen β-C(sp3)-Methylation of Alcohols. *ACS Catal.* **2019**, *9*, 8575–8580. [CrossRef]

33. Coufourier, S.; Gaillard, Q.G.; Lohier, J.-F.; Poater, A.; Gaillard, S.; Renaud, J.-L. Hydrogenation of CO2, Hydrogenocarbonate and Carbonate to Formate in Water using Phosphine Free Bifunctional Iron Complexes. *ACS Catal.* **2020**. [CrossRef]

34. Pagnoux-Ozherelyeva, A.; Pannetier, N.; Mbaye, D.M.; Gaillard, S.; Renaud, J.-L. Knölker's Iron Complex: An Efficient In Situ Generated Catalyst for Reductive Amination of Alkyl Aldehydes and Amines. *Angew. Chem. Int. Ed.* **2012**, *51*, 4976–4980. [CrossRef] [PubMed]

35. Moulin, S.; Dentel, H.; Gaillard, S.; Poater, A.; Cavallo, L.; Lohier, J.-F.; Renaud, J.-L.; Pagnoux-Ozherelyeva, A. Bifunctional (Cyclopentadienone)Iron-Tricarbonyl Complexes: Synthesis, Computational Studies and Application in Reductive Amination. *Chem. A Eur. J.* **2013**, *19*, 17881–17890. [CrossRef] [PubMed]

36. Luh, T.-Y. Trimethylamine N-oxide—A versatile reagent for organometallic chemistry. *Coord. Chem. Rev.* **1984**, *60*, 255–276. [CrossRef]

37. Moyer, S.A.; Funk, T.W. Air-Stable iron catalyse for the oppenauer type oxidation of alcohols. *Tetrahedron Lett.* **2010**, *51*, 5430–5433. [CrossRef]

38. Johnson, T.C.; Clarkson, G.J.; Wills, M. (Cyclopentadienone)iron Shvo Complexes: Synthesis and applications to hydrogen transfer reactions. *Organometallics* **2011**, *30*, 1859–1868. [CrossRef]

39. Plank, T.N.; Drake, J.L.; Kim, D.K.; Funk, T.W. Air-Stable, Nitrile-Ligated (Cyclopentadienone)iron Dicarbonyl Compounds as Transfer Reduction and Oxidation Catalysts. *Adv. Synth. Catal.* **2012**, *354*, 597–601. [CrossRef]

40. Knölker, H.-J.; Heber, J. Transition Metal-Diene Complexes in Organic Synthesis, Part 18.1 Iron-Mediated [2 + 2 + 1] Cycloadditions of Diynes and Carbon Monoxide: Selective Demetalation Reactions. *Synlett* **1993**, *12*, 924–926. [CrossRef]

41. Knölker, H.-J.; Baum, E.; Heber, J. Transition Metal-Diene Complexes in Organic Synthesis, Part 25.1 Cycloadditions of Annulated 2,5-Bis(trimethylsilyl)cyclopentadienones. *Tetrahedron Lett.* **1995**, *36*, 7647–7650. [CrossRef]

42. Knölker, H. Trimethylamine N-Oxide—A useful oxidizing reagent. *J. Für Prakt. Chem.* **1996**, *338*, 190–192. [CrossRef]

43. Knölker, H.-J.; Goesmann, H.; Klauss, R. A Novel Method for the Demetalation of Tricarbonyliron-Diene Complexes by a Photolytically Induced Ligand Exchange Reaction with Acetonitrile. *Angew. Chem. Int. Ed.* **1999**, *38*, 702–705. [CrossRef]

44. Knölker, H.-J.; Baum, E.; Goesmann, H.; Klauss, R. Demetalation of Tricarbonyl(cyclopentadienone)iron Complexes Initiated by a Ligand Exchange Reaction with NaOH—X-Ray Analysis of a Complex with Nearly Square-Planar Coordinated Sodium. *Angew. Chem. Int. Ed.* **1999**, *38*, 2064–2066. [CrossRef]

45. Abdel-Magid, A.F.; Mehrman, S.J. A Review on the Use of Sodium Triacetoxyborohydride in the Reductive Amination of Ketones and Aldehydes. *Org. Process. Res. Dev.* **2006**, *10*, 971–1031. [CrossRef]

46. Nugent, T.C.; El-Shazly, M. Chiral amine synthesis—Recent developments and trends for enamide reduction, reductive amination, and imine reduction. *Adv. Synth. Catal.* **2010**, *352*, 753–819. [CrossRef]

47. Alinezhad, H.; Yavari, H.; Salehian, F. Recent Advances in Reductive Amination Catalysis and Its Applications. *Curr. Org. Chem.* **2015**, *19*, 1021–1049. [CrossRef]

48. Gusak, K.N.; Ignatovich, Z.V.; Koroleva, E.V. New potential of the reductive alkylation of amines. *Russ. Chem. Rev.* **2015**, *84*, 288–309. [CrossRef]
49. Fleischer, S.; Zhou, S.; Junge, K.; Beller, M. An Easy and General Iron-catalyzed Reductive Amination of Aldehydes and Ketones with Anilines. *Chem. Asian J.* **2011**, *6*, 2240–2245. [CrossRef]
50. Lator, A.; Gaillard, Q.G.; Mérel, D.S.; Lohier, J.-F.; Gaillard, S.; Poater, A.; Renaud, J.-L. Room-Temperature Chemoselective Reductive Alkylation of Amines Catalyzed by a Well-Defined Iron(II) Complex Using Hydrogen. *J. Org. Chem.* **2019**, *84*, 6813–6829. [CrossRef]

Sample Availability: Samples of the compounds are available from the authors.

Article

Computational Prediction of Chiral Iron Complexes for Asymmetric Transfer Hydrogenation of Pyruvic Acid to Lactic Acid

Wan Wang [1,2,3] **and Xinzheng Yang** [1,2,3,*]

[1]   State Key Laboratory for Structural Chemistry of Unstable and Stable Species,
      Institute of Chemistry Chinese Academy of Sciences, Beijing 100190, China; ww2016@iccas.ac.cn
[2]   University of Chinese Academy of Sciences, Beijing 100049, China
[3]   Beijing National Laboratory for Molecular Sciences, CAS Research/Education Center for Excellence in
      Molecular Sciences, Beijing 100190, China
*   Correspondence: xyang@iccas.ac.cn

Academic Editor: Hans-Joachim Knölker
Received: 29 February 2020; Accepted: 13 April 2020; Published: 20 April 2020

Abstract: Density functional theory calculations reveal a formic acid-assisted proton transfer mechanism for asymmetric transfer hydrogenation of pyruvic acid catalyzed by a chiral Fe complex, FeH[(R,R)-BESNCH(Ph)CH(Ph)NH$_2$](η^6-p-cymene), with formic acid as the hydrogen provider. The rate-determining step is the hydride transfer from formate anion to Fe for the formation and dissociation of CO_2 with a total free energy barrier of 28.0 kcal mol^{-1}. A series of new bifunctional iron complexes with η^6-p-cymene replaced by different arene and sulfonyl groups were built and computationally screened as potential catalysts. Among the proposed complexes, we found **1$_g$** with η^6-p-cymene replaced by 4-isopropyl biphenyl had the lowest free energy barrier of 26.2 kcal mol^{-1} and excellent chiral selectivity of 98.5% *ee*.

Keywords: asymmetric transfer hydrogenation; density functional theory; bifunctional catalyst

---

1. Introduction

The asymmetric hydrogenation of prochiral unsaturated ketones is an important process for the synthesis of enantiopure alcohols for use in the pharmaceutical, agrochemical, fragrance, and flavor industries [1–3]. As a promising alternative to reduce the use of gaseous hydrogen for the production of enantiopure alcohols, asymmetric transfer hydrogenation (ATH) from low-cost and safe hydrogen donors, usually NH moieties, to ketones catalyzed by organometallic complexes has attracted increasing attention in the past two decades [4–8]. A series of particularly efficient catalytic systems with high reactivity and enantioselectivity for ATH are Noyori and co-workers' chiral bifunctional Ru and Ir complexes, which are stabilized by soft ligands (η^6-arene [9,10] or phosphine [11,12]) and contain at least one hard NH electron donor that activates or directs the ketone toward hydride attack. However, most of the reported efficient catalysts for this reaction were based on noble metals such as Ru [13,14], Os [2,15], Rh [16,17], or Ir [6,18]. Base-metal catalysts, especially the cheaper and less-toxic iron catalyst with comparable activity, would be desirable in asymmetric catalysis. In 2008, Morris' group pioneered the field of Fe(II) catalysis with tetradentate PNNP ligands for the asymmetric hydrogenation of polar bonds and attained an excellent turnover frequency (TOF) of 907 h^{-1} but less activity on the enantioselectivity (29% *ee*) [19]. In 2015, Gao and co-workers [20] reviewed a series of iron, cobalt, and nickel catalysts containing novel chiral aminophosphine ligands for ATH or AH of ketones. Among those base metal catalysts, the P$_2$N$_4$ macrocycles in connection with Fe(0) precursors exhibited high yield (up to 96%) with extraordinary enantioselectivities (up to 99% *ee*) at 65 °C [20]. Mezzetti's group showed that stable, diamagnetic Fe(II) complexes bearing $(NH)_2P_2$ macrocyclic ligands catalyzed the ATH of a broad scope of ketones with up to 99.5% yield and excellent enantioselectivity (up to 98% *ee*) at 75 °C [21,22]. The molecules containing base-labile stereocenters would easily racemize under basic conditions [23]. They furthermore exploited the dicationic $(NH)_2P_2$ macrocyclic ligands to develop the corresponding iron hydride complex, which possessed the active H-Fe-N-H motif for the base-free ATH of ketones in a yield up to 99% with an enantioselectivity up to 99% at 50 °C (Scheme 1a) [24].

Scheme 1. Mezzetti's N_2P_2 macrocyclic iron complex (**a**) and the asymmetric transfer hydrogenation (ATH) of pyruvic acid to lactic acid with HCOOH as hydrogen donor catalyzed by iron hydride complex in this work (**b**).

Although promising progress in the iron-catalyzed enantioselective reduction of ketones has been achieved, most of the reported catalysts contained air and moisture-sensitive phosphine ligands, which require rather rigid conditions in the synthesis of catalyst. The development of highly active phosphine-free iron catalyst is therefore highly desirable. Herein, we proposed a series of chiral Fe complexes based on previously reported Ru and Os complexes with similar arene and Tsdiamine ligands [2,15,25] and computationally evaluated their catalytic activities and enantioselectivities for ATH of pyruvic acid (PA) to lactic acid using the density functional theory (DFT). As shown in Scheme 1b, the ATH reaction catalyzed by $FeH[(R,R)\text{-BESNCH(Ph)CH(Ph)NH}_2](\eta^6\text{-}p\text{-cymene})$(BES, benzenesulfonyl) [26] had formic acid (FA) as the hydrogen donor.

2. Results and Discussion

As illustrated in Figure 1, our DFT calculations revealed a proton-coupled hydride transfer mechanism for the ATH of PA catalyzed by $FeH[(R,R)\text{-BESNCH(Ph)CH(Ph)NH}_2]\text{-}(\eta^6\text{-}p\text{-cymene})$ (**1**) with formic acid as the hydrogen source. The optimized structures of key intermediates and transition states are displayed in Figure 2.

The catalytic cycle began with the approach of a PA molecule to **1** from its *Re*-face for the formation of a 2.7 kcal mol^{-1} less-stable prochiral intermediate ***pro*-2$_R$**. The iron hydride and FA proton in ***pro*-2$_R$** simultaneously transferred to the carbonyl group in PA via transition state **TS$_{2,3\text{-}R}$** to form a D-lactic acid (D-LA) molecule in **3$_R$**, which is a stable solvent-shared ion pair [27] between the cationic Fe complex and the formate anion mediated by D-LA. The C-H$\cdots$Fe distance of 2.73 Å in **3$_R$** suggested a very weak interaction between Fe and the transferred hydride. This process was 26.3 kcal mol^{-1} exothermic and accompanied by the moving of the Fe-N-N plane to a position perpendicular to the arene ring on top of Fe. The dissociation of D-LA and formate anion from **3$_R$** for the formation of cationic Fe complex **4** was a 4.7 kcal mol^{-1} uphill step. The formation of L-lactic acid (L-LA) went through a similar process but had slightly different relative energies. The enantio-determining step (EDS) for the formation of L-LA (**TS$_{2,3\text{-}S}$**) was 3.2 kcal mol^{-1} higher than that of **TS$_{2,3\text{-}R}$**, which suggested a chiral selectivity of 99.1% *ee*. The dissociated formate then came back to **4** and formed a slightly more stable intermediate **5** with a hydride coordinated to the Fe center. The transition state for hydride transfer from formate to Fe (**TS$_{5,1}$**) for the formation and release of CO_2 was 18.5 kcal mol^{-1} higher than that of **5**. Alternatively, the dissociated formate anion could have also coordinated to the Fe center with one of its oxygen atoms and formed a much more stable complex **5′** with a strong Fe-O bond (2.21 Å). According to the energy span model [28], **5′** and the **TS$_{5,1}$** were the rate-determining states (RDS) with a free energy barrier of 28.0 kcal mol^{-1}.

Noyori's work further shows that the application of ligand ArSO$_2$ group [29] and the electron donor alkyl group (CH$_3$, *i*-C$_3$H$_7$) on the η^6-arene [30] are important for maintaining catalytic activity and enantioselectivity. In the hope of discovering more active and enantioselective iron catalysts, we investigated the influence of substituents on the sulfonyl and arene groups for catalytic ability and enantioselectivity. As shown in Scheme 2, we built a series of asymmetric iron complexes **1$_a$-1$_g$** (Table 1) by replacing the phenyl ligand in the sulfonyl group with methyl (**R$_{1a}$**), 4-methylphenyl (**R$_{1b}$**), 4-nitrophenyl (**R$_{1c}$**), and 4-fluorophenyl (**R$_{1d}$**) and replacing the η^6-*p*-cymene ligand with phenyl (**R$_{2e}$**), biphenyl (**R$_{2f}$**), and 4-isopropyl biphenyl (**R$_{2g}$**). Then, we computationally examined the catalytic activities and enantioselectivities of complexes **1$_a$-1$_g$** for ATH of PA by comparing the relative free energies of their EDSs (**TS$_{2,3-R}$**→**TS$_{2,3-S}$**) and RDSs (**5′**→**TS$_{5,1}$**).

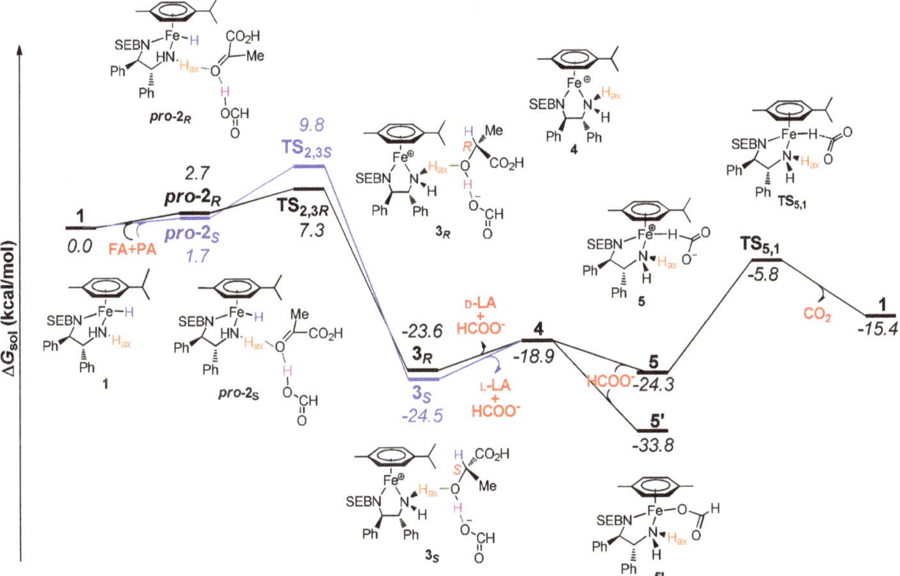

Figure 1. Free energy profile for the ATH of pyruvic acid (PA) catalyzed by **1** with the participation of one formic acid (FA) molecule. The spin-states of all intermediates and transition states are triplet.

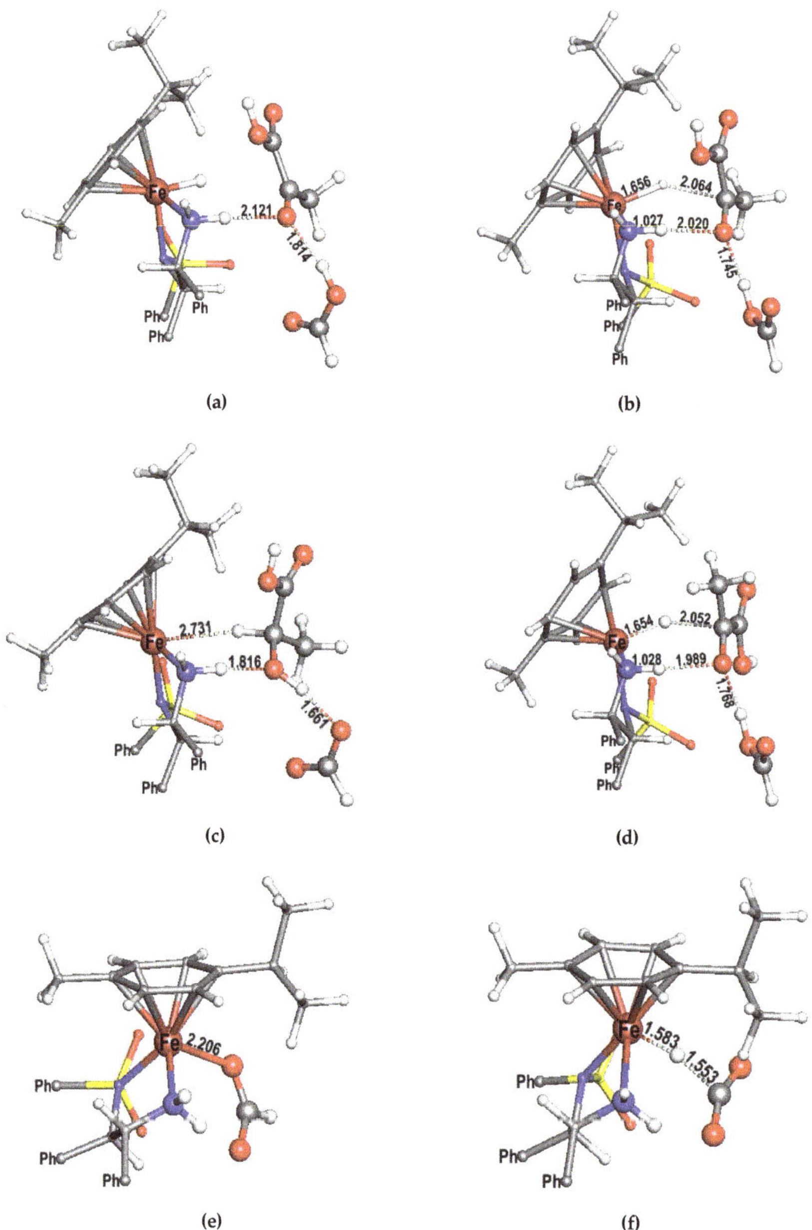

Figure 2. Optimized structures of (**a**) *pro*-**2**$_R$, (**b**) **TS$_{2,3\text{-}R}$** (227i cm^{-1}), (**c**) **3**$_R$, (**d**) **TS$_{2,3\text{-}S}$** (140i cm^{-1}), (**e**) **5′**, and (**f**) **TS$_{5,1}$** (520i cm^{-1}). Bond lengths are in Å. Phenyl groups are omitted for clarity.

$X =$ ⬡ , $\overset{|}{C}H_3$, ⬡–CH_3 , ⬡–NO_2^- , ⬡–F .

Arene

$X-\overset{O}{\underset{O}{S}}-N\diagdown Fe \diagup NH$

R_1 R_{1a} R_{1b} R_{1c} R_{1d}

Arene =

R_2 R_{2e} R_{2f} R_{2g}

Scheme 2. New Fe complexes with methyl or various arene ligands.

Table 1. Relative free energies between EDSs to lactic acid enantiomers ($TS_{2,3\text{-}R}$ and $TS_{2,3\text{-}S}$) and the rate-determining states (RDSs) ($5'$ and $TS_{5,1}$) in catalytic cycles.

Complexes	X	Arene	$TS_{2,3\text{-}R} \rightarrow TS_{2,3\text{-}S}$	$5' \rightarrow TS_{5,1}$
			ΔG (kcal mol^{-1})	
1	R_1	R_2	3.2	28.0
1$_a$	R_{1a}	R_2	0.6	27.5
1$_b$	R_{1b}	R_2	3.8	27.9
1$_c$	R_{1c}	R_2	3.0	28.8
1$_d$	R_{1d}	R_2	4.1	27.5
1$_e$	R_1	R_{2e}	4.9	27.8
1$_f$	R_1	R_{2f}	3.7	28.5
1$_g$	R_1	R_{2g}	2.9	26.2

As is shown in Table 1, the rate-determining barriers of newly proposed Fe complexes **1$_a$**, **1$_b$**, **1$_d$**, **1$_e$**, and **1$_g$** were slightly lower than that of **1**. The lowest barrier belonged to **1$_g$** at 26.2 kcal mol^{-1} while maintaining an excellent enantioselectivity of 2.9 kcal mol^{-1}. Complexes **1$_b$**, **1$_d$**, and **1$_e$**, with higher enantioselectivities of 3.8, 4.1, and 4.9 kcal mol^{-1} respectively, also had lower RDS barriers than those of **1**. The above results suggest that the substituents on both the sulfonyl group and the arene group had significant impact on the catalytic activities and enantioselectivities of proposed iron complexes for asymmetric hydrogenation of PA.

The enantio-determining transfers of the hydride on iron and the proton on FA to the carbonyl group of PA proceeded through an outer-sphere mechanism. In order to understand the weak interaction between unsaturated compound, catalyst, and solvent molecules, we performed a noncovalent interaction (NCI) analysis [31] for transition states $TS_{2,3\text{-}R}$ and $TS_{2,3\text{-}S}$. As shown in Figure 3, we found a stronger attraction between C-H protons of the η^6-p-cymene group and the carboxyl group of PA in $TS_{2,3\text{-}R}$, which is highlighted in position 1 with a zoomed-in view and suggests a stronger C-H$\cdots\pi$ hydrogen bond. In addition, strong hydrogen bonds between the oxygen of carbonyl group in PA and the hydroxyl hydrogen in FA at $TS_{2,3\text{-}R}$ and $TS_{2,3\text{-}S}$ are highlighted as insets labeled 2 with a zoomed-in view. We also compared the NCI plots of the transition states of the proposed complexes with better selectivity or activity, including $TS_{2,3e\text{-}R}$, $TS_{2,3e\text{-}S}$, $TS_{2,3g\text{-}R}$, and $TS_{2,3g\text{-}S}$. As a significant stabilizing factor, the C-H$\cdots\pi$ attraction in $TS_{2,3e\text{-}R}$ was much stronger than that in $TS_{2,3e\text{-}S}$ and thus led to better selectivity.

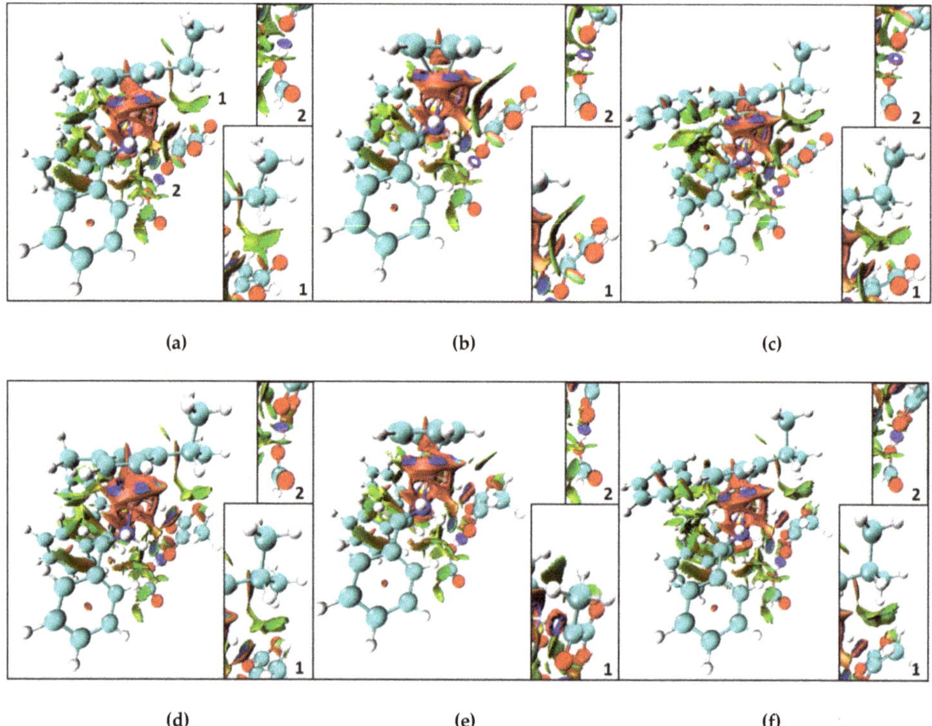

Figure 3. NCI plots for transition states (**a**) $TS_{2,3-R}$, (**b**) $TS_{2,3e-R}$, (**c**) $TS_{2,3g-R}$, (**d**) $TS_{2,3-S}$, (**e**) $TS_{2,3e-S}$, and (**f**) $TS_{2,3g-S}$. The important stabilizing interactions are marked with numbers and highlighted in the box with a zoomed-in view. Color code: red, repulsive; green, weak attractive; blue, attractive.

3. Computational Methods

All DFT calculations were performed using the Gaussian 09 suite of ab initio programs [32] for a hybrid meta-GGA level M06 functional [33] in conjugation with all-electron 6-31G(d) basis set [34] for H, C, N, O, S, and F atoms. Stuttgart relativistic effective core potential basis sets were used for the Fe (ECP10MDF) atom [35]. All structures were fully optimized in formic acid ($\varepsilon = 51.1$) using the integral equation formalism polarizable continuum solvation model (IEFPCM) [36] with the SMD radii [37] for solvent effect corrections. An ultrafine grid (99,590) was used for numerical integrations. The ground states of intermediates were confirmed as triplets through comparison with their low-spin analogs. Thermal corrections were calculated within the harmonic potential approximation on optimized structures under 298.15 K and 1 atm pressure. Unless otherwise noted, the relative energies reported in the text are Gibbs free energies with solvent effect corrections. The optimized structures were confirmed to have no imaginary vibrational mode for all equilibrium structures and only one imaginary vibrational mode for each transition state. Atomic coordinates of all optimized structures (XYZ) and Evaluation of basis sets can be found in Supplementary Materials. Transition states were further characterized by intrinsic reaction coordinate (IRC) calculations to affirm that the correct stationary points were connected. The 3D molecular structures were drawn using the JIMP2 molecular visualizing and manipulating program [38]. NCI analysis was performed using the Multiwfn program [39] for electronic structure analysis and the Virtual Molecular Dynamics (VMD) program [40] for visualization. Computationally enantiomeric ratio was obtained based on Equation (1) [41]:

$$E = \frac{e^{-\frac{\Delta\Delta G}{RT}} - 1}{e^{-\frac{\Delta\Delta G}{RT}} + 1} \times 100\%$$

(1)

4. Evaluation of Ground State Spin Multiplicity

In order to find out correct spin multiplicities of the ground states of the Fe complexes in the catalytic reaction, we optimized the structures of all intermediates with singlet and triplet spins using the methods described above. As shown in Table 2, all singlet states were significantly more stable than the corresponding triplet states. Therefore, we believe the ATH of PA reaction catalyzed by **1** went through a high-spin pathway.

Table 2. Absolute and relative free energies of singlet and triplet states of the intermediates in the ATH of pyruvic acid catalyzed by **1**.

Complexes	Singlet (a.u.)	Triplet (a.u.)	ΔG (kcal mol^{-1})
1	−1944.102697	−1944.118152	−9.7
2$_R$	−2475.926296	−2475.940345	−8.8
2$_S$	−2475.924832	−2475.941943	−10.7
3$_R$	−2475.943887	−2475.982223	−24.1
3$_S$	−2475.946197	−2475.983624	−23.5
4	−1943.384036	−1943.421081	−23.2
5	−2132.591160	−2132.625404	−21.5
5′	−2132.620516	−2132.640419	−12.5

5. Conclusions

In summary, our DFT study of the asymmetric transfer hydrogenation of pyruvic acid to lactic acid enantiomers catalyzed by bifunctional Fe complex revealed a formic acid-assisted proton-coupled hydride transfer mechanism. The formation of D-lactic acid was 2.5 kcal mol^{-1} more favorable than the formation of L-lactic acid with complex **1** as the catalyst. The rate-determining step in the reaction catalyst by **1** was the regeneration of the catalyst with the formation and release of CO_2 with a total free energy barrier of 28.0 kcal mol^{-1}. We also built and computationally examined a series of chiral Fe complexes as potential catalysts for ATH of PA and found that Fe complexes with substituents 4-methylphenyl (**1**$_b$) or 4-fluorophenylphenyl (**1**$_d$) on the sulfonyl group, as well as with substituents 4-isopropyl biphenyl (**1**$_e$) instead of η^6-p-cymene, had potentially better performance than **1** in both catalytic activity and chiral selectivity. Among the proposed iron complexes, **1**$_g$ had the lowest free energy barrier of 26.2 kcal mol^{-1} and computationally chiral selectivity of 98.5% *ee*. Our findings not only provide promising iron complexes for transfer hydrogenation of pyruvic acid, they also pave the way for developing efficient asymmetric catalysts for ATH reactions, as the participation of solvent molecules might be critical for low-barrier proton and hydride transfers.

Supplementary Materials: The following are available online. Atomic coordinates of all optimized structures (XYZ) and Evaluation of basis sets.

Author Contributions: W.W. and X.Y. proposed the research and performed formal analysis. W.W. did calculations and wrote the manuscript. X.Y. reviewed and edited the manuscript. All authors have read and agreed to the published version of the manuscript.

Funding: This work is supported by the National Natural Science Foundation of China (21703256, 21873107, and 21673250).

Conflicts of Interest: The authors declare no conflict of interest.

References

1. Virtanen, P.; Salminen, E.; Mäki-Arvela, P.; Mikkola, J.-P. Selective Hydrogenation for Fine Chemical Synthesis. *Supported Ionic Liquids: Fundam. Appl.* **2014**, 251–262. [CrossRef]
2. Coverdale, J.; Romero-Canelon, I.; Sanchez-Cano, C.; Clarkson, G.J.; Habtemariam, A.; Wills, M.; Sadler, P.J. Asymmetric transfer hydrogenation by synthetic catalysts in cancer cells. *Nat. Chem.* **2018**, *10*, 347–354. [CrossRef]
3. Liu, Y.; Yue, X.; Luo, C.; Zhang, L.; Lei, M. Mechanisms of Ketone/Imine Hydrogenation Catalyzed by Transition-Metal Complexes. *Energy Environ. Mater.* **2019**, *2*, 292–312. [CrossRef]
4. Zassinovich, G.; Mestroni, G.; Gladiali, S. Asymmetric hydrogen transfer reactions promoted by homogeneous transition metal catalysts. *Chem. Rev.* **1992**, *92*, 1051–1069. [CrossRef]
5. Gladiali, S.; Alberico, E. Asymmetric transfer hydrogenation: Chiral ligands and applications. *Chem. Soc. Rev.* **2006**, *35*, 226–236. [CrossRef]
6. Samec, J.S.M.; Bäckvall, J.-E.; Andersson, P.G.; Brandt, P. Mechanistic aspects of transition metal-catalyzed hydrogen transfer reactions. *Chem. Soc. Rev.* **2006**, *35*, 237–248. [CrossRef] [PubMed]
7. Morris, R.H. Asymmetric hydrogenation, transfer hydrogenation and hydrosilylation of ketones catalyzed by iron complexes. *Chem. Soc. Rev.* **2009**, *38*, 2282–2291. [CrossRef]
8. Ager, D.J.; De Vries, A.H.M.; De Vries, J.G. Asymmetric homogeneous hydrogenations at scale. *Chem. Soc. Rev.* **2012**, *41*, 3340. [CrossRef]
9. Fujii, A.; Hashiguchi, S.; Uematsu, N.; Ikariya, T.; Noyori, R. Ruthenium(II)-Catalyzed Asymmetric Transfer Hydrogenation of Ketones Using a Formic Acid–Triethylamine Mixture. *J. Am. Chem. Soc.* **1996**, *118*, 2521–2522. [CrossRef]
10. Uematsu, N.; Fujii, A.; Hashiguchi, S.; Ikariya, T.; Noyori, R. Asymmetric Transfer Hydrogenation of Imines. *J. Am. Chem. Soc.* **1996**, *118*, 4916–4917. [CrossRef]
11. Gao, J.-X.; Ikariya, T.; Noyori, R. A Ruthenium(II) Complex with aC2-Symmetric Diphosphine/Diamine Tetradentate Ligand for Asymmetric Transfer Hydrogenation of Aromatic Ketones†. *Organometallics* **1996**, *15*, 1087–1089. [CrossRef]
12. Haack, K.-J.; Hashiguchi, S.; Fujii, A.; Ikariya, T.; Noyori, R. The catalyst precursor, catalyst, and intermediate in the ruii-promoted asymmetric hydrogen transfer between alcohols and ketones. *Angew. Chem. Int. Ed.* **1997**, *36*, 285–288. [CrossRef]
13. Knighton, R.C.; Vyas, V.K.; Mailey, L.H.; Bhanage, B.M.; Wills, M. Asymmetric transfer hydrogenation of acetophenone derivatives using 2-benzyl-tethered ruthenium (II)/TsDPEN complexes bearing η6-(p-OR) (R = H, iPr, Bn, Ph) ligands. *J. Organomet. Chem.* **2018**, *875*, 72–79. [CrossRef]
14. Dub, P.A.; Ikariya, T. Quantum Chemical Calculations with the Inclusion of Nonspecific and Specific Solvation: Asymmetric Transfer Hydrogenation with Bifunctional Ruthenium Catalysts. *J. Am. Chem. Soc.* **2013**, *135*, 2604–2619. [CrossRef] [PubMed]
15. Wang, W.; Yang, X. Mechanistic insights into asymmetric transfer hydrogenation of pyruvic acid catalysed by chiral osmium complexes with formic acid assisted proton transfer. *Chem. Commun.* **2019**, *55*, 9633–9636. [CrossRef]
16. Cortez, N.A.; Aguirre, G.; Parra-Hake, M.; Somanathan, R.; Cortez-Lemus, N.A. Ruthenium(II) and rhodium(III) catalyzed asymmetric transfer hydrogenation (ATH) of acetophenone in isopropanol and in aqueous sodium formate using new chiral substituted aromatic monosulfonamide ligands derived from (1R,2R)-diaminocyclohexane. *Tetrahedron: Asymmetry* **2008**, *19*, 1304–1309. [CrossRef]
17. Ikariya, T.; Murata, K.; Noyori, R. Bifunctional transition metal-based molecular catalysts for asymmetric syntheses. *Org. Biomol. Chem.* **2006**, *4*, 393–406. [CrossRef]
18. Xie, J.-H.; Liu, X.-Y.; Wang, L.-X.; Zhou, Q.-L. An Additional Coordination Group Leads to Extremely Efficient Chiral Iridium Catalysts for Asymmetric Hydrogenation of Ketones. *Angew. Chem. Int. Ed.* **2011**, *50*, 7329–7332. [CrossRef]
19. Sui-Seng, C.; Freutel, F.; Lough, A.J.; Morris, R.H. Highly efficient catalyst systems using iron complexes with a tetradentate PNNP ligand for the asymmetric hydrogenation of polar bonds. *Angew. Chem. Int. Ed.* **2008**, *47*, 940–943. [CrossRef]
20. Li, Y.-Y.; Yu, S.-L.; Shen, W.-Y.; Gao, J.-X. Iron-, Cobalt-, and Nickel-Catalyzed Asymmetric Transfer Hydrogenation and Asymmetric Hydrogenation of Ketones. *Acc. Chem. Res.* **2015**, *48*, 2587–2598. [CrossRef]

21. Bigler, R.; Huber, R.; Mezzetti, A. Highly enantioselective transfer hydrogenation of ketones with chiral (NH)2P2 macrocyclic iron(II) complexes. *Angew. Chem. Int. Ed.* **2015**, *54*, 5171–5174. [CrossRef] [PubMed]

22. Bigler, R.; Mezzetti, A. Isonitrile Iron(II) Complexes with Chiral N2P2 Macrocycles in the Enantioselective Transfer Hydrogenation of Ketones. *Org. Lett.* **2014**, *16*, 6460–6463. [CrossRef] [PubMed]

23. Corberan, R.; Peris, E.V. An Unusual Example of Base-Free Catalyzed Reduction of C=O and C=NR Bonds by Transfer Hydrogenation and Some Useful Implications. *Organometallics* **2008**, *27*, 1954–1958. [CrossRef]

24. De Luca, L.; Passera, A.; Mezzetti, A. Asymmetric Transfer Hydrogenation with a Bifunctional Iron(II) Hydride: Experiment Meets Computation. *J. Am. Chem. Soc.* **2019**, *141*, 2545–2556. [CrossRef] [PubMed]

25. Coverdale, J.; Sanchez-Cano, C.; Clarkson, G.J.; Soni, R.; Wills, M.; Sadler, P.J. Easy To Synthesize, Robust Organo-osmium Asymmetric Transfer Hydrogenation Catalysts. *Chem. A Eur. J.* **2015**, *21*, 8043–8046. [CrossRef]

26. Maeda, H.; Yamamoto, K.; Kohno, I.; Hafsi, L.; Itoh, N.; Nakagawa, S.; Kanagawa, N.; Suzuki, K.; Uno, T. Design of a Practical Fluorescent Probe for Superoxide Based on Protection–Deprotection Chemistry of Fluoresceins with Benzenesulfonyl Protecting Groups. *Chem. A Eur. J.* **2007**, *13*, 1946–1954. [CrossRef]

27. Macchioni, A. Ion pairing in transition-metal organometallic chemistry. *Chem. Rev.* **2005**, *105*, 2039–2074. [CrossRef]

28. Kozuch, S.; Shaik, S. How to Conceptualize Catalytic Cycles? The Energetic Span Model. *Acc. Chem. Res.* **2011**, *44*, 101–110. [CrossRef]

29. Noyori, R.; Hashiguchi, S. Asymmetric Transfer Hydrogenation Catalyzed by Chiral Ruthenium Complexes. *Acc. Chem. Res.* **1997**, *30*, 97–102. [CrossRef]

30. Masashi, Y.; Issaku, Y.; Ryoji, N. CH/π attraction: The origin of enantioselectivity in transfer hydrogenation of aromatic carbonyl compounds catalyzed by chiral η6-arene-ruthenium(II) complexes. *Angew. Chem. Int. Ed.* **2001**, *40*, 2818–2821.

31. Johnson, E.R.; Keinan, S.; Mori-Sánchez, P.; Contreras-Garcia, J.; Cohen, A.; Yang, W. Revealing Noncovalent Interactions. *J. Am. Chem. Soc.* **2010**, *132*, 6498–6506. [CrossRef] [PubMed]

32. Frisch, M.J.; Trucks, G.W.; Schlegel, H.B.; Scuseria, G.E.; Robb, M.A.; Cheeseman, J.R.; Scalmani, G.; Barone, V.; Mennucci, B.; Petersson, G.A.; et al. *Gaussian 09, Revision E.01*; Gaussian, Inc.: Wallingford, CT, USA, 2013.

33. Zhao, Y.; Truhlar, D. The M06 suite of density functionals for main group thermochemistry, thermochemical kinetics, noncovalent interactions, excited states, and transition elements: two new functionals and systematic testing of four M06 functionals and 12 other functionals. *Theor. Chem. Acc.* **2008**, *119*, 525. [CrossRef]

34. Feller, D. The role of databases in support of computational chemistry calculations. *J. Comput. Chem.* **1996**, *17*, 1571–1586. [CrossRef]

35. Dolg, M.; Wedig, U.; Stoll, H.; Preuss, H. Energy-adjusted abinitio pseudopotentials for the first row transition elements. *J. Chem. Phys.* **1987**, *86*, 866–872. [CrossRef]

36. Tomasi, J.; Mennucci, B.; Cammi, R. Quantum mechanical continuum solvation models. *Chem. Rev.* **2005**, *105*, 2999–3094. [CrossRef]

37. Marenich, A.V.; Cramer, C.J.; Truhlar, D. Universal Solvation Model Based on Solute Electron Density and on a Continuum Model of the Solvent Defined by the Bulk Dielectric Constant and Atomic Surface Tensions. *J. Phys. Chem. B* **2009**, *113*, 6378–6396. [CrossRef]

38. Manson, J.; Webster, C.E.; Hall, M.B. *JIMP2, Version 0.091, A Free Program for Visualizing and Manipulating Molecules*; Texas A&M University: College Station, TX, USA, 2006.

39. Lu, T.; Chen, F. Multiwfn: A multifunctional wavefunction analyzer. *J. Comput. Chem.* **2011**, *33*, 580–592. [CrossRef]

40. Humphrey, W.; Dalke, A.; Schulten, K. VMD: Visual molecular dynamics. *J. Mol. Graph.* **1996**, *14*, 33–38. [CrossRef]

41. Straathof, A.; Jongejan, J. The enantiomeric ratio: origin, determination and prediction. *Enzym. Microb. Technol.* **1997**, *21*, 559–571. [CrossRef]

Sample Availability: Samples of the compounds are not available from the authors.

Article

A Simple Iron-Catalyst for Alkenylation of Ketones Using Primary Alcohols

Motahar Sk, Ashish Kumar, Jagadish Das and Debasis Banerjee *

Department of Chemistry, Laboratory of Catalysis and Organic Synthesis, Indian Institute of Technology Roorkee, Roorkee-247667, India; msk@cy.iitr.ac.in (M.S.); akumar5@cy.iitr.ac.in (A.K.); dasjagadish82@gmail.com (J.D.)

* Correspondence: debasis.banerjee@cy.iitr.ac.in; Tel.: +91-133228-4847

Academic Editor: Diego A. Alonso
Received: 2 February 2020; Accepted: 6 March 2020; Published: 30 March 2020

Abstract: Herein, we developed a simple iron-catalyzed system for the α-alkenylation of ketones using primary alcohols. Such acceptor-less dehydrogenative coupling (ADC) of alcohols resulted in the synthesis of a series of important α,β-unsaturated functionalized ketones, having aryl, heteroaryl, alkyl, nitro, nitrile and trifluoro-methyl, as well as halogen moieties, with excellent yields and selectivity. Initial mechanistic studies, including deuterium labeling experiments, determination of rate and order of the reaction, and quantitative determination of H_2 gas, were performed. The overall transformations produce water and dihydrogen as byproducts.

Keywords: α-alkenylation; iron; alcohols; dehydrogenative coupling; sustainability

1. Introduction

α,β-unsaturated ketones are ubiquitous in various pharmaceutically important plants, and are extensively used as life-saving drugs (choleretic, antiulcer and muco-protective) [1]. Such compounds have wide applications as food additives, pesticides, solar-creams, as well as in materials science [2,3]. Owing to its great demand, synthesis of α,β-unsaturated ketones attracts potential attention in chemical research. In general, Aldol condensation of aldehydes with a suitable ketone as the coupling partner is utilized for such α,β-unsaturated ketones in combination with a stoichiometric amount of a strong base [4–7]. However, such practice not only produces stoichiometric amounts of waste, also raises major concerns about the uses of expensive and highly susceptible aldehydes (Scheme 1a). Though, the Rh-catalyzed protocol having a combination of ketones with internal alkynes was reported for such α,β-unsaturated ketones; this required higher catalyst loading and stoichiometric amounts of ligands for successful transformations [4–7]. Since the last decade, the utilization of biomass-derived renewable alcohols has attained significant attention in catalytic research following hydrogen-borrowing strategy. Alcohols are radially available, earth abundant, easy to store, and often generate water as their sole byproduct, when used as one of the coupling partners in organic transformations [8–15]. Therefore, recently, a series of transition metal-based catalysts, such as Ru [16], Pd [17,18], Pt [19], Au [20], Ti [21], Nb [22] and Cr [23] were established for the synthesis of α,β-unsaturated ketones using alcohols. Importantly, a couple of heterogeneous catalysts based on CeO_2 [24], Bi_2WO_6 [25] and mesoporous Pd-nanoparticles [26], were also reported for such alkenylation of ketones with alcohols (Scheme 1b).

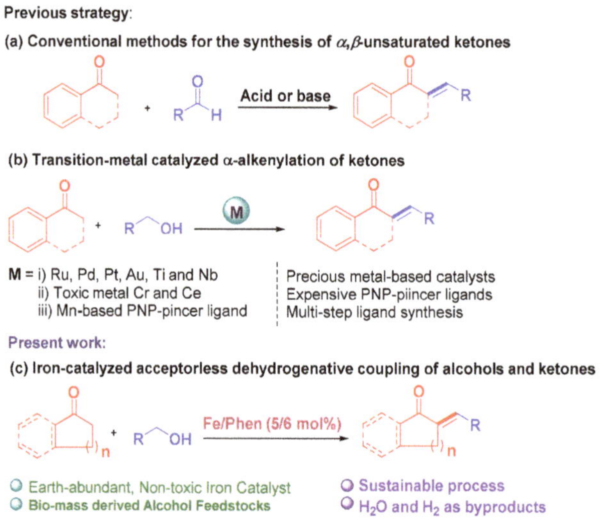

Previous strategy:

(a) Conventional methods for the synthesis of α,β-unsaturated ketones

(b) Transition-metal catalyzed α-alkenylation of ketones

M = i) Ru, Pd, Pt, Au, Ti and Nb ⋮ Precious metal-based catalysts
   ii) Toxic metal Cr and Ce ⋮ Expensive PNP-piincer ligands
   iii) Mn-based PNP-pincer ligand ⋮ Multi-step ligand synthesis

Present work:

(c) Iron-catalyzed acceptorless dehydrogenative coupling of alcohols and ketones

○ Earth-abundant, Non-toxic Iron Catalyst   ○ Sustainable process
○ Bio-mass derived Alcohol Feedstocks   ○ H_2O and H_2 as byproducts

Scheme 1. Approaches for the synthesis of α,β-unsaturated ketones. (**a**) Conventional methods for the synthesis of α,β-unsaturated ketones; (**b**)Transition-metal catalyzed α-alkenylation of ketones; (**c**) Iron-catalyzed acceptorless dehydrogenative coupling of alcohols and ketones.

Recently, much effort has been devoted towards the replacement of the precious-metal-based catalysts with earth abundant and non-precious metals, such as, Fe-, Mn-, Ni-, Co-, etc., for such transformations using alcohols. In this direction, only one report using a manganese-based complex is known (Scheme 1b).

However, such Mn-based catalyst bearing the PNP-pincer ligand is the key for successful transformations and selectivity [27]. Importantly, such PNP-ligands require a multistep synthesis process, and are quite a lot more expensive than the non-precious metals. Therefore, there is still a need to develop a simple, efficient and cost-economic catalyst system for the synthesis of α,β-unsaturated ketones.

Recently, we have started a program for the applications of non-precious, metal-based catalysts for sustainable organic transformations using nickel [28–30], manganese [31] and iron-based-catalysts. Most recently, we successfully developed the acceptor-less dehydrogenative coupling of alcohols with alkyl *N*-heteroaromatics using a simple iron-catalyst [32,33]. Iron is the most earth-abundant, biocompatible, and exists in variable oxidation states. Therefore, application of Fe-based catalysts in organic synthesis attract significant attention [34–40]. Interestingly, iron catalysts were extensively used in cross-coupling reactions, [41–45] hydrogenations and transfer hydrogenations, [46–49], hydrosilylations [50–52], as well as in hydroboration processes [53,54]. However, the application of a Knölker-type complex for the alkylation and methylation of ketones and alcohols is noteworthy [55–59]. Nevertheless, Knölker-type iron complexes require a tedious synthesis procedure and are expensive in nature. Therefore, the recent surge is to develop a simple and inexpensive iron catalyst system in combination with a nitrogen ligand for the dehydrogenative coupling of alcohols with ketones to α, β-unsaturated ketones. To the best of our knowledge, to date, no such iron-catalyzed dehydrogenative coupling of alcohols and ketones to α,β-unsaturated ketones is known (Scheme 1c).

2. Results and Discussions

In our initial investigation for the α-alkenylation of ketones, we choose α-tetralone (**1a**) and 4-methoxy benzyl alcohol (**2a**) as our model substrates. Iron catalysts having oxidation states 0, II and III were examined in combination with 1,10 phenanthroline **L1** (6 mol%), *t*-BuONa (0.25 equiv.) in toluene at 110 °C equipped with a nitrogen balloon. To our delight, iron (II) acetate resulted in the formation of α,β-unsaturated ketone **3a** in 75% isolated yield with excellent selectivity (>25:1, Table 1, entry 1 and SI Table S1). Thereafter, a variety of aromatic and aliphatic nitrogen-based ligands, along with triphenyl phosphine **L2–L6**, were screened, and no increment in product yield was observed (Table 1, entries 4–8 and SI Table S2). Nevertheless, the application of other alkoxide, carbonate and phosphate bases proved less efficient for such transformation (Table 1, entries 9, 10 and SI Table S3).

Thereafter, the replacement of toluene by different nonpolar and polar solvents (*p*-xylene, 1,4-dioxane, DMA and *t*-amyl alcohol) resulted in up to 45% conversion to **3a** (Table 1, entry 11 and SI Table S4). α-alkenylation could also be performed using lower catalyst loading (2.5 mol%), however, the product conversion decreases to 53% with (>26:1) selectivity (Table 1, entry 12). Notably, this reaction could also be performed at lower temperature, and resulted up to 58% conversion to the desired product (Table 1, entry 13 and SI Table S6). Catalytic alkenylation for 9 h gave only moderate product conversion along with unreacted starting substrate (Table 1, entry 14 and SI Table S7). Control experiments in the absence of catalyst, ligand and base establish the potential role of the individual component for the α-alkenylation (Table 1, entries 15, 16 and SI Scheme S4 and Scheme S5). Notably, we have detected benzaldehyde **2a′** and the alkylated product **3a′** in the gas chromatography–mass spectrometry (GC-MS) analysis of the crude reaction mixture.

Table 1. Optimization table for Fe-catalyzed α-alkenylation of ketone [a, b, c].

Entry	Deviations from the above	Conv. 3a (%) [b]	3a/3a′
1	-	**77(75)**	**>25:1**
2	Fe(acac)$_3$	68	17:1
3	Fe$_2$(CO)$_9$	64	16:1
4	**L2** instead of **L1**	68	17:1
5	**L3** instead of **L1**	54	>10:1
6	**L4** instead of **L1**	67	>11:1
7	**L5** instead of **L1**	70	4:1
8	**L6** instead of **L1**	60	2:1
9	*t*-BuOK	58	2:1
10	Na$_2$CO$_3$, K$_2$CO$_3$, Cs$_2$CO$_3$,	12–30	-
11	*p*-xylene, 1,4-dioxane, *t*-amyl-alcohol	25–45	>10
12	Fe(OAc)$_2$ (2.5 mol%), **L1** (3.0 mol%)	53	>26:1
13	Reaction at 80 °C	58	-
14	9 h reaction time	62	>31:1
15	no Fe(OAc)$_2$, no ligand	20	-
16	No base	0	0

Reaction conditions: [a] Unless specified otherwise, the reaction was carried out with **1a** (0.375 mmol), **2a** (0.25 mmol), Fe-cat. (0.0125 mmol), ligand (0.015 mmol), and *t*-BuONa (0.0625 mmol) under an N$_2$ balloon at 110 °C (oil bath) in toluene (1.0 mL) for 12 h in a Schlenk tube. [b] Conversion was determined by gas chromatography–mass spectrometry (GC-MS) (isolated yield in parentheses). [c] = tri-phenyl phosphine (10 mol%) was used. **L1** = 1,10-phenanthroline, **L2** = 2,9-dimethyl-1,10-phenanthroline, **L3** = 2,2′-biquinoline, **L4** = 2,2′-bipyridine, **L5** = triphenyl-phosphine. **L6** = tetramethylethylenediamine (TMEDA).

Next, the optimized conditions of Table 1 were successfully employed for the α-alkenylation using a series of primary alcohols and ketones, and resulted in moderate to excellent yields (Scheme 2). Initially, we explored the reactivity of electronically different aryl, heteroaryl and alkyl alcohols with α-tetralone for the α-alkenylation process. Interestingly, benzyl alcohols, as well as electron-rich *p*-methyl and *p*-*iso*-propyl-substituted benzyl alcohols yielded the α,β-unsaturated ketones **3b–3d** in 55–89% isolated yields (Scheme 2). Importantly, sterically hindered *o*-methyl benzyl alcohol **2e** and α-naphthyl methanol **2k** participated efficiently, and resulted the desired products **3e** and **3k** in moderate yields (Scheme 2). To our delight, F, and Cl-substituted benzyl-alcohols were well tolerated under the reaction conditions, and yielded up to 64% of product (**3f** and **3g**), and we did not observe any de-halogenated product.

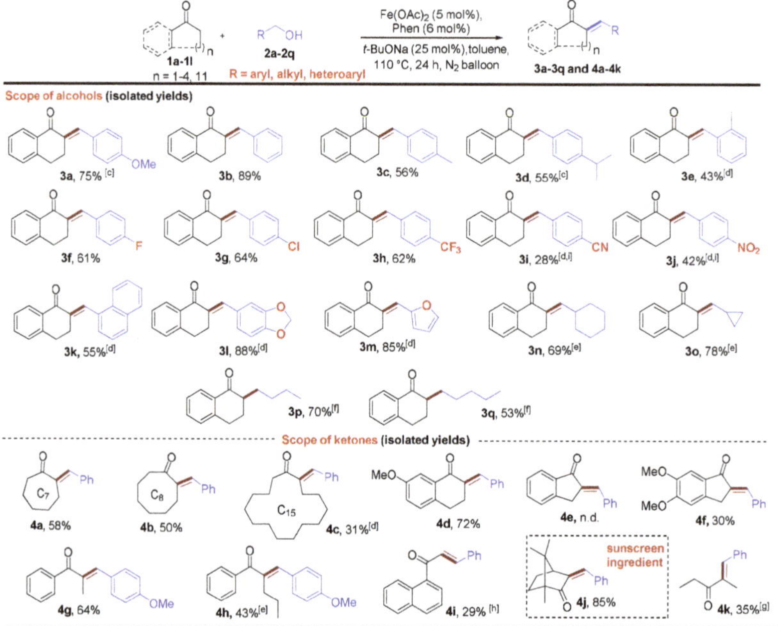

Scheme 2. Substrate scope for the Fe-catalyzed α-alkenylation of ketones and alcohols [a, b]. Reaction conditions: [a] Unless specified otherwise, the reaction was carried out with **1a** (0.375 mmol), **2a** (0.25 mmol), Fe(OAc)$_2$ (0.0125 mmol), ligand (0.015 mmol) and *t*-BuONa (0.0625 mmol) under an N$_2$ balloon at 110 °C (oil bath) in toluene (1.0 mL) for 24 h in a Schlenk tube. [b] The isolated yield reported. [c] The reaction time 12 h. [d] Here, *t*-BuONa (0.125 mmol) was used. [e] In this case, *t*-BuONa (0.25 mmol) was used. [f] For this, α-tetralone (0.25 mmol), alcohols (0.5 mmol) and *t*-BuONa (0.25 mmol) were used. [g] In this case, benzyl alcohol (0.25 mmol), 3-pentanone (0.75 mmol) and *t*-BuONa (0.125 mmol) were used. [h] GC-MS yield. [i] Protonic nuclear magnetic resonance (^{1}H-NMR) yield reported.

Notably, benzyl-alcohols bearing strong electron withdrawing groups, such as, CF$_3$, CN and NO$_2$, also efficiently participated for the α-alkenylation reaction, and the desired products **3h–3j** were obtained in up to 62% yield (Scheme 2). More interestingly, heteroaryl alcohols, such as 1,3-dioxolone-substituted benzyl alcohol **2l** and 2-furfurylmethanol **2m** efficiently reacted with α-tetralone to **3l–3m** in excellent yields (85–88%, Scheme 2). Thereafter, we explored the reactivity with more challenging alkyl alcohols, such as, cyclohexyl-methanol **2n** and cyclopropyl-methanol **2o**, with **1a**, and the desired products **3n** and **3o** were obtained in 69–78% yield (Scheme 2). However, in the case of acyclic alkyl alcohols, such as, *n*-butanol **2p** and *n*-pentanol **2q**, we observed the formation of the reduced products **3p** and **3q** in up to 70% yield (Scheme 2).

Thereafter, to establish the generality of the α-alkenylation process, we further explored the reactivity of more challenging cyclic and acyclic ketones with alcohols (Scheme 2). To our delight C7, C8 and C15 membered cyclic ketones underwent the reaction smoothly and afforded selective mono-benzylated products **4a–4c** in up to 58% isolated yields. Again, 7-methoxy tetralone **1e** and 5,6 dimethoxy indanone **1g** efficiently converted to the α,β-unsaturated ketones **4d** and **4f** in up to 72% isolated yields. Nevertheless, in case of indanone **1f**, we did not observe any desired product (Scheme 2).

Notably, propiophenone **1h**, valerophenone **1i** and 1-acetyl naphthalene **1j** reacted smoothly with 4-methoxy benzyl alcohol **2a** and benzyl alcohol **2b**, and transformed into the desired α,β-unsaturated ketones **4g–4i** in moderate to good yields (Scheme 2). Interestingly, the reaction of more challenging, 3-pentanone **1l** was sluggish, and the product **4k** was obtained in acceptable yield. Additionally, we attempted the reaction of benzyl alcohol with more challenging and highly sterically hindered camphor **1k.** The desired α, β-unsaturated ketone **4j**, extensively used as a sunscreen ingredient, was obtained in an 85% yield (Scheme 2).

Notably, the established catalytic protocol is tolerant to a series of cyclic and acyclic ketones, as well as benzyl alcohols having sensitive functional moieties, such as, nitro, nitrile, trifluoromethyl, fluoro, chloro, 1,3-dioxolonc, alkyl and alkoxy, including furan functionality. As a highlight, we have demonstrated the synthesis of the sunscreen ingredient **4j**.

After having successfully explored the reactivity of various alcohols and ketones for the alkenylation process, next we focused upon the preliminary mechanistic investigations. To establish the participation of 4-methoxybenzaldehyde **2a'**, an in-situ ^{1}H-NMR study was performed, and we detected the formation of aldehyde in ^{1}H-NMR, as well as in the GC-MS analysis of the crude reaction mixture. Again, to demonstrate the potential role of Fe-catalyst for the hydrogen-borrowing process, as well as for the formation of the C–C bond, the reaction of α-tetralone **1a** and 4-methoxy benzyl-aldehyde **2a'** in the presence and absence of our Fe-catalyst was performed. The desired product was obtained in up to 93% yield (Scheme 3a,b). Next, to make evident the involvement of the benzylic C–H/D bond for the alkenylation process, a deuterium labeling experiment was carried out using deuterated benzyl-alcohol **2b–d2** (92% D) with α-tetralone **1a**, and **3b–d** was obtained having 81% deuterium incorporation at the vinyl position (Scheme 3c). However, when the reaction of **1a–d2** (93% D) with **2a** was performed, we did not observe any deuterium incorporation in the product **3a** (Scheme 3d). These experiments are in agreement for the micro-reversible transformation in the hydrogen-borrowing process.

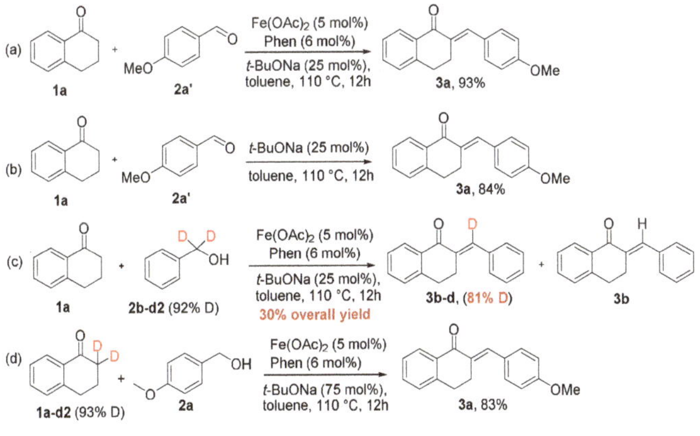

Scheme 3. Preliminary mechanistic study for the α-alkenylation of ketone

Next, to determine the rate and order of the reaction, we performed two different sets of experiments having a variable concentration of ketones and alcohols. The reaction follows first order kinetics with respect to alcohol, considering the steady state approximation for ketone (Figure 1 and SI Scheme S2). Further, we have studied the time-dependent reaction profile for the model reaction of Table 1, and revealed that the concentration of aldehyde is almost constant during the progress of the reaction, whereas, the formation of the reduced product **3a'** increases with prolonged time (Figure 2). Nevertheless, as expected, hydrogen gas is generated during the dehydrogenation process, so for the quantitative determination of the hydrogen gas, we choose the model reaction of Table 1, and it was connected to the gas burette, as shown in Scheme S3. Gratifyingly, the quantitative determination of hydrogen gas produced in the alkenylation process was calculated (SI Scheme S3).

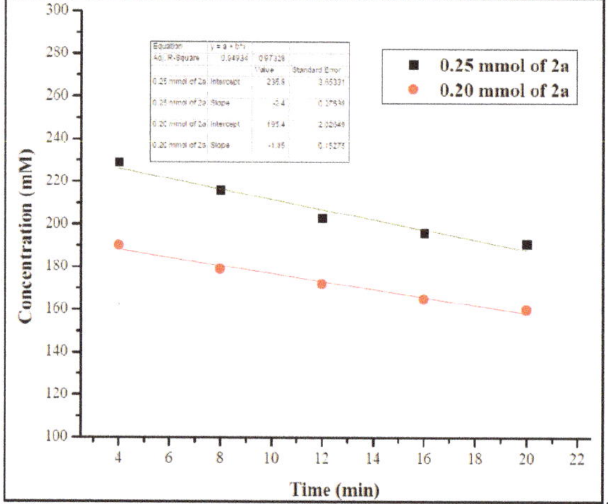

Figure 1. Plot of [**2a**] vs time.

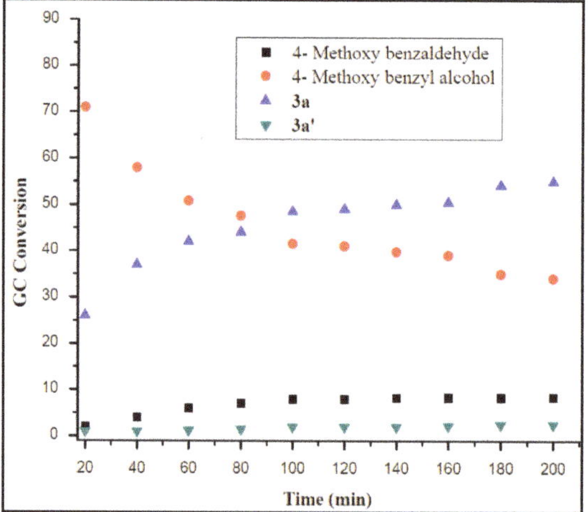

Figure 2. Reaction profile.

Based on these experimental findings, herein we proposed a plausible mechanistic cycle for the α-alkenylation of ketones presented in Scheme 4. Primarily, Fe-catalyzed dehydrogenation of alcohol resulted in the formation of aldehyde, and transient iron-hydride is formed. Next, a base-mediated condensation of aldehyde with ketone affords the desired α, β-unsaturated ketone, releasing water as the byproduct. During the process, the active iron catalyst regenerated via the release of dihydrogen gas.

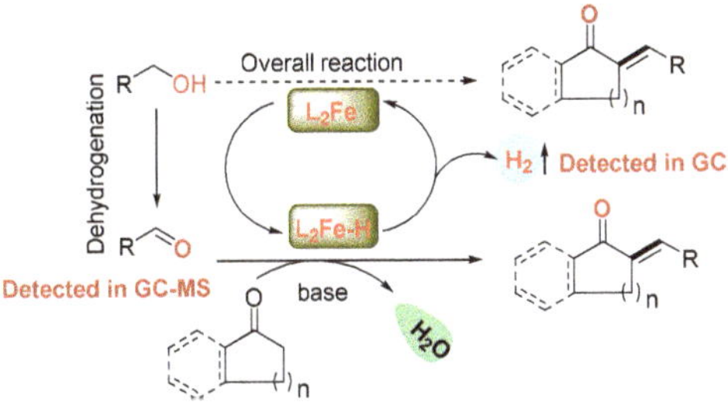

Scheme 4. Plausible mechanistic cycle.

3. Materials and Methods

3.1. General Experimental Details

All solvents and reagents were used, as received from the suppliers. Thin-layer chromatography (TLC) was performed on Merck Kiesel gel 60, F_{254} plates with a layer thickness of 0.25 mm. Column chromatography was performed on silica gel (100–200 mesh) using a gradient of ethyl acetate and hexane as the mobile phase.

Protonic nuclear magnetic resonance (^{1}H NMR) spectral data were collected at 400 MHz (JEOL), 500 MHz (Bruker), and caron-13 nuclear magnetic resonance (^{13}C NMR) values were recorded at 100 MHz. ^{1}H NMR spectral data are given as chemical shifts in ppm followed by multiplicity (*s*- singlet; *d*- doublet; *t*- triplet; *q*- quartet; *m*- multiplet), the number of protons, and the coupling constants. ^{13}C NMR chemical shifts are expressed in ppm. High resolution mass spectrometry (HRMS) electrospray ionization (ESI) spectral data were collected using a Bruker High Resolution Mass Spectrometer. GC-MS was recorded using Agilent Gas Chromatography Mass Spectrometry. Elemental analysis data were recorded using the Vario Micro Cube elemental analyzer. All the reactions were performed in a closed system using a Schlenk tube. $Fe(OAc)_2$ was purchased from TCI Chemicals (India) Pvt. Ltd. (Purity->90%, CAS No: 3094-87-9, Product Number: I0765). 1, 10-Phenanthroline was purchased from Sigma-Aldrich ((Assay- >99%; CAS Number- 66-71-7; EC Number 200-629-2; Pack Size- 131377-25G). Sodium *tert*-butoxide was purchased from Avra Synthesis Pvt. Ltd., India. (Purity-98%, CAS No: **865-48-5**, Catalog No- ASS2615).

3.2. General Procedure for Iron-Catalyzed Alkenylation of Ketones with Alcohols:

In a 15 mL oven-dried Schlenk tube, alcohols (0.25 mmol), *t*-BuONa (0.0625 mmol), $Fe(OAc)_2$ (0.0125 mmol), phen (0.015 mmol) and ketones (0.375 mmol, 1.5 equiv.) were added followed by toluene 1.0 mL under an atmosphere of an N_2 balloon, and the reaction mixture was refluxed at 110 °C for 24 h in a closed system. The reaction mixture was cooled to room temperature, and 3.0 mL of ethyl acetate was added and concentrated in vacuo. The residue was purified by column chromatography, using a gradient of hexane and ethyl acetate (eluent system) to afford the pure product.

(E)-2-(4-methoxybenzylidene)-3,4-dihydronaphthalen-1(2H)-one (Scheme 2, **3a**) [27], Yellow solid (50 mg, 75% yield); ^{1}H NMR (400 MHz, CDCl₃) δ 8.05 (d, *J* = 8.7 Hz, 1H), 7.78 (s, 1H), 7.43–7.34 (m, 3H), 7.29 (t, *J* = 7.5 Hz, 1H), 7.17 (d, *J* = 8.8 Hz, 1H), 6.88 (d, *J* = 8.8 Hz, 2H), 3.78 (s, 3H), 3.09–3.05 (m, 2H), 2.89–2.86 (m, 2H); ^{13}C NMR (100 MHz, CDCl₃) δ 186.9, 158.9, 142.0, 135.7, 132.6, 132.5, 132.1, 130.7, 127.4, 127.1, 127.0, 126.0, 112.9, 54.3, 27.8, 26.2.

(E)-2-benzylidene-3,4-dihydronaphthalen-1(2H)-one (Scheme 2, **3b**) [27], Colorless solid (52 mg, 89% yield); ^{1}H NMR (500 MHz, CDCl₃) δ 8.06 (dd, *J* = 7.8, 0.9 Hz, 1H), 7.80 (s, 1H), 7.43–7.40 (m, 1H), 7.39–7.31 (m, 4H), 7.29–7.26 (m, 2H), 7.19–7.15 (m, 1H), 3.07–3.04 (m, 2H), 2.88–2.83 (m, 2H); ^{13}C NMR (125 MHz, CDCl₃) δ 186.9, 142.2, 135.6, 134.8, 134.5, 132.5, 132.2, 128.9, 127.5, 127.4, 127.2, 127.1, 126.0, 27.9, 26.2.

(E)-2-(4-methylbenzylidene)-3,4-dihydronaphthalen-1(2H)-one (Scheme 2, **3c**) [27], Colorless solid (35 mg, 56% yield); ^{1}H NMR (500 MHz, CDCl$_3$) δ 8.06 (d, *J* = 7.8 Hz, 1H), 7.78 (s, 1H), 7.43–7.40 (m, 1H), 7.29 (dd, *J* = 7.3, 5.2 Hz, 2H), 7.19–7.15 (m, 4H), 3.08–3.05 (m, 2H), 2.89–2.86 (m, 2H), 2.32 (s, 3H); ^{13}C NMR (125 MHz, CDCl$_3$) δ 187.9, 143.2, 138.8, 136.8, 134.7, 133.6, 133.2, 133.0, 130.0, 129.2, 128.2, 128.1, 127.0, 28.9, 27.2, 21.4.

(E)-2-(4-isopropylbenzylidene)-3,4-dihydronaphthalen-1(2H)-one (Scheme 2, **3d**) [27], Yellow oil (40 mg, 55% yield); ^{1}H NMR (500 MHz, CDCl$_3$) δ 8.05 (dd, *J* = 7.8, 1.0 Hz, 1H), 7.79 (s, 1H), 7.42–7.39 (m, 1H), 7.34–7.26 (m, 3H), 7.21 (d, *J* = 8.2 Hz, 2H), 7.17 (d, *J* = 8.1 Hz, 1H), 3.09–3.06 (m, 2H), 2.88–2.85 (m, 3H), 1.20 (d, *J* = 6.9 Hz, 6H); ^{13}C NMR (125 MHz, CDCl$_3$) δ 187.9, 149.7, 143.2, 136.8, 134.7, 133.6, 133.4, 133.2, 130.1, 128.2, 128.1, 127.0, 126.6, 34.0, 28.9, 27.3, 23.9.

(E)-2-(2-methylbenzylidene)-3,4-dihydronaphthalen-1(2H)-one (Scheme 2, **3e**) [27], Yellow oil (27 mg, 43% yield); ^{1}H NMR (500 MHz, CDCl$_3$) δ 8.19 (dd, *J* = 7.8, 1.2 Hz, 1H), 7.94 (s, 1H), 7.54–7.51 (m, 1H), 7.40 (t, *J* = 7.1 Hz, 1H), 7.29–7.26 (m, 5H), 3.02–2.99 (m, 2H), 2.98–2.95 (m, 2H), 2.37 (s, 3H); ^{13}C NMR (125 MHz, CDCl$_3$) δ 188.1, 143.6, 137.9, 136.1, 135.8, 135.2, 133.6, 133.4, 130.4, 129.0, 128.6, 128.4, 128.3, 127.2, 125.6, 29.4, 27.4, 20.2.

(E)-2-(4-fluorobenzylidene)-3,4-dihydronaphthalen-1(2H)-one (Scheme 2, **3f**) [60], Yellow solid (38 mg, 61% yield); ^{1}H NMR (500 MHz, CDCl$_3$) δ 8.06 (dd, *J* = 7.8, 0.9 Hz, 1H), 7.76 (s, 1H), 7.45–7.42 (m, 1H), 7.36 (dd, *J* = 8.5, 5.5 Hz, 2H), 7.31 (t, *J* = 7.6 Hz, 1H), 7.19 (d, *J* = 7.3 Hz, 1H), 7.05 (t, *J* = 8.7 Hz, 2H), 3.06–3.03 (m, 2H), 2.91–2.88 (m, 2H); ^{13}C NMR (125 MHz, CDCl$_3$) δ 186.7, 161.6 (d, *J*$_{C-F}$ = 249.9 Hz), 142.1, 134.5, 134.3, 132.4, 132.3, 130.9, 130.8, 130.7, 127.2 (d, *J*$_{C-F}$ = 9.2 Hz), 126.1, 114.6 (d, *J*$_{C-F}$ = 21.7 Hz), 27.7, 26.1.

(E)-2-(4-chlorobenzylidene)-3,4-dihydronaphthalen-1(2H)-one (Scheme 2, **3g**) [60], Yellow solid (44 mg, 64% yield); ^{1}H NMR (500 MHz, CDCl$_3$) δ 8.06 (d, *J* = 7.8 Hz, 1H), 7.73 (s, 1H), 7.44–7.38 (m, 1H), 7.30 (dd, *J* = 6.7, 4.1 Hz, 4H), 7.19 (dd, *J* = 7.5, 4.9 Hz, 2H), 3.04–3.01 (m, 2H), 2.90–2.87 (m, 2H); ^{13}C NMR (125 MHz, CDCl$_3$) δ 186.6, 142.1, 135.0, 134.2, 133.4, 133.3, 132.4, 130.1, 127.7, 127.3, 127.2, 126.1, 114.0, 27.8, 26.2.

(E)-2-(4-(trifluoromethyl)benzylidene)-3,4-dihydronaphthalen-1(2H)-one (Scheme 2, **3h**) [61], Yellow solid (47 mg, 62% yield); ^{1}H NMR (500 MHz, CDCl$_3$) δ 8.07 (d, *J* = 7.8 Hz, 1H), 7.77 (s, 1H), 7.60 (d, *J* = 8.2 Hz, 2H), 7.46–7.42 (m, 3H), 7.31 (t, *J* = 7.5 Hz, 1H), 7.19 (d, *J* = 7.7 Hz, 1H), 3.03 (td, *J* = 6.4, 1.6 Hz, 2H), 2.91–2.88 (m, 2H); ^{13}C NMR (125 MHz, CDCl$_3$) δ 186.5, 142.2, 138.4, 136.4, 133.7, 132.5, 132.2, 129.3, 129.0, 128.9, 128.4, 127.3–127.2 (d, *J*$_{C-F}$ = 7.3 Hz), 126.1, 124.4–124.3 (q, *J*$_{C-F}$ = 3.8 Hz), 124.0, 121.9, 27.8, 26.1.

(E)-4-((1-oxo-3,4-dihydronaphthalen-2(1H)-ylidene)methyl)benzonitrile (Scheme 2, **3i**) [62], ^{1}H NMR (500 MHz, CDCl$_3$) δ 8.12 (d, *J* = 7.7 Hz, 1H), 7.83 (s, 1H), 7.69 (d, *J* = 8.3 Hz, 2H), 7.46 (td, *J* = 7.5, 1.5 Hz, 3H), 7.35 (d, *J* = 3.2 Hz, 1H), 7.19 (d, *J* = 3.9 Hz, 1H), 3.10–3.04 (t, 2H), 2.95 (t, *J* = 6.1 Hz, 2H); ^{13}C NMR (125 MHz, CDCl$_3$) δ 187.4, 143.1, 140.5, 138.2, 138.1, 134.1, 133.4, 132.3, 132.2, 128.8, 128.4, 128.3, 127.3, 126.7, 111.8, 111.1, 28.7, 23.3.

(E)-2-(4-nitrobenzylidene)-3,4-dihydronaphthalen-1(2H)-one (Scheme 2, **3j**) [60], ^{1}H NMR (500 MHz, CDCl$_3$) δ 8.30 (d, *J* = 1.9 Hz, 2H), 8.14 (d, *J* = 1.3 Hz, 1H), 7.85 (s, 1H), 7.59–7.55 (m, 3H), 7.40–7.38 (m, 1H), 7.30 (s, 1H), 3.15–3.07 (m, 2H), 3.00–2.96 (m, 2H); ^{13}C NMR (125 MHz, CDCl$_3$) δ 187.3, 147.4, 144.7, 138.6, 137.3, 133.8, 133.5, 132.7, 130.5, 128.8, 127.3, 126.7, 123.8, 28.8, 23.4.

(E)-2-(naphthalen-1-ylmethylene)-3,4-dihydronaphthalen-1(2H)-one (Scheme 2, **3k**) [27], Yellow solid (39 mg, 55% yield); ^{1}H NMR (500 MHz, CDCl$_3$) δ 8.30 (s, 1H), 8.14 (dd, *J* = 7.8, 1.1 Hz, 1H), 7.95–7.93 (m, 1H), 7.83–7.79 (m, 2H), 7.49–7.42 (m, 4H), 7.36–7.31 (m, 2H), 7.18 (t, *J* = 3.7 Hz, 1H), 2.93–2.91 (m, 2H), 2.85 (t, *J* = 6.5 Hz, 2H); ^{13}C NMR (125 MHz, CDCl$_3$) δ 186.8, 142.6, 136.5, 133.7, 132.6, 132.5, 132.4, 132.1, 131.0, 127.9, 127.5, 127.4, 127.3, 126.0, 125.8, 125.4, 125.2, 124.1, 123.9, 28.2, 26.7.

(E)-2-(benzo [d] [1,3]dioxol-5-ylmethylene)-3,4-dihydronaphthalen-1(2H)-one (Scheme 2, **3l**) [27], Yellow solid (61 mg, 88% yield); ^{1}H NMR (500 MHz, CDCl$_3$) δ 8.11 (dd, J = 7.8, 0.9 Hz, 1H), 7.79 (s, 1H), 7.49–7.46 (m, 1H), 7.35 (t, J = 7.5 Hz, 1H), 7.27–7.23 (m, 1H), 6.99–6.95 (m, 2H), 6.86 (d, J = 8.0 Hz, 1H), 6.00 (s, 2H), 3.13–3.11 (m, 2H), 2.96–2.93 (m, 2H); ^{13}C NMR (125 MHz, CDCl$_3$) δ 187.8, 148.0, 147.8, 143.1, 136.7, 134.0, 133.6, 133.2, 129.9, 128.2, 127.0, 125.1, 115.0, 109.8, 108.5, 101.4, 28.8, 27.3.

(E)-2-(furan-2-ylmethylene)-3,4-dihydronaphthalen-1(2H)-one (Scheme 2, **3m**) [27], Yellow solid (48 mg, 85% yield); ^{1}H NMR (500 MHz, CDCl$_3$) δ 8.11 (dd, J = 7.8, 1.3 Hz, 1H), 7.60–7.56 (m, 2H), 7.50–7.46 (m, 1H), 7.37–7.33 (m, 1H), 7.27 (d, J = 7.5 Hz, 1H), 6.71 (d, J = 3.5 Hz, 1H), 6.52 (dd, J = 3.4, 1.8 Hz, 1H), 3.36–3.31 (m, 2H), 3.03–2.99 (m, 2H); ^{13}C NMR (125 MHz, CDCl$_3$) δ 187.4, 152.5, 144.3, 143.5, 133.6, 133.1, 131.9, 128.1, 127.0, 122.8, 116.6, 115.0, 112.2, 28.4, 26.7.

(E)-2-(cyclohexylmethylene)-3,4-dihydronaphthalen-1(2H)-one (Scheme 2, **3n**), Yellow solid (41 mg, 69% yield); ^{1}H NMR (500 MHz, CDCl$_3$) δ 8.02 (dd, J = 7.8, 1.0 Hz, 1H), 7.40–7.36 (m, 1H), 7.25 (t, J = 7.5 Hz, 1H), 7.16 (d, J = 7.5 Hz, 1H), 6.70 (d, J = 9.7 Hz, 1H), 2.89–2.86 (m, 2H), 2.74–2.71 (m, 2H), 2.36–2.27 (m, 1H), 1.72–1.68 (m, 2H), 1.63–1.58 (m, 3H), 1.28–1.14 (m, 5H); ^{13}C NMR (125 MHz, CDCl$_3$) δ 186.9, 144.1, 142.7, 132.7, 132.3, 131.9, 127.2, 127.1, 125.8, 36.2, 31.2, 28.3, 24.9, 24.8, 24.6. HRMS (ESI): Calculated for [C$_{17}$H$_{21}$O]$^+$ 241.1587; Found 241.1589.

(E)-2-(cyclopropylmethylene)-3,4-dihydronaphthalen-1(2H)-one (Scheme 2, **3o**) [63], White solid (39 mg, 78% yield); ^{1}H NMR (500 MHz, CDCl$_3$) δ 8.01 (dd, J = 7.8, 1.1 Hz, 1H), 7.39–7.37 (m, 1H), 7.26 (t, J = 7.5 Hz, 1H), 7.18 (d, J = 7.6 Hz, 1H), 6.27 (d, J = 10.9 Hz, 1H), 2.93–2.86 (m, 2H), 2.86–2.83 (m, 2H), 1.67–1.60 (m, 1H), 0.97–0.93 (m, 2H), 0.68–0.65 (m, 2H); ^{13}C NMR (125 MHz, CDCl$_3$) δ 186.7, 145.6, 143.6, 133.7, 132.8, 132.7, 128.2, 128.1, 126.8, 29.0, 25.5, 11.8, 9.0.

2-butyl-3,4-dihydronaphthalen-1(2H)-one (Scheme 2, **3p**) [64], Pale yellow liquid (36 mg, 70% yield); ^{1}H NMR (500 MHz, CDCl$_3$) δ 7.96 (dd, J = 7.8, 1.0 Hz, 1H), 7.40–7.36 (m, 1H), 7.23 (d, J = 7.6 Hz, 1H), 7.16 (d, J = 7.6 Hz, 1H), 2.94–2.86 (m, 3H), 2.42–2.38 (m, 1H), 2.20–2.14(m, 1H), 1.90–1.80 (m, 2H), 1.44–1.40 (m, 1H), 1.36–1.28 (m, 3H), 0.85 (t, J = 7.0 Hz, 3H); ^{13}C NMR (125 MHz, CDCl$_3$) δ 199.5, 143.0, 132.0, 131.6, 127.6, 126.4, 125.5, 46.5, 28.2, 28.1, 27.3, 27.2, 21.8, 13.0.

2-pentyl-3,4-dihydronaphthalen-1(2H)-one (Scheme 2, **3q**) [65], Pale yellow liquid (29 mg, 53% yield); ^{1}H NMR (400 MHz, CDCl$_3$) δ 7.96 (d, J = 7.8 Hz, 1H), 7.40–7.36 (m, 1H), 7.23 (t, J = 7.5 Hz, 1H), 7.16 (d, J = 7.5 Hz, 1H), 2.98–2.84 (m, 3H), 2.44–2.37 (m, 1H), 2.20–2.13 (m, 1H), 1.91–1.78 (m, 2H), 1.35–1.21 (m, 6H), 0.83 (t, J = 6.6 Hz, 3H); ^{13}C NMR (100 MHz, CDCl$_3$) δ 199.5, 142.9, 132.0, 131.5, 127.6, 126.4, 125.5, 46.5, 30.9, 28.3, 27.3, 27.1, 25.7, 21.6, 13.1.

(E)-2-benzylidenecycloheptan-1-one (Scheme 2, **4a**) [66], Colorless liquid (29 mg, 58% yield); ^{1}H NMR (400 MHz, CDCl$_3$) δ 7.45 (s, 1H), 7.34–7.29 (m, 2H), 7.28–7.22 (m, 3H), 2.66–2.61 (m, 4H), 1.77–1.68 (m, 6H); ^{13}C NMR (100 MHz, CDCl$_3$) δ 205.0, 140.8, 136.1, 135.7, 129.5, 128.5, 43.5, 31.4, 30.0, 27.7, 25.5.

(E)-2-benzylidenecyclooctanone (Scheme 2, **4b**) [66], Colorless liquid (27 mg, 50% yield); ^{1}H NMR (500 MHz, CDCl$_3$) δ 7.40 (s, 1H), 7.31 (dd, J = 10.4, 2.8 Hz, 4H), 7.28–7.24 (m, 1H), 2.76–2.74 (m, 2H), 2.65 (dd, J = 7.3, 5.5 Hz, 2H), 1.81–1.76 (m, 2H), 1.71–1.66 (m, 2H), 1.60–1.55 (m, 2H), 1.48–1.43 (m, 2H); ^{13}C NMR (125 MHz, CDCl$_3$) δ 206.6, 139.5, 135.5, 135.1, 128.6, 127.4, 127.4, 38.3, 29.1, 28.6, 25.5, 24.9, 24.5.

(E)-2-benzylidenecyclopentadecanone (Scheme 2, **4c**), Colorless liquid (24 mg, 31% yield); ^{1}H NMR (500 MHz, CDCl$_3$) δ 7.43–7.41 (m, 5H), 7.37–7.34 (m, 1H), 2.84–2.81 (m, 2H), 2.65 (t, J = 7.0 Hz, 2H), 2.44 (t, J = 6.7 Hz, 1H), 1.78–1.75 (m, 2H), 1.49 (dd, J = 9.1, 6.2 Hz, 2H), 1.39–1.29 (m, 17H); ^{13}C NMR (125 MHz, CDCl$_3$) δ 204.0, 143.3, 137.2, 136.1, 129.2, 128.5, 128.2, 42.1, 37.6, 28.6, 27.8, 27.6, 27.5, 26.8, 26.7, 26.5, 26.4, 26.2, 26.1, 24.3, 23.5. HRMS (ESI): Calculated for [C$_{22}$H$_{33}$O]$^+$ 313.2526; Found 241.2528.

(E)-2-benzylidene-7-methoxy-3,4-dihydronaphthalen-1(2H)-one (Scheme 2, **4d**) [66], Yellow solid (48 mg, 72% yield); ^{1}H NMR (500 MHz, CDCl$_3$) δ 7.89 (s, 1H), 7.66 (d, J = 2.8 Hz, 1H), 7.48–7.43 (m, 4H), 7.40–7.37 (m, 1H), 7.19 (d, J = 8.3 Hz, 1H), 7.10 (dd, J = 8.3, 2.8 Hz, 1H), 3.90 (s, 3H), 3.15–3.13 (m, 2H),

2.93–2.42 (m, 2H); ^{13}C NMR (125 MHz, CDCl$_3$) δ 188.0, 158.9, 136.9, 136.2, 136.1, 135.7, 134.5, 130.0, 129.6, 128.7, 128.6, 121.7, 110.5, 55.7, 28.2, 27.6.

(E)-2-benzylidene-5,6-dimethoxy-2,3-dihydro-1H-inden-1-one (Scheme 2, **4f**) [67], Yellow solid (21 mg, 30% yield); ^{1}H NMR (500 MHz, CDCl$_3$) δ 7.69 (d, *J* = 7.3 Hz, 2H), 7.63 (t, *J* = 1.8 Hz, 1H), 7.48 (t, *J* = 7.5 Hz, 2H), 7.42 (t, *J* = 2.0 Hz, 1H), 7.38 (s, 1H), 7.02 (s, 1H), 4.03 (s, 3H), 4.01 (d, *J* = 1.6 Hz, 2H), 3.98 (s, 3H); ^{13}C NMR (125 MHz, CDCl$_3$) δ 206.5, 155.6, 148.9, 141.8, 139.8, 139.7, 135.4, 132.4, 128.9, 128.8, 128.0, 115.0, 107.4, 56.2, 56.1, 31.9.

(E)-3-(4-methoxyphenyl)-2-methyl-1-phenylprop-2-en-1-one (Scheme 2, **4g**) [27], Colorless oil (40 mg, 64% yield); ^{1}H NMR (500 MHz, CDCl$_3$) δ 7.74 (dd, *J* = 8.2, 1.3 Hz, 2H), 7.57–7.54 (m, 1H), 7.49–7.46 (m, 2H), 7.44–7.42 (m, 2H), 7.18 (s, 1H), 6.97–6.95 (m, 2H), 3.87 (s, 3H), 2.31 (d, *J* = 1.3 Hz, 3H); ^{13}C NMR (125 MHz, CDCl$_3$) δ 199.6, 160.0, 142.6, 139.0, 134.9, 131.6, 131.4, 129.4, 128.4, 128.1, 115.0, 114.0, 55.4, 14.4.

(E)-2-(4-methoxybenzylidene)-1-phenylpentan-1-one (Scheme 2, **4h**) [68], Colorless oil liquid. (26 mg, 43% yield); ^{1}H NMR (500 MHz, CDCl$_3$) δ 7.77 (dd, *J* = 8.3, 1.3 Hz, 2H), 7.59–7.53 (m, 1H), 7.47 (dd, *J* = 9.7, 4.0 Hz, 2H), 7.39–7.33 (m, 3H), 6.95 (d, *J* = 8.8 Hz, 2H), 3.87 (s, 3H), 2.78–2.74 (m, 2H), 1.67–1.62 (m, 2H), 1.05 (t, *J* = 7.4 Hz, 3H); ^{13}C NMR (125 MHz, CDCl$_3$) δ 199.5, 159.9, 141.4, 140.4, 139.1, 133.2, 131.0, 129.5, 128.5, 128.2, 114.0, 55.3, 29.6, 22.0, 14.4.

(1S,4S)-3-(benzylidene)-1,7,7-trimethylbicyclo [2.2.1]heptan-2-one (Scheme 2, **4j**) [69], Colorless liquid (51 mg, 85% yield); ^{1}H NMR (500 MHz, CDCl$_3$) δ 7.51–7.50 (m, 2H), 7.43–7.40 (m, 2H), 7.38–7.34 (m, 1H), 7.27 (s, 1H), 3.14 (d, *J* = 4.2 Hz, 1H), 2.24–2.18 (m, 1H), 1.82–1.79 (m, 1H), 1.63–1.53 (m, 2H), 1.06 (s, 3H), 1.03 (s, 3H), 0.83 (s, 3H); ^{13}C NMR (125 MHz, CDCl$_3$) δ 208.2, 142.1, 135.7, 129.8, 128.7, 128.6, 127.5, 57.1, 49.2, 46.7, 30.7, 26.0, 20.6, 18.3, 9.3.

2-methyl-1-phenylpent-1-en-3-one (Scheme 2, **4k**) [70], Colorless liquid (15 mg, 35% yield); ^{1}H NMR (500 MHz, CDCl$_3$) δ 7.46 (d, *J* = 1.2 Hz, 1H), 7.34 (d, *J* = 4.4 Hz, 3H), 7.28 (d, *J* = 4.1 Hz, 1H), 7.19 (s, 1H), 2.78 (q, *J* = 7.3 Hz, 2H), 2.00 (s, 3H), 1.11 (t, *J* = 7.3 Hz, 3H); ^{13}C NMR (125 MHz, CDCl$_3$) δ 203.0, 138.2, 137.2, 136.1, 129.7, 128.4, 128.4, 30.8, 13.2, 8.9.

4. Conclusions

In conclusion, we have established a simple, cost-efficient and commercially available iron catalyst system for the dehydrogenative coupling of alcohols and ketones to α, β-unsaturated ketone. The process is highly selective (>25:1), and a variety of benzyl alcohols bearing nitro, nitrile, trifluoro-methyl fluoro, chloro and 1,3-dioxolone-functionalities, 2-furfurylmethanol, as well as cyclic and acyclic alkyl alcohols, were well tolerated under the optimized reaction conditions. Preliminary mechanistic investigations establish the hydrogen-borrowing process. As a highlight, we have demonstrated the synthesis of a sunscreen ingredient **4j** (Scheme 2).

Supplementary Materials: The following are available online. 1H-NMR, 13C-NMR, HRMS data and deuterium labeling experiments.

Author Contributions: M.S. and A.K. conducted experimental work and analyzed the data; M.S., A.K., J.D., and D.B. initiated the project, and designed experiments to develop this reaction. J.D. and D.B. wrote the paper. All of the authors have read and agreed to the published version of the manuscript.

Funding: This research has been funded by DAE-BRNS, India (Young Scientist Research Award to D.B., 37(2)/20/33/2016-BRNS).

Acknowledgments: The authors thank IIT Roorkee, SMILE-32, for providing the infrastructure and instrumentation facilities. DST (FIST) is gratefully acknowledged for the HRMS. M.S. thanks DST/2018/IF180079, A.K. thanks UGC, and J.D. thanks IIT-R for financial support.

Conflicts of Interest: The authors declare no conflict of interest.

References

1. Zhuang, C.; Zhang, W.; Sheng, C.; Zhang, W.; Xing, C.; Miao, Z. Chalcone: A Privileged Structure in Medicinal Chemistry. *Chem. Rev.* **2017**, *117*, 7762–7810. [CrossRef] [PubMed]

2. Climent, M.J.; Corma, A.; Iborra, S.; Velty, A. Activated Hydrotalcites as Catalysts for the Synthesis of Chalcones of Pharmaceutical Interest. *J. Catal.* **2004**, *221*, 474–482. [CrossRef]

3. Li, R.; Kenyon, G.L.; Cohen, F.E.; Chen, X.; Gong, B.; Dominguez, J.N.; Davidson, E.; Kurzban, G.; Miller, R.E.; Nuzman, E.O. Vitro Antimalarial Activity of Chalcones and Their Derivatives. *J. Med. Chem.* **1995**, *38*, 5031–5037. [CrossRef]

4. Dhar, D.N. *Chemistry of Chalcones and Related Compounds*; Wiley: New York, NY, USA, 1981.

5. Harbone, J.B.; Mabry, T.J. *The Flavonoids: Advances in Research*; Chapman & Hall: New York, NY, USA, 1982.

6. Micheli, F.; Degiorgis, F.; Feriani, A.; Paio, A.; Pozzan, A.; Zarantonello, P.; Seneci, P.A. Combinatorial Approach to [1,5]- Benzothiazepine Derivatives as Potential Antibacterial Agents. *J. Comb. Chem.* **2001**, *3*, 224–228. [CrossRef] [PubMed]

7. Sinisterra, J.V.; Garcia-Raso, J.V.; Cabello, J.A.; Marinas, J.M. An Improved Procedure for the Claisen-Schmidt Reaction. *Synthesis* **1984**, *1984*, 502–504. [CrossRef]

8. Crabtree, R.H. Homogeneous Transition Metal Catalysis of Acceptorless Dehydrogenative Alcohol Oxidation: Applications in Hydrogen Storage and to Heterocycle Synthesis. *Chem. Rev.* **2017**, *117*, 9228–9246. [CrossRef]

9. Corma, A.; Navas, J.; Sabater, M.J. Advances in One-Pot Synthesis through Borrowing Hydrogen Catalysis. *Chem. Rev.* **2018**, *118*, 1410–1459. [CrossRef]

10. Irrgang, T.; Kempe, R. 3d-Metal Catalyzed N- and C–Alkylation Reactions via Borrowing Hydrogen or Hydrogen Autotransfer. *Chem. Rev.* **2019**, *119*, 2524–2549. [CrossRef]

11. Yang, Q.; Wanga, Q.; Yu, Z. Substitution of alcohols by N-nucleophiles via transition metal-catalyzed dehydrogenation. *Chem. Soc. Rev.* **2015**, *44*, 2305–2329. [CrossRef]

12. Reed-Berendt, B.G.; Polidano, K.; Morrill, L.C. Recent Advances in Homogeneous Borrowing Hydrogen Catalysis using Earth-abundant First Row Transition Metals. *Org. Biomol. Chem.* **2019**, *17*, 1595–1607. [CrossRef]

13. Barta, K.; Ford, P.C. Catalytic Conversion of Nonfood Woody Biomass Solids to Organic Liquids. *Acc. Chem. Res.* **2014**, *47*, 1503–1512. [CrossRef] [PubMed]

14. Vispute, T.P.; Zhang, H.; Sanna, A.; Xiao, R.; Huber, G.W. Renewable Chemical Commodity Feedstocks from Integrated Catalytic Processing of Pyrolysis Oils. *Science* **2010**, *330*, 1222–1227. [CrossRef] [PubMed]

15. Tuck, C.O.; Pérez, E.; Horváth, I.T.; Sheldon, R.A.; Poliakoff, M. Valorization of Biomass: Deriving More Value from Waste. *Science* **2012**, *337*, 695–699. [CrossRef] [PubMed]

16. Martínez, R.; Ramón, D.J.; Yus, M. Easy α-Alkylation of Ketones with Alcohols Through a Hydrogen Autotransfer Process Catalyzed by $RuCl_2(DMSO)_4$. *Tetrahedron* **2006**, *62*, 8988–9001. [CrossRef]

17. Kwon, M.S.; Kim, N.; Seo, S.H.; Park, I.S.; Cheedrala, R.K.; Park, J. Recyclable Palladium Catalyst for Highly Selective α-Alkylation of Ketones with Alcohols. *Angew. Chem.* **2005**, *117*, 7073–7075. [CrossRef]

18. Jana, S.K.; Kubota, Y.; Tatsumi, T. Selective α-Alkylation of Ketones with Alcohols Catalyzed by Highly Active Mesoporous $Pd/MgO-Al_2O_3$ Type Basic Solid Derived from Pd-Supported Mg Al-Hydrotalcite Stud. *Surf. Sci. Catal.* **2007**, *165*, 701–704.

19. Chaudhari, C.; Siddiki, S.M.A.H.; Kon, K.; Tomita, A.; Tai, Y.; Shimizu, K. C-3 Alkylation of Oxindole with Alcohols by Pt/CeO_2 Catalyst in Additive-Free Conditions. *Catal. Sci. Technol.* **2014**, *4*, 1064–1069. [CrossRef]

20. Kim, S.; Bae, S.W.; Lee, J.S.; Park, J. Recyclable Gold Nanoparticle Catalyst for the Aerobic Alcohol Oxidation and C–C Bond Forming Reaction Between Primary Alcohols and Ketones Under Ambient Conditions. *Tetrahedron* **2009**, *65*, 1461–1466. [CrossRef]

21. Fischer, A.; Makowski, P.; Müller, J.-O.; Antonietti, M.; Thomas, A.; Goettmann, F. High Surface Area TiO_2 and TiN as Catalysts for the C–C Coupling of Alcohols and Ketones. *Chem Sus Chem* **2008**, *1*, 444–449. [CrossRef]

22. Yao, W.; Makowski, P.; Giordano, C.; Goettmann, F. Synthesis of Early-Transition-Metal Carbide and Nitride Nanoparticles Through the Urea Route anTheir Use as Alkylation Catalysts. *Chem.-Eur. J.* **2009**, *15*, 11999–12004. [CrossRef]

23. Li, Y.; Chen, D. Novel and Efficient One Pot Condensation Reactions between Ketones and Aromatic Alcohols in the Presence of CrO$_3$ Producing α, β-Unsaturated Carbonyl Compounds. *Chin. J. Chem.* **2011**, *29*, 2086–2090. [CrossRef]

24. Zhang, Z.; Wang, Y.; Wang, M.; Lu, J.; Zhang, C.; Li, L.; Jiang, J.; Wang, F. The Cascade Synthesis of α, β-Unsaturated Ketones via Oxidative C–C Coupling of Ketones and Primary Alcohols Over a Ceria Catalyst. *Catal. Sci. Technol.* **2016**, *6*, 1693–1700. [CrossRef]

25. Liu, X.; Li, H.; Ma, J.; Yu, X.; Wang, Y.; Li, J. Preparation of a Bi$_2$WO$_6$ catalyst and its catalytic performance in an alpha alkylation reaction under visible light irradiation. *Mol. Catal.* **2019**, *466*, 157–166. [CrossRef]

26. Zou, H.; Daib, J.; Wang, R. Encapsulating mesoporous metal nanoparticles: Towards a highly active and stable nanoreactor for oxidative coupling reactions in water. *Chem. Commun.* **2019**, *55*, 5898–5901. [CrossRef] [PubMed]

27. Gawali, S.S.; Pandia, B.K.; Gunanathan, C. Manganese (I)-Catalyzed α–Alkenylation of Ketones Using Primary Alcohols. *Org. Lett.* **2019**, *21*, 3842–3847. [CrossRef]

28. Das, J.; Singh, K.; Vellakkaran, M.; Banerjee, D. Nickel-Catalyzed Hydrogen-Borrowing Strategy for α-Alkylation of Ketones with Alcohols: A New Route to Branched gem-Bis(alkyl) Ketones. *Org. Lett.* **2018**, *20*, 5587–5591. [CrossRef]

29. Das, J.; Vellakkaran, M.; Banerjee, D. Nickel-Catalyzed Alkylation of Ketone Enolates: Synthesis of Monoselective Linear Ketones. *J. Org. Chem.* **2019**, *84*, 769–779. [CrossRef]

30. Das, J.; Vellakkaran, M.; Banerjee, D. Nickel-catalysed direct α-olefination of alkyl substituted N-heteroarenes with alcohols. *Chem. Commun.* **2019**, *55*, 7530–7533. [CrossRef]

31. Kabadwal, L.M.; Das, J.; Banerjee, D. Mn(II)-catalysed alkylation of methylene ketones with alcohols: Direct access to functionalised branched products. *Chem. Commun.* **2018**, *54*, 14069–14072. [CrossRef]

32. Das, J.; Vellakkaran, M.; Sk, M.; Banerjee, D. Iron-Catalyzed Coupling of Methyl N-Heteroarenes with Primary Alcohols: Direct Access to E-Selective Olefins. *Org. Lett.* **2019**, *21*, 7514–7518. [CrossRef]

33. Alanthadka, A.; Bera, S.; Banerjee, D. Iron-Catalyzed Ligand Free α-Alkylation of Methylene Ketones and β-Alkylation of Secondary Alcohols Using Primary Alcohols. *J. Org. Chem.* **2019**, *84*, 11676–11686. [CrossRef] [PubMed]

34. Morris, R.H. Exploiting Metal–Ligand Bifunctional Reactions in the Design of Iron Asymmetric Hydrogenation Catalysts. *Acc. Chem. Res.* **2015**, *48*, 1494–1502. [CrossRef] [PubMed]

35. Chirik, P.J. Iron- and Cobalt-Catalyzed Alkene Hydrogenation: Catalysis with Both Redox-Active and Strong Field Ligands. *Acc. Chem. Res.* **2015**, *48*, 1687–1695. [CrossRef] [PubMed]

36. Bauer, I.; Knolker, H.J. Iron Catalysis in Organic Synthesis. *Chem. Rev.* **2015**, *115*, 3170–3387. [CrossRef]

37. Filonenko, G.A.; Putten, R.V.; Hensen, E.J.M.; Pidko, E.A. Catalytic (de)hydrogenation promoted by non-precious metals–Co, Fe and Mn: Recent advances in an emerging field. *Chem. Soc. Rev.* **2018**, *47*, 1459–1483. [CrossRef]

38. Wei, D.; Darcel, C. Iron Catalysis in Reduction and Hydrometalation Reactions. *Chem. Rev.* **2019**, *119*, 2550–2610. [CrossRef]

39. Balaraman, E.; Nandakumar, A.; Jaiswalab, G.; Sahoo, M.K. Iron-catalyzed dehydrogenation reactions and their applications in sustainable energy and catalysis. *Catal. Sci. Technol.* **2017**, *7*, 3177–3195. [CrossRef]

40. Wei, D.; Netkaew, C.; Darcel, C. Multi-Step Reactions Involving Iron-Catalysed Reduction and Hydrogen Borrowing Reactions. *Eur. J. Inorg. Chem.* **2019**, 2471–2487. [CrossRef]

41. Kessler, S.N.; Bäckvall, J.E. Iron-Catalyzed Cross-Coupling of Propargyl Carboxylates and Grignard Reagents: Synthesis of Substituted Allenes. *Angew. Chem. Int. Ed.* **2016**, *55*, 3734–3738. [CrossRef]

42. Fürstner, A.; Leitner, A.; Mendez, M.; Krause, H. Iron-Catalyzed Cross-Coupling Reactions. *J. Am. Chem. Soc.* **2002**, *124*, 13856–13863. [CrossRef]

43. Sui-Seng, C.; Freutel, F.; Lough, A.J.; Morris, R.H. Highly Efficient Catalyst Systems Using Iron Complexes with a Tetradentate PNNP Ligand for the Asymmetric Hydrogenation of Polar Bonds. *Angew. Chem. Int. Ed.* **2008**, *47*, 940–943. [CrossRef] [PubMed]

44. Zuo, W.; Lough, A.J.; Li, Y.F.; Morris, R.H. Amine (imine)-diphosphine Iron Catalysts for Asymmetric Transfer Hydrogenation of Ketones and Imines. *Science* **2013**, *342*, 1080–1083. [CrossRef]

45. Kessler, S.N.; Hundemer, F.; Bäckvall, J.E. A Synthesis of Substituted α-Allenols via Iron-Catalyzed Cross-Coupling of Propargyl Carboxylates with Grignard Reagents. 2016, 6, 7448–7451. *ACS Catal.* **2016**, *6*, 7448–7451. [CrossRef] [PubMed]

46. Yan, T.; Feringa, B.L.; Barta, K. Iron Catalysed Direct Alkylation of Amines with Alcohols. *Nat. Commun.* **2014**, *5*, 5602. [CrossRef]

47. Polidano, K.; Allen, B.D.W.; Williams, J.M.J.; Morrill, L.C. Iron- Catalyzed Methylation Using the Borrowing Hydrogen Approach. *ACS Catal.* **2018**, *8*, 6440–6445. [CrossRef]

48. Brown, T.J.; Cumbes, M.; Diorazio, L.J.; Clarkson, G.J.; Wills, M. Use of (Cyclopentadienone)iron Tricarbonyl Complexes for C−N Bond Formation Reactions between Amines and Alcohols. *J. Org. Chem.* **2017**, *82*, 10489–10503. [CrossRef]

49. Chakraborty, S.; Lagaditis, P.O.; Förster, M.; Bielinski, A.E.; Hazari, N.; Holthausen, M.C.; Jones, W.D.; Schneider, S. Well-Defined Iron Catalysts for the Acceptorless Reversible dehydrogenation-Hydrogenation of Alcohols and Ketones. *ACS Catal.* **2014**, *4*, 3994–4003. [CrossRef]

50. Shaikh, N.S.; Enthaler, S.; Junge, K.; Beller, M. Iron-Catalyzed Enantioselective Hydrosilylation of Ketones. *Angew. Chem. Int. Ed.* **2008**, *47*, 2497–2501. [CrossRef]

51. Jiang, F.; Bézier, D.; Sortais, J.-B.; Darcel, C. N-Heterocyclic Carbene Piano-Stool Iron Complexes as Efficient Catalysts for Hydrosilylation of Carbonyl Derivatives. *Adv. Synth. Catal.* **2011**, *353*, 239–244. [CrossRef]

52. Tondreau, A.M.; Atienza, C.C.H.; Weller, K.J.; Nye, S.A.; Lewis, K.M.; Delis, J.G.; Chirik, P.J. Iron Catalysts for Selective Anti-Markovnikov Alkene Hydrosilylation Using Tertiary Silanes. *Science* **2012**, *335*, 567–570. [CrossRef]

53. Gorgas, N.; Alves, L.G.; Stoger, B.; Martins, A.M.; Veiros, L.F.; Kirchner, K. Stable, Yet Highly Reactive Nonclassical Iron (II) Polyhydride Pincer Complexes: Z-Selective Dimerization and Hydroboration of Terminal Alkynes. *J. Am. Chem. Soc.* **2017**, *139*, 8130–8133. [CrossRef] [PubMed]

54. Zhang, F.; Song, H.; Zhuang, X.; Tung, C.-H.; Wang, W. Iron-Catalyzed 1,2-Selective Hydroboration of N-Heteroarenes. *J. Am. Chem. Soc.* **2017**, *139*, 17775–17778. [CrossRef] [PubMed]

55. Bettoni, L.; Seck, C.; Mbaye, M.D.; Gaillard, S.; Renaud, J.L. Iron-Catalyzed Tandem Three-Component Alkylation: Access to α-Methylated Substituted Ketones. *Org. Lett.* **2019**, *21*, 3057–3061. [CrossRef] [PubMed]

56. Elangovan, S.; Sortais, J.B.; Beller, M.; Darcel, C. Iron-Catalyzed α-Alkylation of Ketones with Alcohols. *Angew. Chem. Int. Ed.* **2015**, *54*, 14483–14486. [CrossRef] [PubMed]

57. Seck, C.; Mbaye, M.D.; Coufourier, S.; Lator, A.; Lohier, J.-F.; Poater, A.; Ward, T.R.; Gaillard, S.; Renaud, J.-L. Alkylation of Ketones Catalyzed by Bifunctional Iron Complexes: From Mechanistic Understanding to Application. *ChemCatChem* **2017**, *9*, 4410–4416. [CrossRef]

58. Polidano, K.; Williams, M.J.J.; Morrill, L.C. Iron-Catalyzed Borrowing Hydrogen β—C (sp^3) -Methylation of Alcohols. *ACS Catal.* **2019**, *9*, 8575–8580. [CrossRef]

59. Dambatta, M.B.; Polidano, K.; Northey, A.D.; Williams, M.J.J.; Morrill, L.C. Iron-Catalyzed Borrowing Hydrogen C-Alkylation of Oxindoles with Alcohols. *ChemSusChem* **2019**, *12*, 2345–2349. [CrossRef]

60. Ceylan, M.; Kocyigit, U.M.; Usta, N.C.; Gürbüzlü, B.; Temel, Y.; Alwasel, S.H.; Gülçin, Y. Synthesis, carbonic anhydrase I and II isoenzymes inhibition properties, and antibacterial activities of novel tetralone-based 1,4-benzothiazepine derivatives. *J. Biochem. Mol. Toxicol.* **2016**, *31*, 1–11. [CrossRef]

61. Feng, L.; Lanfranchi, D.A.; Cotos, L.; Rodo, E.C.; Ehrhardt, K.; Alice-Anne Goetz, A.A.; Zimmermann, H.; Fenaille, F.; Blandin, S.A.; Charvet, D.E. Synthesis of plasmodione metabolites and 13C-enriched plasmodione as chemical tools for drug metabolism investigation. *Org. Biomol. Chem.* **2018**, *16*, 2647–2665. [CrossRef] [PubMed]

62. Wagner, G.; Horn, H.; Eppner, B.; Kuehmstedt, H. Synthesis of 2-(3- and 4-amidinobenzylidene) Derivatives of 1-indanone, 1-tetralone and benzosuberone. *Pharmazie* **1979**, *34*, 56.

63. Trost, B.M.; Stambuli, J.P.; Silverman, S.M.; Schworer, U. Palladium-Catalyzed Asymmetric [3 + 2] Trimethylenemethane Cycloaddition Reactions. *J. Am. Chem. Soc.* **2006**, *128*, 13328–13329. [CrossRef]

64. Yokota, M.; Fujita, D.; Ichikawa, J. Activation of 1,1-Difluoro-1-alkenes with a Transition-Metal Complex: Palladium(II)-Catalyzed Friedel Crafts Type Cyclization of 4,4-(Difluorohomoallyl)arenes. *Org. Lett.* **2007**, *9*, 4639–4642. [CrossRef] [PubMed]

65. Dhakshinamoorthy, A.; Alvaro, M.; Garciaa, H. Claisen–Schmidt Condensation Catalyzed by Metal-Organic Frameworks. *Adv. Synth. Catal.* **2010**, *352*, 711–717. [CrossRef]

66. Chu, G.H.; Li, P.K. Synthesis of Naphthalenic Melatonin Receptor Ligands. *Synth. Commun.* **2001**, *31*, 621–629. [CrossRef]

67. Lantaño, B.; Aguirre, J.M.; Drago, E.V.; Bollini, M.; Faba, D.J.; Mufato, J.D. Synthesis of Benzylidenecycloalkan-1-ones and 1,5-diketones under Claisen–Schmidt Reaction: Influence of the Temperature and Electronic Nature of Arylaldehydes. *Synth. Commun.* **2017**, *47*, 2202–2214. [CrossRef]
68. Zhou, X.; Zhao, Y.; Cao, Y.; He, L. Catalytic Efficient Nazarov Reaction of Unactivated Aryl Vinyl Ketones via a Bidentate Diiron Lewis Acid Activation Strategy. *Adv. Synth. Catal.* **2017**, *359*, 3325–3331. [CrossRef]
69. Rahman, A.F.M.; Ali, R.; Jahng, Y.; Kadi, A.A. A Facile Solvent Free Claisen-Schmidt Reaction: Synthesis of α, α'-*bis*-(Substituted-benzylidene) cycloalkanones and α, α'-*bis*-(Substituted-alkylidene)cycloalkanones. *Molecules* **2012**, *17*, 571–583. [CrossRef]
70. Bechara, W.S.; Pelletier, G.; Charette, A.B. Chemoselective Synthesis of Ketones and Ketimines by Addition of Organometallic Reagents to Secondary Amides. *Nat. Chem.* **2012**, *4*, 228–234. [CrossRef]

Sample Availability: Samples of the compounds are not available from the authors.

 *molecules*

Article

Revealing the Iron-Catalyzed β-Methyl Scission of *tert*-Butoxyl Radicals via the Mechanistic Studies of Carboazidation of Alkenes

Mong-Feng Chiou [1], Haigen Xiong [1,2], Yajun Li [1], Hongli Bao [1,2,*] and Xinhao Zhang [3,*]

[1] Key Laboratory of Coal to Ethylene Glycol and Its Related Technology, State Key Laboratory of Structural Chemistry, Center for Excellence in Molecular Synthesis, Fujian Institute of Research on the Structure of Matter, Chinese Academy of Sciences, 155 Yangqiao Road West, Fuzhou 350002, Fujian, China; qiumengfeng@fjirsm.ac.cn (M.-F.C.); xionghaigen@fjirsm.ac.cn (H.X.); liyajun@fjirsm.ac.cn (Y.L.)

[2] School of Chemistry and Chemical Engineering of University of Chinese Academy of Sciences, Beijing 100049, China

[3] Lab of Computational Chemistry and Drug Design, Key Laboratory of Chemical Genomics, Peking University Shenzhen Graduate School, Shenzhen 518055, China

* Correspondence: hlbao@fjirsm.ac.cn (H.B.); zhangxh@pkusz.edu.cn (X.Z.); Tel.: +86-0591-63179307 (H.B.); +86-0755-26037219 (X.Z.)

Academic Editor: Hans-Joachim Knölker
Received: 26 December 2019; Accepted: 5 March 2020; Published: 9 March 2020

Abstract: We describe here a mechanistic study of the iron-catalyzed carboazidation of alkenes involving an intriguing metal-assisted β-methyl scission process. Although t-BuO radical has frequently been observed in experiments, the β-methyl scission from a t-BuO radical into a methyl radical and acetone is still broadly believed to be thermodynamically spontaneous and difficult to control. An iron-catalyzed β-methyl scission of t-BuO is investigated in this work. Compared to a free t-BuO radical, the coordination at the iron atom reduces the activation energy for the scission from 9.3 to 3.9 ~ 5.2 kcal/mol. The low activation energy makes the iron-catalyzed β-methyl scission of t-BuO radicals almost an incomparably facile process and explains the selective formation of methyl radicals at low temperature in the presence of some iron catalysts. In addition, a radical relay process and an outer-sphere radical azidation process in the iron-catalyzed carboazidation of alkenes are suggested by density functional theory (DFT) calculations.

Keywords: iron-catalysis; carboazidation; β-methyl scission; radical; DFT

1. Introduction

The carboazidation of alkenes, a powerful and promising method for the synthesis of amino acid precursors and other useful building blocks, has attracted much attention recently [1–8]. Iron-catalyzed carboazidation of alkenes has recently been developed by our group in which *tert*-butyl peroxybenzoate (TBPB) was employed as the initiator (Scheme 1a) [9].

tert-Butoxy-containing peroxides, including di-*tert*-butyl peroxide (DTBP), [10,11] *tert*-butyl hydroperoxide (TBHP), [12–21] and *tert*-butyl peroxybenzoate (TBPB), [22–32] have versatile roles in organic synthesis and have been proven to be good sources of t-BuO radical. However, these peroxides can also occasionally serve as a source of methyl radicals (Scheme 1b) [33–37]. The β-methyl scission of alkoxy radicals which is a common fragmentation process forming corresponding alkyl radicals was discovered more than fifty years ago [38], and is described in organic chemical textbooks [39]. The β-methyl scission from a t-BuO radical accordingly is believed to be an easily spontaneous process, [40–44] and offers a facile pathway to methyl radicals; however, it is inconsistent with the common experimental observation of t-BuO radical [25–32].

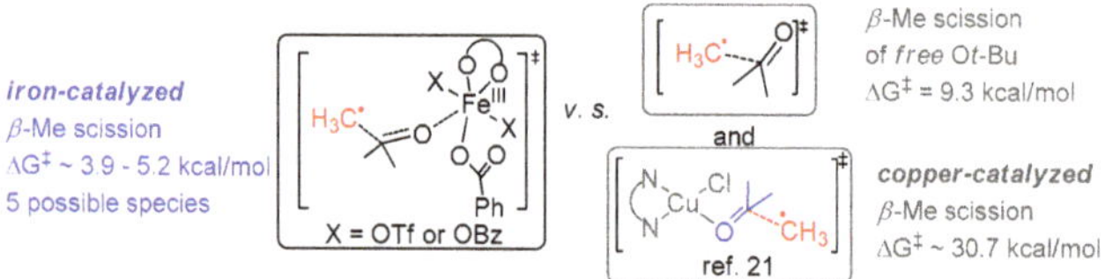

Scheme 1. (**a**) Carboazidation of alkenes in previous study, [9] (**b**) selective formation of methyl radicals [33–37] or t-BuO radical, [10–32] and (**c**) mechanistic studies.

Although β-methyl scission from a t-BuO radical can afford methyl radical, the factors which determine the selective formation of methyl radicals or retaining as t-BuO radical are still unclear. To the best of our knowledge, no further investigation of t-BuO radical splitting has been reported. Very recently, our studies suggested that a copper catalyst may not assist the β-methyl scission (Scheme 1c, 30.7 kcal/mol) and the t-BuO radical can become untethered which serves as the radical initiator and does not proceed β-methyl scission [21]. On the other hand, the selective or dominant formation of methyl radicals in iron-catalyzed reactions has frequently been observed in our previous work [33–35]. It is questionable why the t-BuO radical behaves very differently when catalyzed by iron.

Herein, the crucial factors for the selective formation of methyl radicals have been investigated and a rare iron-catalyzed β-methyl scission was revealed (Scheme 1c). In addition, experimental and theoretical investigations were conducted to support a radical relay mechanism for carboazidation reactions [9].

2. Results and Discussion

Experiments exploring the carboazidation reactions of alkenes were conducted to probe the mechanism. First, the reaction was conducted in the absence of alkyl iodide, as expected, a methyl adduct **2**, (1-azidopropyl)benzene, was obtained in 52% yield (Scheme 2a). The reaction employing *tert*-butyl ethaneperoxoate as initiator instead of TBPB also delivers the desired product (**2**) in 33% yield demonstrating that the methyl radical can be easily generated at room temperature under these conditions. Because of the absence of diazidation product which can indirectly prove the existence of t-BuO radial [45] in all cases under standard conditions or these two conditions, the formation of methyl radical can be regarded as highly selective. Next, (2-phenylcyclopropyl)styrene (**3**), a radical clock compound, for which the rate constant of the ring opening step is approximately 10^8 s^{-1}, was used and afforded a ring-opened product (**4**) in 42% yield (Scheme 2b) [46]. A ring closure reaction was also conducted with 1,6-heptadiene **5** and the ring-closure products (**6** and **7**) and non-ring-closed product (**8**) were obtained in 80% and 18% yields, respectively (Scheme 2c). This result suggests that the azidation step is quite fast and is comparable to 5-exo-cyclization of 5-hexenyl radical (~10^5/s). Besides, radical scavengers, 2,6-di-tert-butyl-4-methylphenol (BHT) or hydroquinone, can interrupt the standard reaction to reduce the yield of product (See Preliminary mechanistic studies in supporting information). These results are consistent with a radical mechanism [47,48].

Interestingly, the acetone and methyl iodide formed under standard conditions could be observed by GC-MS (See supporting information), implying that the radical relay process possibly begins with a methyl radical.

(a)

Ph + TBPB + TMSN$_3$ $\xrightarrow[\text{rt, 30 min}]{\text{Fe(OTf)}_2\ (5\ \text{mol%})\quad\text{DME (2 mL)}}$ **2**

1a, 0.5 mmol   2.0 equiv   2.0 equiv

52%(33%[a])

(b)

3, 0.5 mmol + *n*-C$_4$F$_9$I + TMSN$_3$ $\xrightarrow[\text{DME (2 mL)}\ \text{rt, 30 min}]{\text{Fe(OTf)}_2\ (5\ \text{mol %})\quad\text{TBPB (1.5 equiv)}}$ **4**, 42%, *E/Z* = 17:1

0.65 mmol   0.7 mmol

(c)

5, 0. 5 mmol + C$_8$F$_{17}$I + TMSN$_3$ $\xrightarrow[\text{total yield}\ 98\%]{\text{Fe(OTf)}_2\ (5\ \text{mol %})\quad\text{TBPB (2 equiv)}}$ **6** + **7** + **8**

1.3 equiv   2.0 equiv

6 31%, dr = 3.8:1   **7** 49%, dr = 5.2:1   **8** 18%

[a]*tert*-butyl ethaneperoxoate (CH$_3$CO$_3$C(CH$_3$)$_3$) was used in place of TBPB.

Scheme 2. Experimental studies. (**a**) Participation of the methyl radical within the caboazidation of alkenes in the absence of further alkyl iodides, (**b**) and (**c**) Ring-opening and ring-closing experiments for exploring the radical relay mechanism.

Next, density functional theory (DFT) calculations on the iron-catalyzed carboazidation of styrene were performed in an attempt to understand the mechanism at the atomic level [49,50]. Figure 1a shows the overall potential energy surface of the iron-catalyzed reaction. According to the systematic computations on the spin states and the conformations of iron species, a quintet state of catalyst Fe(OTf)$_2$ (^{5}**INT1**) coordinated by two 1,2-dimethoxyethane (DME) molecules, was found to have the lowest free energy (Table S1), and thus can be considered as the starting catalyst for first cycle [51,52].

The interaction of TBPB with ^{5}**INT1** yields ^{5}**INT2** by an associative ligand exchange process. An associative intermediate, ^{5}**INT12**, with a relative free energy of 8.3 kcal/mol can be considered as a barrier to the ligand exchange (Supplementary Figure S1) [53,54]. Subsequently, a single electron transfer (SET) occurs, breaking the O-O bond of TBPB with an energy barrier of 8.7 kcal/mol and resulting in a septet, (^{7}**INT3**) of the Fe(III) species coordinated by a tethered t-BuO radical with a exergonicity of 5.7 kcal/mol. As displayed in Figure 1b, the oxygen atom of the tethered t-BuO in ^{7}**INT3** acquires an unpaired spin density, indicating that the t-BuO moiety becomes a radical during the SET process. Since the selective formation of methyl radical was observed in this reaction and in our previous work [33], two pathways of methyl radical generation were therefore considered.

With a terminal carbon having a spin density in ^{7}**INT3** (cf. Figure 1b), a transition state ^{7}**TS2**, corresponding to C-C bond cleavage along this coordinate, is located with a quite low barrier, 5.2 kcal/mol, and leads to a sextet (^{6}**INT4**) with a free methyl radical. Surprisingly, ^{7}**TS2-OtBu**, dissociation of an t-BuO radical from ^{7}**INT3** was found to be an unfavorable process, requiring a higher energy barrier (6.7 kcal/mol) to lead to a free t-BuO radical and a sextet ^{6}**INT4**, with only 0.9 kcal/mol small exergonicity (red path in Figure 1a). A much higher barrier of 9.3 kcal/mol is required (^{2}**TS3**) for dissociation of a methyl radical from a free t-BuO radical [21] indicating that a free methyl radical generated directly from ^{7}**INT3** is thermodynamically and kinetically favorable. In addition, since there will be benzoate anions in the system following the occurrence of SET on the TBPB, several possible iron(III) species ligated by different anions with the tethered t-BuO radical were also examined for the possibility that they could assist the β-methyl scission. Figure 2 depicts the free energy profiles of SET and iron-catalysed β-methyl scission processes for four candidate iron complexes, ^{5}**INT1-1** to ^{5}**INT1-4** (Supplementary Table S3). Encouragingly, the energy barriers of SET for these species are in the range of 6.7–11.0 kcal/mol smaller than that for ^{5}**INT1**. Besides, the energy barriers of iron-catalyzed β-methyl scission for these species are in the range of 3.9–4.7 kcal/mol which are all smaller than that of β-methyl scission from the free t-BuO radical (9.3 kcal/mol) and even smaller than that of ^{7}**TS2** (5.2 kcal/mol) indicating that, in the reaction condition, these possible iron-catalyst species can also perform the SET on TBPB and assist the β-methyl scission well after initial catalytic cycle. These results are in good agreement with our experiment results in which no t-BuO radical derivative was observed due to the incomparable process of the generation of a free methyl radical. This implies that the Fe(III) catalyst may assist the β-methyl scission even at room temperature. This study offers a clear image of the whole decomposition process from TBPB to the t-BuO radical and a methyl radical.

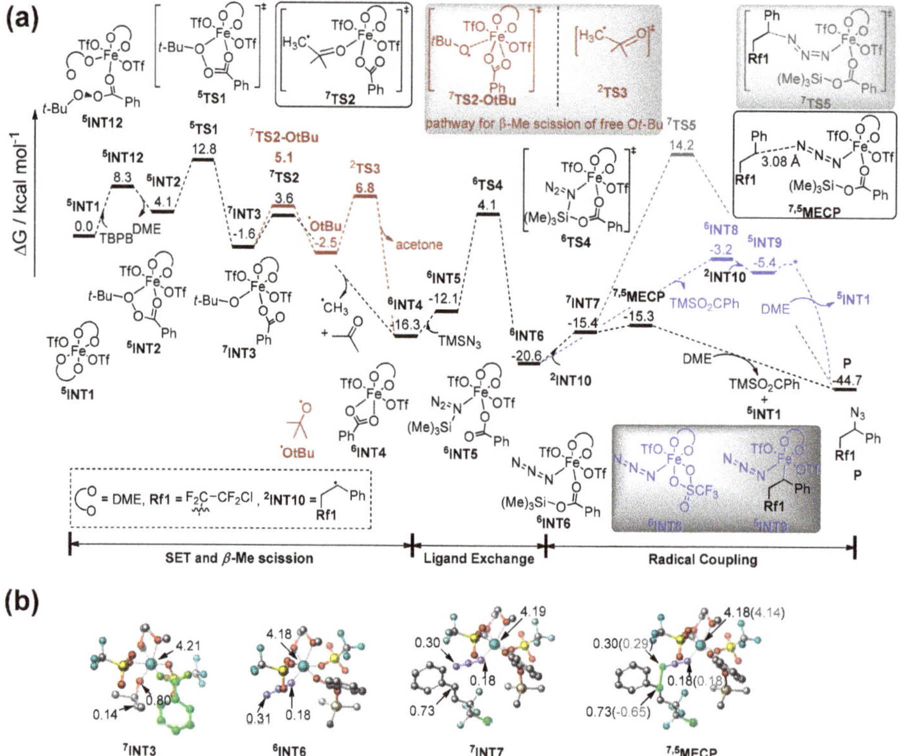

Figure 1. (**a**) The Gibbs free energy profile of the Fe-catalyzed carboazidation of alkenes. The transition state corresponding to reductive elimination from ^{5}**INT9** cannot be explicitly located and is indicated as *. (**b**) Optimized structures of selected intermediates. Spin densities on selected atoms are shown in each structure beside the arrows. For ^{7,5}**MECP**, the spin densities of the quintet state are shown in parenthesis. Hydrogen atoms are omitted for clarity.

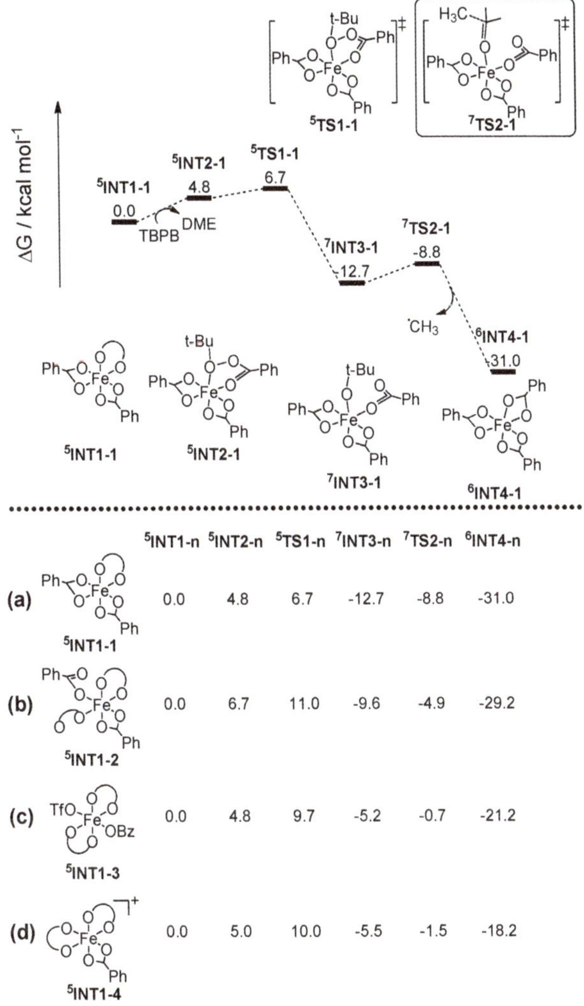

Figure 2. The Gibbs free energy profiles of SET and iron-catalyzed β-methyl scission processes for (**a**) Fe(OBz)$_2$(DME), **^{5}INT1-1**, (**b**) Fe(OBz)$_2$(DME)$_2$, **^{5}INT1-2**, (**c**) Fe(OTf)(OBz)(DME)$_2$, **^{5}INT1-3** and (**d**) cation species [Fe(OBz)(DME)$_2$]$^+$, **^{5}INT1-4**, showing that different possible Fe(II/III) species facilitate the β-scission of t-BuO radical. Relative free energies are in kcal/mol.

Subsequently, a radical relay starting from a free methyl radical and generating the benzyl radical **^{2}INT10** was demonstrated to be a facile process and shown in Figure 3. Herein, 1-chloro-1,1,2,2-tetrafluoro-2-iodoethane is employed as a perhaloalkyl iodide model to conduct the radical relay process. Transition state **^{2}TS6**, corresponding to CH$_3$I and perhaloalkyl radical **Rf1** generations is located with the lowest barrier of 7.2 kcal/mol which is much lower than that of chlorine extraction (**^{2}TS7**, 19.8 kcal/mol). On the other hand, methyl β-addition to styrene [35,55] is also considered; however, a much higher barrier (13.0 kcal/mol of **^{2}TS8**) is found for **^{2}INT11** producing. Although the relative free energy of **^{2}INT11** (−17.8 kcal/mol) is lower than that of **Rf1** (−8.1 kcal/mol), **Rf1** addition to styrene is barrierless to result in a much exergonic **^{2}INT10** (−30.7 kcal/mol). Owing to the flat and long range effective (~3.22 Å) potential energy surface (PES), TS for

2INT10 production cannot be located, suggesting that this radical relay process should be fast (see Supplementary Figure S2).

We then focused on the iron-catalyzed azidation. Formation of **6INT5** by TMSN$_3$ (trimethylsilyl azide) complexing with the Fe in **6INT4** via the internal nitrogen atom, was initially calculated. A trimethylsilyl group migration transition state (**6TS4**) leading to an exergonic azide complex (**6INT6**) was identified [8]. Charge transfer to the iron center from the azide increases gradually during trimethylsilyl group migration. In particular, spin spreads to the internal and terminal nitrogen atom, implying that the azide adopts a radical characteristic even though its net charge is negative (cf. Figure 1b). Three possible pathways leading to the C-N bond coupling via a septet state, a quintet state or a septet-quintet crossing, were considered. In the septet state, the reaction pathway in which the benzyl radical, **2INT10**, couples directly with the terminal nitrogen atom of the azide in **6INT6** was considered due to its larger spin density and the reduced steric hindrance. An outer-sphere azide reaction has also been proposed in Mn catalysis [56], but the transition state **7TS5** has an extremely high barrier of 34.8 kcal/mol.

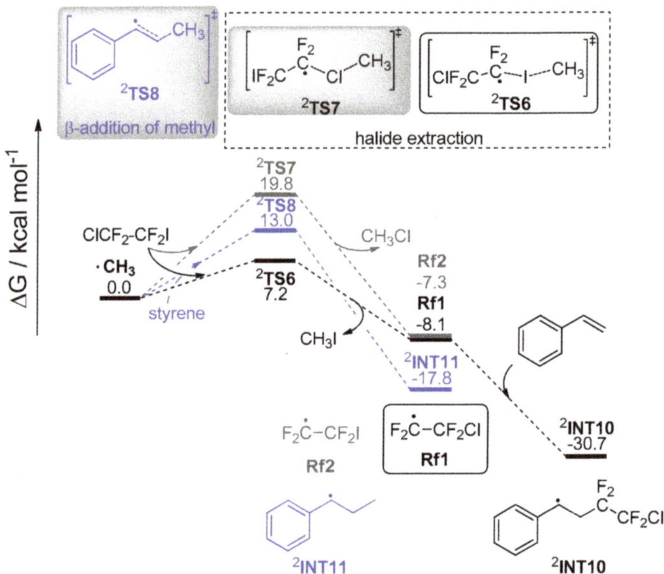

Figure 3. The Gibbs free energy profile for the benzyl radical generation from methyl radical via radical relay pathway. Relative free energies are in kcal/mol.

The inner-sphere pathway via an intermediate with an iron-carbon bond was also considered [57]. A quintet, **5INT9**, formed through the TMSOBz-dissociated sextet **6INT8**, was found to have a much higher relative free energy (−5.4 kcal/mol), suggesting that this pathway involving an intermediate containing a newly formed Fe-C bond, leading to the dissociation of the TMSOBz and association of **2INT10** to **6INT6**, is unfavorable. This result suggests that the oxidation of a Fe(III) by a benzyl radical to form Fe(IV) is an unfavorable pathway in this reaction and this finding is consistent with Gutierrez's study although they investigated different iron species [58].

We then sought a minimum energy crossing point (MECP) crossing between the septet and the quintet states. Intermediate **7INT7**, approaching an **2INT10** to **6INT6**, was located with the relative free energy only 5.2 kcal/mol higher than that of **6INT6**, and the MECP was found at a distance $d(C_b\text{-}N_t)$, between carbon and the terminal nitrogen atom of 3.08 Å (cf. 3.23 Å in **7INT7**, Figure 1b). The electronic energy of the MECP was estimated to be slightly higher (~0.1 kcal/mol) than that of **7INT7** [59]. After spin state crossing, no transition state relevant to product formation can be located due to the flat potential energy surface corresponding to $d(\text{Fe-N}_i)$ elongation around $d(\text{C}_b\text{-N}_t)$ of *ca.* 3.08 Å (Figure S3). The potential energy surface corresponding to $d(\text{Fe-N}_i)$ elongation accompanying the $d(\text{C}_b\text{-N}_t)$ shortening shows no barrier and can proceed downhill to product formation. This result is analogous to the halogenations of carbon centered radicals with iron(III)-halide species, [60,61] and suggests that only 5.2~5.3 kcal/mol is required to conduct spin state crossing, after which product formation is spontaneous.

A septet intermediate, [7]**INT9**, has been calculated with the energy of 4.8 kcal/mol higher than [5]**INT9**, (see Supplementary Table S4). A transition state between [6]**INT8** and [5]**INT9** for inner-sphere radical coupling may exist but cannot be located. On the other hand, much effort was made to locate the transition state after the inner-sphere spin crossing point, but was unsuccessful. This result may be regarded as a barrierless reductive elimination for a high-valent metal complex, [62,63] but the absence of a transition state for the inner-sphere pathway does not affect the conclusion of a favorable outer-sphere pathway since the free energy of [5]**INT9** is much higher than that of [7]**INT7**. Moreover, such azidation processes with another possible Fe(III)N$_3$ species, [6]**INT6-1**, has also been calculated, and as expected, the outer-sphere pathway remains the favorable route (see Supplementary Figure S4), indicating that the catalytic cycle can perform after first cycle as well as the Fe(III)N$_3$ species with OTf$^-$ anion.

In view of the results of these mechanistic studies, a radical relay-involved catalytic cycle is proposed and is shown in Scheme 3. A SET between iron catalyst and TBPB initiates the reaction by generating a methyl radical (**A**), an Fe(III) species and acetone. A radical relay process then occurs between the methyl radical and the alkyl iodide affording a new carbon radical (**B**) and methyl iodide. This carbon radical adds to the olefin, generating an internal radical (**C**). Azidotrimethylsilane as a ligand delivers an Fe(III)N$_3$ species, [8] which ultimately reacts with the radical (**C**) to deliver the desired alkylazidation products, regenerating Fe(II) [(**C**) + Fe(III)N$_3$ → (**C**)-N$_3$ + Fe(II)]. The C-N$_3$ bond formation from alkenes can be facilitated by the Fe(III)N$_3$ species as well as by the putative Mn(III) species [64]. According to the theoretical study, an iron-catalyzed β-methyl scission is an incomparable process for generation of the initial methyl radical; in addition, an outer-sphere radical coupling pathway [56,65,66] is thought to be the more favorable pathway. Similar outer-sphere radical capture for direct C-N bond formation have been reported on the C-H amination of copper(II) anilides [67–69].

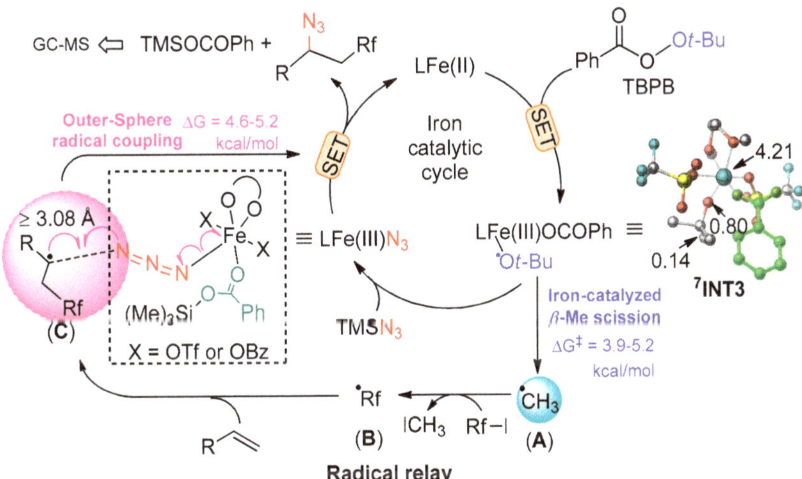

Scheme 3. Proposed mechanism for carboazidation of alkenes. SET, single electron transfer. Optimized structure, [7]**INT3**, are depicted as in Figure 1b.

3. Materials and Methods

3.1. Experimental Section

3.1.1. General Information

All reactions were carried out under an atmosphere of nitrogen in dried glassware with magnetic stirring unless otherwise indicated. Compound **3** in Scheme 2 was synthesized in our lab and other chemicals obtained from commercial suppliers were used without further purification. The purity of iron catalyst between different vendors (Energy Chemical, Bokachem and ®HEOWNS) did not change the yields of products when other batches of iron triflate were purchased. Solvents were dried by Innovative Technology Solvent Purification System. Liquids and solutions were transferred via syringe. All reactions were monitored by thin-layer chromatography. GC and GC-MS data were recorded on Thermo Trace 1300 (Thermo Fisher Scientific, Milan, Italy) and Thermo ISQ QD, respectively. ^{1}H-, ^{19}F-, and ^{13}C-NMR spectra were recorded on Bruker-BioSpin AVANCE III HD-400 Hz (Bruker BioSpin GmbH, Rheinstetten, Germany). Data for ^{1}H-NMR spectra are reported relative to chloroform as an internal standard (7.26 ppm) and are reported as follows: chemical shift (ppm), multiplicity, coupling constant (Hz), and integration. Data for ^{13}C-NMR spectra were reported relative to chloroform as an internal standard (77.00 ppm) and are reported in terms of chemical shift (ppm). IR data were obtained from Bruker VERTEX 70. All melting points were determined on a Beijing Science Instrument Dianguang Instrument (Beijing, China) Factory XT4B melting point apparatus and are uncorrected. HRMS(ESI) data were recorded on Agilent Technologies 6224 TOF LC/MS (Agilent, Palo Alto, CA, USA); HRMS(EI) data were recorded on Waters Micromass GCT Premier (Waters, MMAS, New York, NY, USA).

3.1.2. General Procedure for Confirmation of Methyl Radical

To a dried Schlenk tube equipped with a magnetic bar, Fe(OTf)$_2$ (9 mg, 0.025 mmol) was added, flushed with nitrogen gas (3 times) and maintained the nitrogen atmosphere using the balloon. A thoroughly mixed solution of vinylarene (0.5 mmol), TMSN$_3$ (1.0 mmol) and TBPB (or *tert*-butyl ethaneperoxoate) (1.0 mmol) in DME (2 mL) was added to the catalyst via syringe and stirred vigorously for 30 min at room temperature. The solvent was evaporated, and the residue was purified by flash chromatography on silica gel to give the corresponding product **9** in 52% (33%) yield.

3.2. Computational Method and Details

Density functional theory (DFT) studies on the iron-catalyzed carboazidation of styrene were performed at B3LYP [70,71] -D3 [72,73] /Def2-SVP [74,75] level of theory in gas-phase for geometrical optimizations, thermal energy calculations, and frequency analyses. Transition state structures were searched by simply performing a crude relaxed potential energy surface (RPES) scan connecting reactants and products, and then optimized by the three-structure synchronous transit-guided quasi-Newton (STQN) method, [76,77] and rational function optimization (RFO) method of TS as well [78]. In addition, transition state vibrational frequencies were verified to have one and only one imaginary frequency and confirm the correctness of the imaginary frequency by viewing normal mode vibrational vector. All optimized stationary points were characterized by frequency calculation for identification of minimum points and saddle points. Single point energies based upon the optimized structures were calculated at the B3LYP-D3/Def2-TZVP [74,75] level of theory with SMD solvation model calculation in DME solution, [79] and the reported Gibbs free energy is obtained by adding the solution-phase electronic energy with the gas-phase Gibbs free energy correction for saving the computational time consumption. To verify the reliability of the geometries and thermal corrections obtained in the gas phase, geometrical optimizations as well as the frequency calculations for ^{5}**INT1**, TBPB, DME, ^{5}**INT2**, ^{5}**TS1**, and ^{7}**INT3** were also carried out with SMD solvation model. Figure S5 depicts the free energy profile of pathway from ^{5}**INT1** to ^{7}**INT3** in which the gas-phase thermal energy correction shows well comparative to the solvation thermal energy correction. Other functionals including types of generalized gradient approximation (GGA), meta-GGA and hybrid functional including dispersion were also employed for [5]**INT1** (i.e., Fe(OTf)$_2$(DME)$_2$) optimization on quintet, triplet and singlet spin states to confirm the validity of quintet state. Supplementary Table S2 shows the similar energetic tendency supporting that employing the quintet state [5]**INT1** to initiate studies should be reliable. On the other hand, for radical coupling, minimum energy crossing point (MECP) was also located by using the hybrid approach method of Harvey [80]. All calculations were performed by the Gaussian 09 package (Gaussian, Wallingford, CT, USA) [81].

4. Conclusions

In summary, experimental studies have established the selective formation of methyl radical formation for this iron-catalyzed carboazidation of alkenes. The methyl radical can be identified by GC-MS and be found in the product in the absence of further alkyl iodides indicating that a methyl radical is easily propagated at room temperature under the reaction conditions. Theoretical studies reveal that the methyl radical propagation via β-Me scission of the t-BuO radical will be assisted by the iron catalysis. The energy barrier of methyl radical release from a coordinated t-BuO radical is far lower than that of untethered one. In addition, the formed methyl radical has a lower barrier to abstract an iodine atom from the alkyl iodide instead of reacting with the styrene which explains the outcome of products through the radical relay process. In the end, the iron-catalyzed carboazidation of alkenes may undergo an outer-sphere radical coupling via an Fe(III)N$_3$ intermediate to form products. This study may shed some light on the metal-catalyzed SET reactions of peroxides and may offer a partial explanation of the formation of methyl radical [21,36,37,44].

Supplementary Materials: The following are available online, Supporting Information including all NMR spectroscopic analysis, characterization data, GC-MS analysis, Figure S1–S5, Tables S1–S4 and Cartesian coordinates of all optimized structures.

Author Contributions: Conceptualization, M.-F.C. and H.B.; methodology, M.-F.C., X.Z. and H.B.; formal analysis, M.-F.C., H.X. and Y.L.; investigation, M.-F.C. and H.X.; resources, H.B. and X.Z.; data curation, M.-F.C., Y.L. and H.B.; writing—original draft preparation, M.-F.C.; writing—review and editing, H.B. and X.Z.; visualization, M.-F.C.; supervision, H.B. and X.Z.; project administration, H.B. and X.Z.; funding acquisition, H.B. and X.Z. All authors have read and agreed to the published version of the manuscript.

Funding: This research was funded by the National Key R&D Program of China (Grant No. 2017YFA0700103), the NSFC (Grant Nos. 21672213, 21871258), the Strategic Priority Research Program of the Chinese Academy of Sciences (Grant No. XDB20000000), the Haixi Institute of CAS (Grant No. CXZX-2017-P01), the Shenzhen STIC (Grant JCYJ20170412150343516) and the Shenzhen San-Ming Project (SZSM201809085) for financial support. APC was sponsored by MDPI.

Acknowledgments: We acknowledge the National Supercomputer Centre in Guangzhou (NSCC-GZ) for supercomputer supporting.

Conflicts of Interest: The authors declare no conflict of interest.

References

1. Huang, W.-Y.; Lü, L. The reaction of perfluoroalkanesulfinates VII. Fenton reagent-initiated addition of sodium perfluoroalkanesulfinates to alkenes. *Chin. J. Chem.* **1992**, *10*, 365–372. [CrossRef]

2. Renaud, P.; Ollivier, C.; Panchaud, P. Radical Carboazidation of Alkenes: An Efficient Tool for the Preparation of Pyrrolidinone Derivatives. *Angew. Chem. Int. Ed.* **2002**, *41*, 3460–3462. [CrossRef]

3. Weidner, K.; Giroult, A.; Panchaud, P.; Renaud, P. Efficient carboazidation of alkenes using a radical desulfonylative azide transfer process. *J. Am. Chem. Soc.* **2010**, *132*, 17511–17515. [CrossRef]

4. Wang, F.; Qi, X.; Liang, Z.; Chen, P.; Liu, G. Copper-Catalyzed Intermolecular Trifluoromethylazidation of Alkenes: Convenient Access to CF3-Containing Alkyl Azides. *Angew. Chem. Int. Ed.* **2014**, *53*, 1881–1886. [CrossRef]

5. Dagousset, G.; Carboni, A.; Magnier, E.; Masson, G. Photoredox-induced three-component azido- and aminotrifluoromethylation of alkenes. *Org. Lett.* **2014**, *16*, 4340–4343. [CrossRef]

6. Bunescu, A.; Ha, T.M.; Wang, Q.; Zhu, J. Copper-Catalyzed Three-Component Carboazidation of Alkenes with Acetonitrile and Sodium Azide. *Angew. Chem. Int. Ed.* **2017**, *56*, 10555–10558. [CrossRef]

7. Geng, X.; Lin, F.; Wang, X.; Jiao, N. Azidofluoroalkylation of Alkenes with Simple Fluoroalkyl Iodides Enabled by Photoredox Catalysis. *Org. Lett.* **2017**, *19*, 4738–4741. [CrossRef]

8. Zhu, C.-L.; Wang, C.; Qin, Q.-X.; Yruegas, S.; Martin, C.D.; Xu, H. Iron(II)-Catalyzed Azidotrifluoromethylation of Olefins and N-Heterocycles for Expedient Vicinal Trifluoromethyl Amine Synthesis. *ACS Catal.* **2018**, *8*, 5032–5037. [CrossRef]

9. Xiong, H.; Ramkumar, N.; Chiou, M.-F.; Jian, W.; Li, Y.; Su, J.-H.; Zhang, X.; Bao, H. Iron-catalyzed carboazidation of alkenes and alkynes. *Nat. Commun.* **2019**, *10*, 122. [CrossRef]

10. Li, Z.; Cao, L.; Li, C.-J. FeCl2-Catalyzed Selective C–C Bond Formation by Oxidative Activation of a Benzylic C–H Bond. *Angew. Chem. Int. Ed.* **2007**, *46*, 6505–6507. [CrossRef]

11. Lv, L.; Lu, S.; Guo, Q.; Shen, B.; Li, Z. Iron-Catalyzed Acylation-Oxygenation of Terminal Alkenes for the Synthesis of Dihydrofurans Bearing a Quaternary Carbon. *J. Org. Chem.* **2015**, *80*, 698–704. [CrossRef]

12. Pan, S.; Liu, J.; Li, H.; Wang, Z.; Guo, X.; Li, Z. Iron-Catalyzed N-Alkylation of Azoles via Oxidation of C–H Bond Adjacent to an Oxygen Atom. *Org. Lett.* **2010**, *12*, 1932–1935. [CrossRef]

13. Liu, W.; Li, Y.; Liu, K.; Li, Z. Iron-Catalyzed Carbonylation-Peroxidation of Alkenes with Aldehydes and Hydroperoxides. *J. Am. Chem. Soc.* **2011**, *133*, 10756–10759. [CrossRef]

14. Boess, E.; Schmitz, C.; Klussmann, M. A Comparative Mechanistic Study of Cu-Catalyzed Oxidative Coupling Reactions with N-Phenyltetrahydroisoquinoline. *J. Am. Chem. Soc.* **2012**, *134*, 5317–5325. [CrossRef]

15. Leifert, D.; Daniliuc, C.G.; Studer, A. 6-Aroylated phenanthridines via base promoted homolytic aromatic substitution (BHAS). *Org. Lett.* **2013**, *15*, 6286–6289. [CrossRef]

16. Wang, J.; Liu, C.; Yuan, J.; Lei, A. Copper-Catalyzed Oxidative Coupling of Alkenes with Aldehydes: Direct Access to α,β-Unsaturated Ketones. *Angew. Chem. Int. Ed.* **2013**, *52*, 2256–2259. [CrossRef]

17. Wei, W.-T.; Zhou, M.-B.; Fan, J.-H.; Liu, W.; Song, R.-J.; Liu, Y.; Hu, M.; Xie, P.; Li, J.-H. Synthesis of Oxindoles by Iron-Catalyzed Oxidative 1,2-Alkylarylation of Activated Alkenes with an Aryl C(sp2)–H Bond and a C(sp3)–H Bond Adjacent to a Heteroatom. *Angew. Chem. Int. Ed.* **2013**, *52*, 3638–3641. [CrossRef]

18. Zhou, M.-B.; Song, R.-J.; Ouyang, X.-H.; Liu, Y.; Wei, W.-T.; Deng, G.-B.; Li, J.-H. Metal-free oxidative tandem coupling of activated alkenes with carbonyl C(sp2)–H bonds and aryl C(sp2)–H bonds using TBHP. *Chem. Sci.* **2013**, *4*, 2690–2694. [CrossRef]

19. Wei, W.-T.; Song, R.-J.; Li, J.-H. Copper-Catalyzed Oxidative α-Alkylation of α-Amino Carbonyl Compounds with Ethers via Dual C(sp3)-H Oxidative Cross-Coupling. *Adv. Synth. Catal.* **2014**, *356*, 1703–1707. [CrossRef]

20. Zhao, J.; Li, P.; Xia, C.; Li, F. Direct N-acylation of azoles via a metal-free catalyzed oxidative cross-coupling strategy. *Chem. Commun.* **2014**, *50*, 4751–4754. [CrossRef]

21. Jiao, Y.; Chiou, M.-F.; Li, Y.; Bao, H. Copper-Catalyzed Radical Acyl-Cyanation of Alkenes with Mechanistic Studies on the tert-Butoxy Radical. *ACS Catal.* **2019**, *9*, 5191–5197. [CrossRef]

22. Kharasch, M.S.; Sosnovsky, G. The Reactions of *t*-Butyl Perbenzoate and Olefins—A Stereospecific Reaction. *J. Am. Chem. Soc.* **1958**, *80*, 756. [CrossRef]

23. Kochi, J.K.; Mains, H.E. Studies on the Mechanism of the Reaction of Peroxides and Alkenes with Copper Salts. *J. Org. Chem.* **1965**, *30*, 1862–1872. [CrossRef]

24. Pryor, W.A.; Hendrickson, W.H. Reaction of nucleophiles with electron acceptors by SN2 or electron transfer (ET) mechanisms: Tert-butyl peroxybenzoate/dimethyl sulfide and benzoyl peroxide/N,N-dimethylaniline systems. *J. Am. Chem. Soc.* **1983**, *105*, 7114–7122. [CrossRef]

25. Sekar, G.; DattaGupta, A.; Singh, V.K. Asymmetric Kharasch Reaction: Catalytic Enantioselective Allylic Oxidation of Olefins Using Chiral Pyridine Bis(diphenyloxazoline)–Copper Complexes and tert-Butyl Perbenzoate. *J. Org. Chem.* **1998**, *63*, 2961–2967. [CrossRef]

26. Zhou, S.-L.; Guo, L.-N.; Wang, H.; Duan, X.-H. Copper-Catalyzed Oxidative Benzylarylation of Acrylamides by Benzylic C–H Bond Functionalization for the Synthesis of Oxindoles. *Chem. Eur. J.* **2013**, *19*, 12970–12973. [CrossRef]

27. Cai, Z.-J.; Lu, X.-M.; Zi, Y.; Yang, C.; Shen, L.-J.; Li, J.; Wang, S.-Y.; Ji, S.-J. I2/TBPB Mediated Oxidative Reaction of N-Tosylhydrazones with Anilines: Practical Construction of 1,4-Disubstituted 1,2,3-Triazoles under Metal-Free and Azide-Free Conditions. *Org. Lett.* **2014**, *16*, 5108–5111. [CrossRef]

28. Chen, C.; Xu, X.-H.; Yang, B.; Qing, F.-L. Copper-Catalyzed Direct Trifluoromethylthiolation of Benzylic C–H Bonds via Nondirected Oxidative C(sp3)–H Activation. *Org. Lett.* **2014**, *16*, 3372–3375. [CrossRef]

29. Tang, S.; Liu, K.; Long, Y.; Gao, X.; Gao, M.; Lei, A. Iodine-Catalyzed Radical Oxidative Annulation for the Construction of Dihydrofurans and Indolizines. *Org. Lett.* **2015**, *17*, 2404–2407. [CrossRef]

30. Chen, W.; Zhang, Y.; Li, P.; Wang, L. tert-Butyl peroxybenzoate mediated formation of 3-alkylated quinolines from N-propargylamines via a cascade radical addition/cyclization reaction. *Org. Chem. Front.* **2018**, *5*, 855–859. [CrossRef]

31. Shen, S.-J.; Zhu, C.-L.; Lu, D.-F.; Xu, H. Iron-Catalyzed Direct Olefin Diazidation via Peroxyester Activation Promoted by Nitrogen-Based Ligands. *ACS Catal.* **2018**, *8*, 4473–4482. [CrossRef]

32. Yu, H.; Li, Z.; Bolm, C. Nondirected Copper-Catalyzed Sulfoxidations of Benzylic C–H Bonds. *Org. Lett.* **2018**, *20*, 2076–2079. [CrossRef]

33. Zhu, N.; Zhao, J.; Bao, H. Iron catalyzed methylation and ethylation of vinyl arenes. *Chem. Sci.* **2017**, *8*, 2081–2085. [CrossRef]

34. Jian, W.; Ge, L.; Jiao, Y.; Qian, B.; Bao, H. Iron-Catalyzed Decarboxylative Alkyl Etherification of Vinylarenes with Aliphatic Acids as the Alkyl Source. *Angew. Chem. Int. Ed.* **2017**, *56*, 3650–3654. [CrossRef]

35. Qian, B.; Chen, S.; Wang, T.; Zhang, X.; Bao, H. Iron-Catalyzed Carboamination of Olefins: Synthesis of Amines and Disubstituted beta-Amino Acids. *J. Am. Chem. Soc.* **2017**, *139*, 13076–13082. [CrossRef]

36. Guo, S.; Wang, Q.; Jiang, Y.; Yu, J.-T. tert-Butyl Peroxybenzoate-Promoted α-Methylation of 1,3-Dicarbonyl Compounds. *J. Org. Chem.* **2014**, *79*, 11285–11289. [CrossRef]

37. Bao, X.; Yokoe, T.; Ha, T.M.; Wang, Q.; Zhu, J. Copper-catalyzed methylative difunctionalization of alkenes. *Nat. Commun.* **2018**, *9*, 3725. [CrossRef]

38. Gray, P.; Williams, A. The Thermochemistry And Reactivity Of Alkoxyl Radicals. *Chem. Rev.* **1959**, *59*, 239–328. [CrossRef]

39. Carey, F.A.; Sundberg, R.J. Free Radical Reactions. In *Advanced Organic Chemistry, Part A: Structure and Mechanisms*, 5th ed.; Springer: New York, NY, USA, 2007; pp. 956–1062.

40. Tran, B.L.; Driess, M.; Hartwig, J.F. Copper-Catalyzed Oxidative Dehydrogenative Carboxylation of Unactivated Alkanes to Allylic Esters via Alkenes. *J. Am. Chem. Soc.* **2014**, *136*, 17292–17301. [CrossRef]

41. Tran, B.L.; Li, B.; Driess, M.; Hartwig, J.F. Copper-Catalyzed Intermolecular Amidation and Imidation of Unactivated Alkanes. *J. Am. Chem. Soc.* **2014**, *136*, 2555–2563. [CrossRef]

42. Bunescu, A.; Wang, Q.; Zhu, J. Synthesis of Functionalized Epoxides by Copper-Catalyzed Alkylative Epoxidation of Allylic Alcohols with Alkyl Nitriles. *Org. Lett.* **2015**, *17*, 1890–1893. [CrossRef] [PubMed]

43. Tang, S.; Wang, P.; Li, H.; Lei, A. Multimetallic catalysed radical oxidative C(sp3)–H/C(sp)–H cross-coupling between unactivated alkanes and terminal alkynes. *Nat. Commun.* **2016**, *7*, 11676. [CrossRef] [PubMed]

44. Wu, X.; Riedel, J.; Dong, V.M. Transforming Olefins into γ,δ-Unsaturated Nitriles through Copper Catalysis. *Angew. Chem. Int. Ed.* **2017**, *56*, 11589–11593. [CrossRef] [PubMed]

45. Zhou, H.; Jian, W.; Qian, B.; Ye, C.; Li, D.; Zhou, J.; Bao, H. Copper-Catalyzed Ligand-Free Diazidation of Olefins with TMSN3 in CH3CN or in H2O. *Org. Lett.* **2017**, *19*, 6120–6123. [CrossRef]

46. Newcomb, M. *Radical Kinetics and Clocks in Encyclopedia of Radicals in Chemistry, Biology and Materials*; Wiley: Chichester, UK, 2012.

47. Pryor, W.A.; Gu, J.T.; Church, D.F. Trapping free radicals formed in the reaction of ozone with simple olefins using 2,6-di-tert-butyl-4-cresol (BHT). *J. Org. Chem.* **1985**, *50*, 185–189. [CrossRef]

48. Thavasi, V.; Leong, L.P.; Bettens, R.P. Investigation of the influence of hydroxy groups on the radical scavenging ability of polyphenols. *J. Phys. Chem. A* **2006**, *110*, 4918–4923. [CrossRef]

49. Cheng, G.-J.; Zhang, X.; Chung, L.W.; Xu, L.; Wu, Y.-D. Computational Organic Chemistry: Bridging Theory and Experiment in Establishing the Mechanisms of Chemical Reactions. *J. Am. Chem. Soc.* **2015**, *137*, 1706–1725. [CrossRef] [PubMed]

50. Plata, R.E.; Singleton, D.A. A Case Study of the Mechanism of Alcohol-Mediated Morita Baylis–Hillman Reactions. The Importance of Experimental Observations. *J. Am. Chem. Soc.* **2015**, *137*, 3811–3826. [CrossRef]

51. Sameera, W.M.; Hatanaka, M.; Kitanosono, T.; Kobayashi, S.; Morokuma, K. The Mechanism of Iron(II)-Catalyzed Asymmetric Mukaiyama Aldol Reaction in Aqueous Media: Density Functional Theory and Artificial Force-Induced Reaction Study. *J. Am. Chem. Soc.* **2015**, *137*, 11085–11094. [CrossRef]

52. Verma, P.; Varga, Z.; Klein, J.; Cramer, C.J.; Que, L.; Truhlar, D.G. Assessment of electronic structure methods for the determination of the ground spin states of Fe(ii), Fe(iii) and Fe(iv) complexes. *Phys. Chem. Chem. Phys.* **2017**, *19*, 13049–13069. [CrossRef]

53. Mazumder, S.; Crandell, D.W.; Lord, R.L.; Baik, M.-H. Switching the Enantioselectivity in Catalytic [4 + 1] Cycloadditions by Changing the Metal Center: Principles of Inverting the Stereochemical Preference of an Asymmetric Catalysis Revealed by DFT Calculations. *J. Am. Chem. Soc.* **2014**, *136*, 9414–9423. [CrossRef]

54. Lee, Y.; Baek, S.-Y.; Park, J.; Kim, S.-T.; Tussupbayev, S.; Kim, J.; Baik, M.-H.; Cho, S.H. Chemoselective Coupling of 1,1-Bis[(pinacolato)boryl]alkanes for the Transition-Metal-Free Borylation of Aryl and Vinyl Halides: A Combined Experimental and Theoretical Investigation. *J. Am. Chem. Soc.* **2017**, *139*, 976–984. [CrossRef]

55. Zhu, N.; Wang, T.; Ge, L.; Li, Y.; Zhang, X.; Bao, H. gamma-Amino Butyric Acid (GABA) Synthesis Enabled by Copper-Catalyzed Carboamination of Alkenes. *Org. Lett.* **2017**, *19*, 4718–4721. [CrossRef]

56. Huang, X.; Bergsten, T.M.; Groves, J.T. Manganese-catalyzed late-stage aliphatic C-H azidation. *J. Am. Chem. Soc.* **2015**, *137*, 5300–5303. [CrossRef]

57. Toriyama, F.; Cornella, J.; Wimmer, L.; Chen, T.G.; Dixon, D.D.; Creech, G.; Baran, P.S. Redox-Active Esters in Fe-Catalyzed C-C Coupling. *J. Am. Chem. Soc.* **2016**, *138*, 11132–11135. [CrossRef]

58. Lee, W.; Zhou, J.; Gutierrez, O. Mechanism of Nakamura's Bisphosphine-Iron-Catalyzed Asymmetric C(sp(2))-C(sp(3)) Cross-Coupling Reaction: The Role of Spin in Controlling Arylation Pathways. *J. Am. Chem. Soc.* **2017**, *139*, 16126–16133. [CrossRef] [PubMed]

59. Harvey, J.N. Spin-forbidden reactions: Computational insight into mechanisms and kinetics. *Wiley Interdiscip. Rev. Comput. Mol. Sci.* **2014**, *4*, 1–14. [CrossRef]

60. Kulik, H.J.; Blasiak, L.C.; Marzari, N.; Drennan, C.L. First-Principles Study of Non-heme Fe(II) Halogenase SyrB2 Reactivity. *J. Am. Chem. Soc.* **2009**, *131*, 14426–14433. [CrossRef] [PubMed]

61. Rana, S.; Biswas, J.P.; Sen, A.; Clémancey, M.; Blondin, G.; Latour, J.-M.; Rajaraman, G.; Maiti, D. Selective C–H halogenation over hydroxylation by non-heme iron(iv)-oxo. *Chem. Sci.* **2018**, *9*, 7843–7858. [CrossRef] [PubMed]

62. Shin, K.; Park, Y.; Baik, M.-H.; Chang, S. Iridium-catalysed arylation of C–H bonds enabled by oxidatively induced reductive elimination. *Nat. Chem.* **2017**, *10*, 218–224. [CrossRef]

63. Chiou, M.-F.; Jayakumar, J.; Cheng, C.-H.; Chuang, S.-C. Impact of the Valence Charge of Transition Metals on the Cobalt- and Rhodium-Catalyzed Synthesis of Indenamines, Indenols, and Isoquinolinium Salts: A Catalytic Cycle Involving MIII/MV M = Co, Rh for 4 + 2 Annulation. *J. Org. Chem.* **2018**, *83*, 7814–7824. [CrossRef] [PubMed]

64. Fu, N.; Sauer, G.S.; Saha, A.; Loo, A.; Lin, S. Metal-catalyzed electrochemical diazidation of alkenes. *Science* **2017**, *357*, 575–579. [CrossRef] [PubMed]

65. Chen, B.; Fang, C.; Liu, P.; Ready, J.M. Rhodium-Catalyzed Enantioselective Radical Addition of CX_4 Reagents to Olefins. *Angew. Chem. Int. Ed.* **2017**, *56*, 8780–8784. [CrossRef] [PubMed]

66. Lo, J.C.; Kim, D.; Pan, C.-M.; Edwards, J.T.; Yabe, Y.; Gui, J.; Qin, T.; Gutierrez, S.; Giacoboni, J.; Smith, M.W.; et al. Fe-Catalyzed C-C Bond Construction from Olefins via Radicals. *J. Am. Chem. Soc.* **2017**, *139*, 2484–2503. [CrossRef] [PubMed]

67. Jang, E.S.; McMullin, C.L.; Käß, M.; Meyer, K.; Cundari, T.R.; Warren, T.H. Copper(II) Anilides in sp3 C-H Amination. *J. Am. Chem. Soc.* **2014**, *136*, 10930–10940. [CrossRef] [PubMed]

68. Gephart Iii, R.T.; Huang, D.L.; Aguila, M.J.B.; Schmidt, G.; Shahu, A.; Warren, T.H. Catalytic C–H Amination with Aromatic Amines. *Angew. Chem. Int. Ed.* **2012**, *51*, 6488–6492. [CrossRef]

69. Carsch, K.M.; DiMucci, I.M.; Iovan, D.A.; Li, A.; Zheng, S.-L.; Titus, C.J.; Lee, S.J.; Irwin, K.D.; Nordlund, D.; Lancaster, K.M.; et al. Synthesis of a copper-supported triplet nitrene complex pertinent to copper-catalyzed amination. *Science* **2019**, *365*, 1138–1143. [CrossRef]

70. Lee, C.; Yang, W.; Parr, R.G. Development of the Colle-Salvetti correlation-energy formula into a functional of the electron density. *Phys. Rev. B* **1988**, *37*, 785–789. [CrossRef]

71. Becke, A.D. Density-functional thermochemistry. III. The role of exact exchange. *J. Chem. Phys.* **1993**, *98*, 5648–5652. [CrossRef]

72. Grimme, S.; Antony, J.; Ehrlich, S.; Krieg, H. A consistent and accurate ab initio parametrization of density functional dispersion correction (DFT-D) for the 94 elements H-Pu. *J. Chem. Phys.* **2010**, *132*, 154104. [CrossRef]

73. Grimme, S.; Ehrlich, S.; Goerigk, L. Effect of the damping function in dispersion corrected density functional theory. *J. Comput. Chem.* **2011**, *32*, 1456–1465. [CrossRef] [PubMed]

74. Weigend, F. Accurate Coulomb-fitting basis sets for H to Rn. *Phys. Chem. Chem. Phys.* **2006**, *8*, 1057–1065. [CrossRef] [PubMed]

75. Weigend, F.; Ahlrichs, R. Balanced basis sets of split valence, triple zeta valence and quadruple zeta valence quality for H to Rn: Design and assessment of accuracy. *Phys. Chem. Chem. Phys.* **2005**, *7*, 3297–3305. [CrossRef] [PubMed]

76. Peng, C.; Bernhard Schlegel, H. Combining Synchronous Transit and Quasi-Newton Methods to Find Transition States. *Isr. J. Chem.* **1993**, *33*, 449–454. [CrossRef]

77. Peng, C.; Ayala, P.Y.; Schlegel, H.B.; Frisch, M.J. Using redundant internal coordinates to optimize equilibrium geometries and transition states. *J. Comput. Chem.* **1996**, *17*, 49–56. [CrossRef]

78. Besalú, E.; Bofill, J.M. On the automatic restricted-step rational-function-optimization method. *Theor. Chem. Acc.* **1998**, *100*, 265–274. [CrossRef]

79. Marenich, A.V.; Cramer, C.J.; Truhlar, D.G. Universal Solvation Model Based on Solute Electron Density and on a Continuum Model of the Solvent Defined by the Bulk Dielectric Constant and Atomic Surface Tensions. *J. Phys. Chem. B* **2009**, *113*, 6378–6396. [CrossRef]

80. Harvey, J.N.; Aschi, M.; Schwarz, H.; Koch, W. The singlet and triplet states of phenyl cation. A hybrid approach for locating minimum energy crossing points between non-interacting potential energy surfaces. *Theor. Chem. Acc.* **1998**, *99*, 95–99. [CrossRef]

81. Frisch, M.J.; Trucks, G.W.; Schlegel, H.B.; Scuseria, G.E.; Robb, M.A.; Cheeseman, J.R.; Scalmani, G.; Barone, V.; Mennucci, B.; Petersson, G.A.; et al. *Gaussian ~09 Revision D.01*; Gaussian, Inc.: Wallingford, CT, USA, 2013.

Sample Availability: Samples of the compounds are not available from the authors.

Article

Iron-Catalysed C(sp^2)-H Borylation Enabled by Carboxylate Activation

Luke Britton [1], Jamie H. Docherty [1,*], Andrew P. Dominey [2] and Stephen P. Thomas [1,*]

[1] EaStCHEM School of Chemistry, University of Edinburgh, Joseph Black Building, David Brewster Road, Edinburgh EH9 3FJ, UK; s1942174@ed.ac.uk
[2] GSK Medicines Research Centre, Gunnels Wood Road, Stevenage, Hertfordshire SG1 2NY, UK; andrew.q.dominey@gsk.com
* Correspondence: jdocher5@ed.ac.uk (J.H.D.); stephen.thomas@ed.ac.uk (S.P.T.)

Received: 30 January 2020; Accepted: 14 February 2020; Published: 18 February 2020

Abstract: Arene C(sp^2)-H bond borylation reactions provide rapid and efficient routes to synthetically versatile boronic esters. While iridium catalysts are well established for this reaction, the discovery and development of methods using Earth-abundant alternatives is limited to just a few examples. Applying an in situ catalyst activation method using air-stable and easily handed reagents, the iron-catalysed C(sp^2)-H borylation reactions of furans and thiophenes under blue light irradiation have been developed. Key reaction intermediates have been prepared and characterised, and suggest two mechanistic pathways are in action involving both C-H metallation and the formation of an iron boryl species.

Keywords: catalysis; borylation; Iron; C-H functionalisation; pinacolborane; photochemistry

---

1. Introduction

The development of sustainable methods for the selective C(sp^2)-H functionalisation of arenes is an area of intense research but is still dominated by the use of 2nd- and 3rd-row transition metals [1–7]. Earth-abundant metals offer low toxicity and inexpensive alternatives, with iron being a leading example [8–12]. Direct C(sp^2)-H borylation offers a simple and efficient route to aryl-boronic esters, which are key platforms for organic synthesis [13–15]. Iridium-based complexes have become a "go-to" for C(sp^2)-H borylation reactions [16–24], while the discovery and development of Earth-abundant alternatives remains comparatively rare [25–37].

Tatsumi and Ohki showed that arenes would undergo thermally promoted C(sp^2)-H borylation using an *N*-heterocyclic carbene cyclopentadienyl iron(II) alkyl complex [NHC(Cp*)FeMe] as a catalyst in the presence of *tert*-butylethylene (Scheme 1a) [35]. Mankad applied heterobimetallic Fe-Cu and Fe-Zn complexes under continuous ultraviolet light irradiation to arene C(sp^2)-H borylation [36]. Similarly, Darcel and co-workers reported the use of a bis(diphosphino) iron(II) dialkyl and dihydride complexes for arene C(sp^2)-H borylation, again under continuous ultraviolet light irradiation [37]. While these landmark reports are highly significant developments, all require the prior synthesis of sensitive inorganic complexes which are synthetically challenging and difficult to handle for the non-specialist practitioner, thus limiting use by the broader synthetic community.

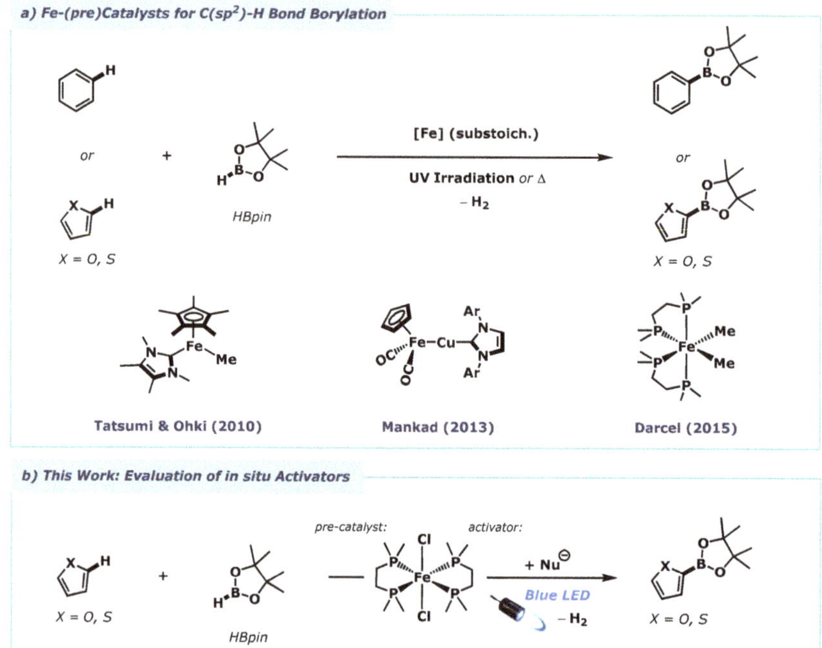

Scheme 1. Iron-catalysed C-H borylation of arenes. (**a**) Prior approaches to iron-catalysed C(*sp*^{2})-H-bond borylation with pinacolborane (HBpin) using organoiron and iron/copper bimetallic catalysts. (**b**) This work: C(*sp*^{2})-H bond borylation using dmpe$_2$FeCl$_2$ as a pre-catalyst, activated by exogenous nucleophiles, under blue light irradiation.

To reduce the synthetic challenges, and need for organometallic reagents, we questioned whether the C(*sp*^{2})-H borylation chemistry reported previously could be simplified by in situ catalyst activation using only bench stable reagents. In the example reported by Darcel and co-workers the bis[1,2-bis(dimethylphosphino)ethane-*P*,*P'*]dimethyliron(II) pre-catalyst (dmpe$_2$FeMe$_2$) was generated by the addition of methyllithium to the corresponding iron(II) dichloride complex (dmpe$_2$FeCl$_2$) [37]. Similarly, the catalytically active bis[1,2-bis(dimethylphosphino)ethane-*P*,*P'*]iron(II) dihydride (dmpe$_2$FeH$_2$) could be accessed using either LiHBEt$_3$ or LiAlH$_4$ [37,38]. Given our previous work on the in situ generation of hydride donors formed by the combination of alkoxide salts and pinacolborane (HBpin) [39], we postulated that the active C(*sp*^{2})-H borylation pre-catalyst, dmpe$_2$FeH$_2$, may be accessible by the same method. Reaction of substoichiometric alkoxide salt with HBpin, the boron source used for this borylation, would generate a hydride reductant in situ to activate the dmpe$_2$FeCl$_2$ pre-catalyst to dmpe$_2$FeH$_2$, the active borylation catalyst, and thus initiate catalysis. Importantly, the dmpe$_2$FeCl$_2$ complex displays much greater air- and moisture stability compared to the dihydride and dialkyl analogues. Herein, we report the in situ activation of dmpe$_2$FeCl$_2$ and application to the C(*sp*^{2})-H borylation reaction of heteroarenes (Scheme 1b).

2. Results

Guided by the work of Darcel and co-workers, we selected 2-methylfuran **2a** as an ideal test substrate for our investigations. Darcel and co-workers showed that dmpe$_2$FeMe$_2$ could be used as a pre-catalyst for the borylation of furan **2a** (3 equiv.) using HBpin (1 equiv.) under continuous ultraviolet light irradiation to give a regioisomeric mixture of 5- and 4-borylated furans, **3a** and **4a** respectively

(67%, **3a:4a** = 82:18) [37]. Using our alkoxide activation strategy we found the use of ultraviolet light for this reaction was not necessary, instead operating with lower energy blue light (Kessil A160 WE, 40 W Blue LED). Additionally, we used an inverted stoichiometry of arene (1 equiv.) and HBpin (1.2 equiv) and a reduced catalyst loading. Using these reaction parameters, we assessed the ability of a selection of potential activators to initiate catalysis alongside the dmpe$_2$FeCl$_2$ **1** pre-catalyst. (Scheme 2).

Any of LiOMe, KOMe, TBAOMe (TBA = tetra-*n*-butylammonium), NaO^{i}Pr, NaO^{t}Bu or KO^{t}Bu triggered pre-catalyst activation and the formation of both furyl boronic ester regioisomers, **3a** and **4a**, albeit in modest yields (17% to 39%) and with varying regioselectivity, after 24 h. The use of carboxylate salts also initiated catalysis; NaO$_2$CH, LiOAc, NaOAc, Na(2-EH) (2-EH = 2-ethylhexanoate), TBA(2-EH), NaO$_2$CPh, NaO$_2$CCF$_3$ all successfully initiated catalysis with varying efficiency (2% to 45%). Na(2-EH) and NaO$_2$CPh outperformed all alkoxide salts, and the yields obtained using these activators could be increased with prolonged reaction times to give a mixture of furyl boronic esters **3a** and **4a** in good yield and regioselectivity (Na(2-EH), 59%, **3a:4a** = 71:29). Control reactions with no catalyst, no added activator, and with no light irradiation showed no reactivity, highlighting the necessity of each reaction component.

Activator

	LiOMe	KOMe	TBAOMe	NaO^{i}Pr	NaO^{t}Bu	KO^{t}Bu	NaO$_2$CH	LiOAc	NaOAc
Yield (%)[a]	17	29	39	37	39	39	8	2	32
(**3a:4a**)	(76:24)	(76:24)	(74:26)	(70:30)	(85:15)	(85:15)	–	–	(72:28)

	NaHCO$_3$	LiAlH$_4$	NaHMDS	NaSEt	TBA(2-EH)	NaO$_2$CCF$_3$	Na(2-EH)	NaO$_2$CPh
	0	0[b]	35[b]	36	27	35	45	45
	–	–	(74:26)	(70:30)	(70:30)	(74:26)	(71:29)	(73:27)

Control reactions:	No [Fe]	No activator	No light				Na(2-EH)	NaO$_2$CPh
	0	0	0				59[c]	49[c]
	–	–	–				(71:29)	(71:29)

Scheme 2. Activator screening for the borylation of 2-methyl furan by dmpe$_2$FeCl$_2$ **1**. [a] Yields determined by ^{1}H-NMR spectroscopy of the crude reaction mixtures using 1,3,5-trimethoxybenzene as an internal standard. Product ratios were determined by ^{1}H-NMR spectroscopy of the crude reaction mixtures. [b] Reaction time = 15 h. [c] Reaction time = 48 h. 2-EH = 2-ethylhexanoate. TBA = tetra-*n*-butylammonium.

2.1. Substrate Scope

With optimised reaction conditions established using dmpe$_2$FeCl$_2$ (4 mol%), Na(2-EH) (8 mol%), arene (1.0 equiv.) and HBpin (1.2 equiv.) in THF under blue light irradiation, we assessed the reactivity of the system by application to a subset of furan and thiophene derivatives (Scheme 3). 2-Methylfuran **2a** underwent efficient borylation to generate a mixture of 5- and 4-borylated regioisomers **3a** and **4a** in good yield and regioselectivity (72%, 81:19). The parent, unsubstituted furan **2b**, also underwent successful borylation but gave a regioisomeric mixture of the 2- and 3-substituted boronic ester regioisomers **3b** and **4b**, and additionally the bis-boryl furans **5b**. Borylation of 2,3-dimethylfuran **2c** gave the corresponding 5-boryl regioisomer **3c** exclusively in good yield. 2-Ethylfuran **2d** reacted similarly to the 2-methylfuran analogue **2a** giving a mixture of 4- and 5-substituted boronic esters **3d** and **4d**. Unfortunately, application to thiophenes demonstrated limited reactivity under the established reaction conditions, giving only low yields of boryl-arenes **3e-g** and **4e-g**, again as a mixture of regioisomers, and bis-borylated product when the parent thiophene was used [40].

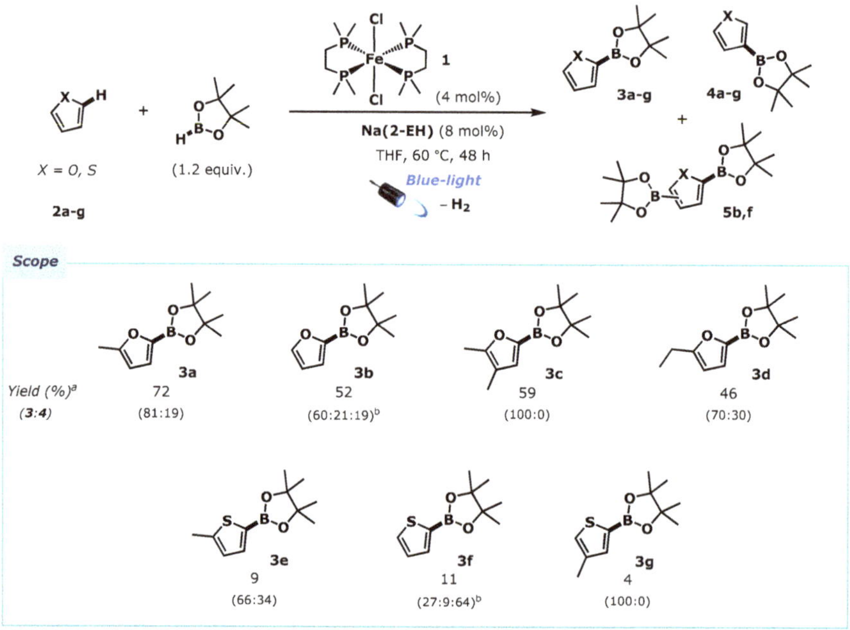

Scheme 3. Na(2-EH) activated borylation of furan and thiophene derivatives using dmpe$_2$FeCl$_2$ **1**. [a] Yields determined by ^{1}H-NMR spectroscopy of the crude reaction mixtures using 1,3,5-trimethoxybenzene as an internal standard. Product ratios were determined by ^{1}H-NMR spectroscopy of the crude reaction mixtures. [b] Values represent the ratio of 2-boryl:3-boryl:bis-boryl products.

2.2. Mechanistic Investigations

On the basis of successful catalysis we presumed that our in situ activation system provided access to the active iron(II) dihydride complex dmpe$_2$FeH$_2$ **7**, which had been shown to be catalytically active by Darcel and co-workers [37]. To support this, we combined each component in the absence of light or arene, i.e., the reaction of dmpe$_2$FeCl$_2$ **1**, Na(2-EH) and HBpin (Scheme 4a). This showed the formation of both the monohydride product dmpe$_2$FeHCl **6** and the expected dihydride, dmpe$_2$FeH$_2$ **7**, as observed by ^{31}P-NMR spectroscopy (see Supplementary Materials, S16). Reaction of the activator, Na(2-EH), and HBpin in the absence of pre-catalyst showed ligand redistribution to a mixture of boron-containing species, including boron "ate" complexes, BH$_3$ and [BH$_4$]$^-$, as observed by ^{11}B-NMR spectroscopy (see Supplementary Materials, S3). This reactivity is in accordance with that when using other nucleophiles such as alkoxide salts [39,41]. Taken together, these observations are indicative of an in situ activation process, whereby the added carboxylate reagent Na(2-EH) triggers hydride transfer from boron to iron to form the dihydride dmpe$_2$FeH$_2$ **7**. Once formed, the iron dihydride dmpe$_2$FeH$_2$ **7** can efficiently catalyse the C(sp^2)-H borylation reaction.

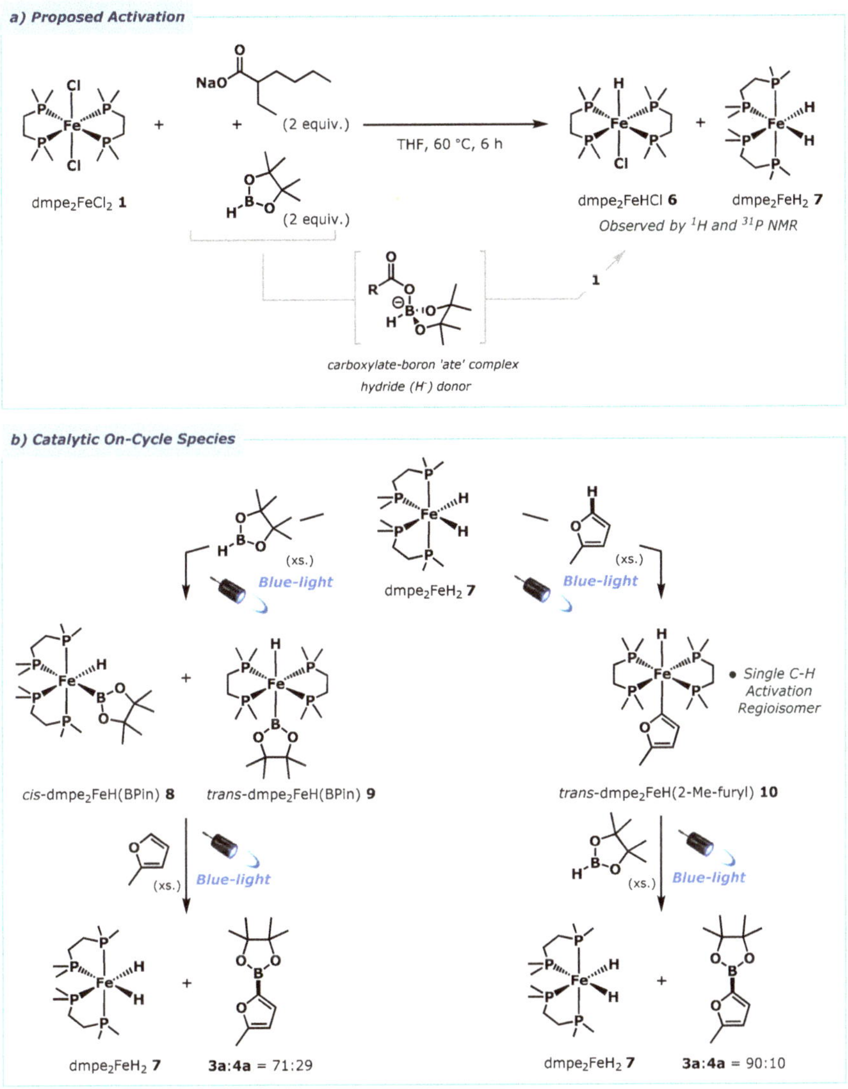

Scheme 4. (**a**) Pre-catalyst activation and hydride formation. (**b**) Mechanistic investigations of dmpe₂FeH₂ **7** produced by hydride transfer from HBpin and Na(2-EH).

As the dihydride complex dmpe₂FeH₂ **7** was readily formed using our in situ hydride transfer method, and was observable by ¹H and ³¹P-NMR spectroscopy, we next investigated the fundamental steps of this borylation reaction with the aim of identifying key reaction intermediates. Reaction of the in situ generated dmpe₂FeH₂ **7** with excess HBpin under blue light irradiation led to the formation of both *cis*-dmpe₂FeH(Bpin) **8** and *trans*-dmpe₂FeH(Bpin) **9** boryl iron complexes, as observed by ¹H, ¹¹B, and ³¹P NMR spectroscopy (see Supplementary Materials, S17-19). These complexes were previously reported by Darcel and co-workers, where they were formed from the reaction of the related dialkyl complex, dmpe₂FeMe₂, with HBpin [37]. Addition of 2-methylfuran **2a** to the mixture of *cis*-dmpe₂FeH(Bpin) **8** and *trans*-dmpe₂FeH(Bpin) **9** under blue light irradiation gave the formation of the regioisomeric furyl boronic esters **3a** and **4a** (**3a:4a** = 71:29), notably in a different ratio to that observed during catalysis (vide supra, **3a:4a** = 81:19).

Blue light irradiation of the dihydride complex dmpe$_2$FeH$_2$ **7** in the presence of excess 2-methylfuran **2a** led to exclusive C(sp^2)-H bond metallation at the 5-position to give *trans*-dmpe$_2$FeH(2-Me-furyl) **10**, as observed by ^{1}H and ^{31}P NMR spectroscopy (see Supplementary Materials, S22-24). Addition of HBpin to *trans*-dmpe$_2$FeH(2-Me-furyl) **10** and irradiation with blue light induced formation of the furyl boronic esters **3a** and **4a** (**3a:4a** = 90:10). Again, in a different ratio to that observed under catalysis. As the ratio of regioisomers observed under catalytic conditions (**3a:4a** = 81:19) appears to be a combination of the ratios observed in the stoichiometric studies (**3a:4a** = 71:29, and 90:10 respectively), it is suggestive that both the C-H metallation and iron boryl pathways are operative. Specifically, the reaction can precede by C(sp^2)-H bond metallation to give dmpe$_2$FeH(2-Me-Furyl) **10**, followed by C(sp^2)-B bond formation, or by direct reaction of arene with the iron boryl species *cis*-dmpe$_2$FeH(Bpin) **8** and *trans*-dmpe$_2$FeH(Bpin) **9**. The relative ratios of the furyl boronic ester regioisomers indicate both pathways are equally accessible for the activated catalyst. (53% by C-H metallation, 47% by the iron boryl species).

3. Conclusions

In summary, we have investigated the applicability of several alkoxide, carboxylate and other, common bench stable reagents towards the in situ activation of an iron(II) pre-catalyst for C(sp^2)-H bond borylation. We found a sodium carboxylate salt Na(2-EH) in combination with HBpin to be a potent pre-catalysts activator generating the iron dihydride dmpe$_2$FeH$_2$ **7** in situ. The validity of this method was demonstrated by the generation of catalytically relevant species that were used as mechanistic probes. These suggest two C-H borylation pathways are operating to give the aryl boronic ester products; C-H metallation followed by borylation, and formation of an iron boryl species followed by arylation.

4. Materials and Methods

4.1. General Information

All compounds reported in the manuscript are commercially available or have been previously described in the literature unless indicated otherwise. All experiments involving iron were performed using standard Schlenk techniques under argon or nitrogen atmosphere. All yields refer to yields determined by ^{1}H-NMR spectroscopy of crude reaction mixtures using an internal standard. All product ratios refer to product ratios determined by ^{1}H-NMR spectroscopy of the crude reaction mixtures. ^{1}H-NMR and ^{13}C-NMR data are given for all compounds when possible in the experimental section for characterisation purposes. Spectroscopic data matched those reported previously.

4.2. Activator Synthesis

Tetra-*n*-butylammonium 2-ethylhexanoate TBA(2-EH)

A suspension of KH (80 mg, 2 mmol) in anhydrous THF (20 mL) was prepared under an N$_2$ atmosphere, 2-ethylhexanoic acid (0.32 mL, 2 mmol) was added dropwise whilst stirring. *n.b.* gas evolution (H$_2$). The solution was stirred for 3 h at room temperature, and the THF removed in vacuo to give an amorphous colourless solid. The solid was re-dissolved in MeOH (20 mL) and tetra-*n*-butylammonium chloride (556 mg, 2 mmol) was added, the solution was stirred for 16 h, filtered through a glass frit and dried in vacuo without further purification to give tetra-*n*-butylammonium 2-ethylhexanoate (0.72 g, 1.86 mmol, 93%) as an amorphous white solid.

^{1}H-NMR (500 MHz, CDCl$_3$) δ 3.51–3.35 (m, 8H), 2.09 (tt, *J* = 8.4, 5.4 Hz, 1H), 1.66 (m, 8H), 1.62–1.54 (m, 2H), 1.48–1.39 (m, 8H), 1.39–1.23 (m, 6H), 0.98 (t, *J* = 7.3 Hz, H), 0.91 (t, *J* = 7.4 Hz, 3H), 0.88–0.82 (m, 3H). ^{13}C-NMR (126 MHz, CDCl$_3$) δ 181.2, 59.0, 51.1, 33.1, 30.5, 26.4, 24.3, 23.2, 19.8, 14.2, 13.7, 12.7.

4.3. Pre-catalyst Synthesis

dmpe$_2$FeCl$_2$ **1** [42]

Anhydrous iron dichloride (0.21 g, 1.67 mmol) was charged to a Schlenk flask and dissolved in anhydrous THF (10 mL), dmpe [(bis(dimethylphosphino)ethane]; 0.50 g, 3.33 mmol) were added to the flask under an Ar atmosphere and the solution left to stir for 48 h at room temperature. The solvent was removed in vacuo, and in an argon-filled glove box, the residue was re-dissolved in dichloromethane (5 mL) and filtered through glass wool. The filtrate was reduced in vacuo to produce a green amorphous solid (0.549 g, 1.29 mmol, 77%).

^{1}H-NMR (400 Hz, d^8-THF) δ 2.18 (s, 8H), 1.42 (s, 24H). ^{31}P-NMR (202 MHz, CDCl$_3$) δ 59.0.

4.4. General Borylation Procedure

In an argon-filled glovebox, dmpe$_2$FeCl$_2$ **1** (8.6 mg, 0.02 mmol), sodium 2-ethylhexanoate (6.6 mg, 0.04 mmol), HBpin (87 μL, 0.6 mmol), substrate (0.5 mmol), and THF (1 mL) were added to a 1.7 mL sample vial and shaken to ensure full dissolution. The vial was placed under blue light radiation for 48 h and then allowed to cool to room temperature. Yields determined by ^{1}H-NMR spectroscopy of the crude reaction mixtures using 1,3,5-trimethoxybenzene as an internal standard [0.5 mL; standard solution = 1,3,5-trimethoxybenzene (0.336 g, 2.0 mmol) in diethyl ether (10 mL)]. Product ratios were determined by ^{1}H-NMR spectroscopy of the crude reaction mixtures.

4.5. Characterisation of Borylated Products

4.5.1. 2-Methylfuran Derivatives

4,4,5,5-Tetramethyl-2-(5-methylfuran-2-yl)-1,3,2-dioxaborolane **3a** [37], 4,4,5,5-tetramethyl-2-(5-methylfuran-3-yl)- 1,3,2-dioxaborolane **4a** [37]

Following the general procedure; 2-methylfuran **2a** (41 mg, 44 μL, 0.5 mmol). Yield = 72%. **3a:4a** = 81:19. ^{1}H-NMR (500 MHz, CDCl$_3$) **3a**: δ 6.99 (d, *J* = 3.2 Hz, 1H), 6.06–6.01 (m, 1H), 2.36 (s, 3H), 1.34 (s, 12H). **4a**: δ 7.62 (d, *J* = 0.9 Hz, 1H), 6.15 (t, *J* = 1.0 Hz, 1H), 2.29 (d, *J* = 1.1 Hz, 3H), 1.31 (s, 12H). ^{13}C-NMR (126 MHz, CDCl$_3$) **3a**: δ 157.8, 124.8, 106.9, 84.0, 24.7, 13.9. **4a**: δ 152.7, 149.7, 108.8, 83.3, 24.9, 13.1. ^{11}B-NMR (160 MHz, CDCl$_3$) **3a**: δ 27.1. **4a**: δ 29.8.

4.5.2. Furan Derivatives

4,4,5,5-Tetramethyl-2-(furanyl-2-yl)-1,3,2-dioxaborolane **3b** [43], 4,4,5,5-tetramethyl-2-(furanyl-3-yl)-1,3,2-dioxaborolane **4b** [44], 2,5-bis(4,4,5,5-tetramethyl-1,3,2-dioxaborolan-2-yl)furan **5ba** [45], 2,4-bis(4,4,5,5-tetramethyl-1,3,2-dioxaborolan-2-yl)furan **5bb** [45]

Following the general procedure; furan **2b** (34 mg, 36 μL, 0.5 mmol). Yield = 52%. **3b:4b:5ba:5bb** 60:21:4:5. ^{1}H-NMR (600 MHz, CDCl$_3$) **3b**: δ 7.65 (d, *J* = 1.7 Hz, 1H), 7.07 (d, *J* = 3.4 Hz, 1H), 6.44 (dd, *J* = 3.4, 1.6 Hz, 1H), 1.35 (s, 12H). **4b**: δ 7.78 (s, 1H), 7.46 (m, *J* = 1.6 Hz, 1H), 6.59 (d, *J* = 1.7 Hz, 1H), 1.32 (s, 1H). **5ba**: δ 7.06 (s, 2H), 1.33 (s, 24H). **5bb**: δ 7.78 (s, 1H), 7.28 (s, 1H), 1.30 (s, 24H). ^{13}C-NMR (126 MHz, CDCl$_3$) **3b**: δ 147.3, 123.2, 110.3, 84.2, 24.8. **4b**: δ 151.2, 142.9, 113.1, 83.5, 24.9. **5ba**: δ 123.2, 83.5, 24.8. **5bb**: δ 151.2, 83.2, 75.1, 24.6. ^{11}B-NMR (160 MHz, CDCl$_3$) **3b**: δ 27.2. **4b**: δ 29.8.

4.5.3. 2.3-Dimethylfuran Derivatives

4,4,5,5-Tetramethyl-2-(4,5-dimethylfuran-2-yl)-1,3,2-dioxaborolane **3c** [46]

Following the general procedure; 2,3-dimethylfuran **2c** (48 mg, 53 µL, 0.5 mmol). Yield = 46%. **3c:4c** = 100:0. ^{1}H-NMR (500 MHz, CDCl$_3$) δ 6.87 (s, 1H), 2.26 (s, 3H), 1.94 (s, 3H), 1.33 (s, 12H). ^{13}C-NMR (126 MHz, CDCl$_3$) δ 153.4, 127.1, 115.2, 83.9, 24.7, 11.8, 9.7. ^{11}B-NMR (160 MHz, CDCl$_3$) δ 27.2.

4.5.4. 2-Ethylfuran Derivatives

4,4,5,5-Tetramethyl-2-(5-ethylfuran-2-yl)-1,3,2-dioxaborolane **3d** [47], 4,4,5,5-tetramethyl-2-(5-methylfuran-3-yl)-1,3,2-dioxaborolane **4d** [47]

Following the general procedure; 2-ethylfuran **2d** (48 mg, 53 µL, 0.5 mmol). Yield = 59%. **3d:3d** = 70:30. ^{1}H-NMR (500 MHz, CDCl$_3$) **3d**: δ 7.01 (d, *J* = 3.3 Hz, 1H), 6.05 (d, *J* = 3.1, 1H), 2.72 (q, *J* = 7.6 Hz, 2H), 1.34 (s, 12H), 1.25 (m, 3H). **4d**: δ 7.64 (d, *J* = 0.8 Hz, 1H), 6.18 (d, *J* = 1.1 Hz, 1H), 2.6 (q, *J* = 7.5, 2H), 1.31 (s, 12H), 1.25 (m, 3H). ^{13}C-NMR (126 MHz, CDCl$_3$) **3d**: δ 163.6, 124.7, 105.2, 84.0, 24.7, 21.6, 12.2. **4d**: δ 163.6, 149.6, 107.2, 83.3, 24.9, 21.1, 12.1. ^{11}B-NMR (160 MHz, CDCl$_3$) **3d**: δ 27.2. **4d**: δ 29.9.

4.5.5. 2-Methylthiophene Derivatives

4,4,5,5-Tetramethyl-2-(5-methylthiophen-2-yl)-1,3,2-dioxaborolane **3e** [47], 4,4,5,5-tetramethyl-2-(5-methylthiophen-3-yl)-1,3,2-dioxaborolane **4e** [47]

Following the general procedure; 2-methylthiophene **2e** (49 mg, 48 µL, 0.5 mmol). Yield = 9%. **3e:4e** = 66:34. ^{1}H-NMR (500 MHz, CDCl$_3$) **3e**: δ 7.45 (d, *J* = 3.4 Hz, 1H), 6.84 (d, *J* = 3.4, 1H), 2.53 (s, 3H), 1.33 (s, 12H). **4e**: δ 7.67 (d, *J* = 1.2 Hz, 1H), 7.04 (s, 1H), 2.49 (d, *J* = 1.1 Hz, 3H), 1.32 (s, 12H). ^{13}C-NMR (126 MHz, CDCl$_3$) **3e**: δ 147.5, 137.6, 127.0, 83.9, 24.9, 15.4. **4e**: not observed. ^{11}B-NMR (160 MHz, CDCl$_3$) δ 28.7.

4.5.6. Thiophene Derivatives

4,4,5,5-Tetramethyl-2-(thiophen-2-yl)-1,3,2-dioxaborolane **3f** [48], 4,4,5,5-tetramethyl-2-(thiophen-3-yl)-1,3,2-dioxaborolane **4f** [49], 2,5-bis(4,4,5,5-tetramethyl-1,3,2-dioxaborolan-2-yl)thiophene **5f** [48]

Following the general procedure; thiophene **2f** (42 mg, 40 µL, 0.5 mmol). Yield = 11%. **3f:4f:5f** = 27:9:64. ^{1}H-NMR (500 MHz, CDCl$_3$) **3f**: δ 7.64 (m, 1H), 7.63 (d, *J* = 4.8 Hz, 1H), 7.19 (d, *J* = 4.8 Hz, 1H), 1.35 (s, 12H). **4f**: δ 7.92 (d, *J* = 2.6 Hz, 1H), 7.41 (d, *J* = 4.9 Hz, 1H), 7.34 (dd, *J* = 4.8, 2.7 Hz, 1H), 1.35 (s, 12H). **5f**: δ 7.66 (s, 2H), 1.34 (s, 24H). ^{13}C-NMR (126 MHz, CDCl$_3$) **3f**: δ 137.2, 132.4, 128.2, 83.2, 24.9. **4f**: δ 136.5, 129.0, 125.3, 83.2, 24.8. **5f**: δ 137.7, 84.1, 24.8. ^{11}B-NMR (160 MHz, CDCl$_3$) δ 29.2.

4.5.7. 3-Methylthiophene Derivatives

4,4,5,5-Tetramethyl-2-(4-methylthiophen-2-yl)-1,3,2-dioxaborolane **3g** [17]

Following the general procedure; 3-methylthiophene **2g** (49 mg, 48 µL, 0.5 mmol). Yield = 4%. **3g:4g** = 100:0. ^{1}H-NMR (500 MHz, CDCl$_3$) δ 7.44 (d, *J* = 1.2 Hz, 1H), 7.19 (t, *J* = 1.1 Hz, 1H), 2.29 (d, *J* = 0.9 Hz, 3H), 1.34 (s, 12H). ^{13}C-NMR (126 MHz, CDCl$_3$) δ 139.5, 139.0, 128.2, 84.0, 24.9, 15.1. ^{11}B-NMR (160 MHz, CDCl$_3$) δ 29.1.

4.6. Mechanistic Investigations

dmpe$_2$FeH$_2$ **7** [37]

dmpe$_2$FeCl$_2$ **1** (10 mg, 0.023 mmol), sodium 2-ethylhexanoate (7.6 mg, 0.046 mmol), and HBpin (7 µL, 0.046 mmol) were added to a Young's NMR tube under an Ar atmosphere and heated at 60 °C for 3 days. ^{1}H NMR (600 MHz, THF) δ −14.38 (m). ^{31}P-NMR (500 MHz, THF) δ 76.9 (t, *J* = 28 Hz), 67.7 (t, *J* = 28 Hz).

cis-dmpe$_2$FeH(Bpin) **8** and *trans*-dmpe$_2$FeH(Bpin) **9** [37]

dmpe$_2$FeCl$_2$ **1** (4.3 mg, 0.001 mmol), sodium 2-ethylhexanoate (3.3 mg, 0.002 mmol), and HBpin (87 µL, 0.6 mmol) were added to a Young's NMR tube under an Ar atmosphere and irradiated with blue light for 16 h. ^{1}H-NMR (500 MHz, THF) δ −13.1 (p, *J* = 43.2 Hz), −14.0 (m). ^{31}P-NMR (500 MHz, THF) δ 77.6(m), 77.2 (m), 59.7 (m), 58.9 (m).

dmpe$_2$FeH(2-Me-furyl) **10**

dmpe$_2$FeCl$_2$ **1** (10 mg, 0.023 mmol), sodium 2-ethylhexanoate (30.4 mg, 0.184 mmol), and HBpin (7 µL, 0.046 mmol) were added to a Young's NMR tube under an Ar atmosphere and warmed at 60 °C for 24 h. 2-methylfuran (8 µL, 0.092 mmol) was added under an Ar atmosphere and the sample irradiated with blue light for 3 h. This complex was observed *in situ*. ^{1}H-NMR (500 MHz, THF) δ -18.93 (q, *J* = 45.8 Hz). ^{31}P-NMR (500 MHz, THF) δ 77.1 (d, *J* = 38.1 Hz). MS: (HRMS – EI$^+$) Found 438.12041 (C$_{17}$H$_{38}$O$_1$^{56}Fe$_1$P$_4$), requires 438.12171.

Supplementary Materials: The following are available online.

Author Contributions: L.B. and J.H.D. carried out the experimental work. J.H.D., A.P.D. and S.P.T. conceived and supervised the project. L.B., J.H.D., and S.P.T. wrote the paper. All authors have read and agreed to the published version of the manuscript.

Funding: This research was funded by The Royal Society, UF130393 and RF191015, and GSK/EPSRC, PIII0002.

Acknowledgments: S.P.T. acknowledges the University of Edinburgh and the Royal Society for a University Research Fellowship. J.H.D. and S.P.T. acknowledge GSK, EPSRC, and the University of Edinburgh for post-doctoral funding. L.B. acknowledges the Royal Society and the University of Edinburgh for a PhD studentship.

Conflicts of Interest: The authors declare no conflict of interest.

References

1. Kakiuchi, F.; Chatani, N. Catalytic Methods for C-H Bond Functionalization: Application in Organic Synthesis. *Adv. Synth. Catal.* **2003**, *345*, 1077–1101. [CrossRef]
2. Godula, K.; Sames, D. C-H bond functionalization in complex organic synthesis. *Science* **2006**, *312*, 67–72. [CrossRef]
3. Lyons, T.W.; Sanford, M.S. Palladium-catalyzed ligand-directed C-H functionalization reactions. *Chem. Rev.* **2010**, *110*, 1147–1169. [CrossRef] [PubMed]
4. McMurray, L.; O'Hara, F.; Gaunt, M.J. Recent developments in natural product synthesis using metal-catalysed C-H bond functionalisation. *Chem. Soc. Rev.* **2011**, *40*, 1885–1898. [CrossRef]
5. Yamaguchi, J.; Yamaguchi, A.D.; Itami, K. C-H bond functionalization: Emerging synthetic tools for natural products and pharmaceuticals. *Angew. Chem. Int. Ed.* **2012**, *51*, 8960–9009. [CrossRef] [PubMed]
6. Davies, H.M.L.; Morton, D. Recent Advances in C-H Functionalization. *J. Org. Chem.* **2016**, *81*, 343–350. [CrossRef] [PubMed]
7. Roudesly, F.; Oble, J.; Poli, G. Metal-catalyzed C-H activation/functionalization: The fundamentals. *J. Mol. Catal. A Chem.* **2017**, *426*, 275–296. [CrossRef]
8. Su, B.; Cao, Z.C.; Shi, Z.J. Exploration of Earth-Abundant Transition Metals (Fe, Co, and Ni) as Catalysts in Unreactive Chemical Bond Activations. *Acc Chem. Res.* **2015**, *48*, 886–896. [CrossRef]
9. Cera, G.; Ackermann, L. Iron-Catalyzed C–H Functionalization Processes. *Top. Curr. Chem.* **2016**, *374*, 57. [CrossRef]
10. Shang, R.; Ilies, L.; Nakamura, E. Iron-Catalyzed C-H Bond Activation. *Chem. Rev.* **2017**, *117*, 9086–9139. [CrossRef]
11. Gandeepan, P.; Müller, T.; Zell, D.; Cera, G.; Warratz, S.; Ackermann, L. 3d Transition Metals for C-H Activation. *Chem. Rev.* **2019**, *119*, 2192–2452. [CrossRef] [PubMed]
12. Loup, J.; Dhawa, U.; Pesciaioli, F.; Wencel-Delord, J.; Ackermann, L. Enantioselective C–H Activation with Earth-Abundant 3d Transition Metals. *Angew. Chem. Int. Ed.* **2019**, *58*, 12803–12818. [CrossRef] [PubMed]
13. Mkhalid, I.A.I.; Barnard, J.H.; Marder, T.B.; Murphy, J.M.; Hartwig, J.F. C-H activation for the construction of C-B bonds. *Chem. Rev.* **2010**, *110*, 890–931. [CrossRef] [PubMed]
14. Fyfe, J.W.B.; Watson, A.J.B. Recent Developments in Organoboron Chemistry: Old Dogs, New Tricks. *Chemistry* **2017**, *3*, 31–55. [CrossRef]

15. Xu, L.; Wang, G.; Zhang, S.; Wang, H.; Wang, L.; Liu, L.; Jiao, J.; Li, P. Recent advances in catalytic C–H borylation reactions. *Tetrahedron* **2017**, *73*, 7123–7157. [CrossRef]

16. Ishiyama, T.; Takagi, J.; Ishida, K.; Miyaura, N.; Anastasi, N.R.; Hartwig, J.F. Mild iridium-catalyzed borylation of arenes. High turnover numbers, room temperature reactions, and isolation of a potential intermediate. *J. Am. Chem. Soc.* **2002**, *124*, 390–391. [CrossRef]

17. Chotana, G.A.; Kallepalli, V.A.; Maleczka, R.E.; Smith, M.R. Iridium-catalyzed borylation of thiophenes: Versatile, synthetic elaboration founded on selective C-H functionalization. *Tetrahedron* **2008**, *64*, 6103–6114. [CrossRef] [PubMed]

18. Hartwig, J.F. Regioselectivity of the borylation of alkanes and arenes. *Chem. Soc. Rev.* **2011**, *40*, 1992–2002. [CrossRef]

19. Hartwig, J.F. Borylation and silylation of C-H bonds: A platform for diverse C-H bond functionalizations. *Acc. Chem. Res.* **2012**, *45*, 864–873. [CrossRef]

20. Preshlock, S.M.; Plattner, D.L.; Maligres, P.E.; Krska, S.W.; Maleczka, R.E.; Smith, M.R. A traceless directing group for C-H borylation. *Angew. Chem. Int. Ed.* **2013**, *52*, 12915–12919. [CrossRef]

21. Larsen, M.A.; Hartwig, J.F. Iridium-catalyzed C-H borylation of heteroarenes: Scope, regioselectivity, application to late-stage functionalization, and mechanism. *J. Am. Chem. Soc.* **2014**, *136*, 4287–4299. [CrossRef] [PubMed]

22. Sadler, S.A.; Tajuddin, H.; Mkhalid, I.A.I.; Batsanov, A.S.; Albesa-Jove, D.; Cheung, M.S.; Maxwell, A.C.; Shukla, L.; Roberts, B.; Blakemore, D.C.; et al. Iridium-catalyzed C-H borylation of pyridines. *Org. Biomol. Chem.* **2014**, *12*, 7318–7327. [CrossRef] [PubMed]

23. Saito, Y.; Segawa, Y.; Itami, K. Para C-H borylation of benzene derivatives by a bulky iridium catalyst. *J. Am. Chem. Soc.* **2015**, *137*, 5193–5198. [CrossRef] [PubMed]

24. Yang, L.; Semba, K.; Nakao, Y. Para-Selective C–H Borylation of (Hetero)Arenes by Cooperative Iridium/Aluminum Catalysis. *Angew. Chem. Int. Ed.* **2017**, *56*, 4853–4857. [CrossRef] [PubMed]

25. Yan, G.; Jiang, Y.; Kuang, C.; Wang, S.; Liu, H.; Zhang, Y.; Wang, J. Nano-Fe_2O_3-catalyzed direct borylation of arenes. *Chem. Commun.* **2010**, *46*, 3170–3172. [CrossRef]

26. Zhang, H.; Hagihara, S.; Itami, K. Aromatic C-H borylation by nickel catalysis. *Chem. Lett.* **2015**, *44*, 779–781. [CrossRef]

27. Furukawa, T.; Tobisu, M.; Chatani, N. Nickel-catalyzed borylation of arenes and indoles via C-H bond cleavage. *Chem. Commun.* **2015**, *51*, 6508–6511. [CrossRef] [PubMed]

28. Mazzacano, T.J.; Mankad, N.P. Thermal C-H borylation using a CO-free iron boryl complex. *Chem. Commun.* **2015**, *51*, 5379–5382. [CrossRef]

29. Obligacion, J.V.; Semproni, S.P.; Pappas, I.; Chirik, P.J. Cobalt-Catalyzed C(sp^2)-H Borylation: Mechanistic Insights Inspire Catalyst Design. *J. Am. Chem. Soc.* **2016**, *138*, 10645–10653. [CrossRef]

30. Léonard, N.G.; Bezdek, M.J.; Chirik, P.J. Cobalt-Catalyzed C(sp^2)-H borylation with an air-stable, readily prepared terpyridine cobalt(II) Bis(acetate) precatalyst. *Organometallics* **2017**, *36*, 142–150. [CrossRef]

31. Yoshigoe, Y.; Kuninobu, Y. Iron-Catalyzed ortho-Selective C-H Borylation of 2-Phenylpyridines and Their Analogs. *Org. Lett.* **2017**, *19*, 3450–3453. [CrossRef] [PubMed]

32. Obligacion, J.V.; Bezdek, M.J.; Chirik, P.J. C(sp^2)-H Borylation of Fluorinated Arenes Using an Air-Stable Cobalt Precatalyst: Electronically Enhanced Site Selectivity Enables Synthetic Opportunities. *J. Am. Chem. Soc.* **2017**, *139*, 2825–2832. [CrossRef] [PubMed]

33. Kamitani, M.; Kusaka, H.; Yuge, H. Iron-catalyzed versatile and efficient C(sp^2)-H borylation. *Chem. Lett.* **2019**, *48*, 898–901. [CrossRef]

34. Agahi, R.; Challinor, A.J.; Dunne, J.; Docherty, J.H.; Carter, N.B.; Thomas, S.P. Regiodivergent hydrosilylation, hydrogenation, [$2\pi + 2\pi$]-cycloaddition and C-H borylation using counterion activated earth-abundant metal catalysis. *Chem. Sci.* **2019**, *10*, 5079–5084. [CrossRef]

35. Hatanaka, T.; Ohki, Y.; Tatsumi, K. C-H bond activation/borylation of furans and thiophenes catalyzed by a half-sandwich iron N-heterocyclic carbene complex. *Chemistry* **2010**, *5*, 1657–1666. [CrossRef]

36. Mazzacano, T.J.; Mankad, N.P. Base metal catalysts for photochemical C-H borylation that utilize metal-metal cooperativity. *J. Am. Chem. Soc.* **2013**, *135*, 17258–17261. [CrossRef]

37. Dombray, T.; Werncke, C.G.; Jiang, S.; Grellier, M.; Vendier, L.; Bontemps, S.; Sortais, J.B.; Sabo-Etienne, S.; Darcel, C. Iron-catalyzed C-H borylation of arenes. *J. Am. Chem. Soc.* **2015**, *137*, 4062–4065. [CrossRef]

38. Allen, O.R.; Dalgarno, S.J.; Field, L.D.; Jensen, P.; Turnbull, A.J.; Willis, A.C. Addition of CO_2 to alkyl iron complexes, Fe(PP)$_2$Me$_2$. *Organometallics* **2008**, *27*, 2092–2098. [CrossRef]

39. Docherty, J.H.; Peng, J.; Dominey, A.P.; Thomas, S.P. Activation and discovery of earth-abundant metal catalysts using sodium *tert*-butoxide. *Nat. Chem.* **2017**, *9*, 595–600. [CrossRef]

40. Buys, I.E.; Field, L.D.; Hambley, T.W.; McQueen, A.E.D.J. Photochemical reactions of [cis-Fe(H)2(Me2PCH2CH2PMe2)2] with thiophenes: Insertion into C–H and C–S bonds. *Chem. Soc. Chem. Commun.* **1994**, *5*, 557–558. [CrossRef]

41. Query, I.P.; Squier, P.A.; Larson, E.M.; Isley, N.A.; Clark, T.B. Alkoxide-catalyzed reduction of ketones with pinacolborane. *J. Org. Chem.* **2011**, *76*, 6452–6456. [CrossRef]

42. Elton, T.E.; Ball, G.E.; Bhadbhade, M.; Field, L.D.; Colbran, S.B. Evaluation of Organic Hydride Donors as Reagents for the Reduction of Carbon Dioxide and Metal-Bound Formates. *Organometallics* **2018**, *37*, 3972–3982. [CrossRef]

43. Wollenburg, M.; Moock, D.; Glorius, F. Hydrogenation of Borylated Arenes. *Angew. Chem. Int. Ed.* **2019**, *58*, 6549–6553. [CrossRef] [PubMed]

44. Zhang, L.; Jiao, L. Pyridine-Catalyzed Radical Borylation of Aryl Halides. *J. Am. Chem. Soc.* **2017**, *139*, 607–610. [CrossRef] [PubMed]

45. Miyaura, N.; Ishiyama, T. Process for the Production of Hetereoaryl-type Boron Compounds with Iridium Catalyst. European Patent EP1481978A1, 1 December 2004.

46. Xue, C.; Luo, Y.; Teng, H.; Ma, Y.; Nishiura, M.; Hou, Z. Ortho-Selective C–H Borylation of Aromatic Ethers with Pinacol-borane by Organo Rare-Earth Catalysts. *ACS Catalysis.* **2018**, *8*, 5017–5022. [CrossRef]

47. Kato, T.; Kuriyama, S.; Nakajima, K.; Nishibayashi, Y. Catalytic C–H Borylation Using Iron Complexes Bearing 4,5,6,7-Tetrahydroisoindol-2-ide-Based PNP-Type Pincer Ligand. *Chem Asian J.* **2019**, *14*, 2097–2101. [CrossRef]

48. Tobisu, M.; Igarashi, T.; Chatani, N. Iridium/N-heterocyclic carbene-catalyzed C–H borylation of arenes by diisopropylaminoborane. *Beilstein J. Org. Chem.* **2016**, *12*, 654–661. [CrossRef]

49. Ratniyom, J.; Dechnarong, N.; Yotphan, S.; Kiatisevi, S. Convenient Synthesis of Arylboronates through a Synergistic Pd/Cu-Catalyzed Miyaura Borylation Reaction under Atmospheric Conditions. *Eur. J. Org. Chem.* **2014**, *7*, 1381–1385. [CrossRef]

Sample Availability: Samples of the pre-catalyst **1** are available from the authors.

Communication

Fe-catalyzed Decarbonylative Alkylative Spirocyclization of *N*-Arylcinnamamides: Access to Alkylated 1-Azaspirocyclohexadienones

Xiang Peng [†], Ren-Xiang Liu [†], Xiang-Yan Xiao and Luo Yang *

Key Laboratory for Environmentally Friendly Chemistry and Application of the Ministry of Education, Key Laboratory for Green Organic Synthesis and Application of Hunan Province, College of Chemistry, Xiangtan University, Hunan 411105, China; pxsb123456@163.com (X.P.); lrx201821531602@163.com (R.-X.L.); xxy0707@163.com (X.-Y.X.)

* Correspondence: yangluo@xtu.edu.cn; Tel.: +86-0731-59282229; Fax: +86-0731-58292251

† These authors contributed equally to this work.

Received: 29 December 2019; Accepted: 16 January 2020; Published: 21 January 2020

Abstract: For the convenient introduction of simple linear/branched alkyl groups into biologically important azaspirocyclohexadienones, a practical Fe-catalyzed decarbonylative cascade spiro-cyclization of *N*-aryl cinnamamides with aliphatic aldehydes to provide alkylated 1-azaspiro-cyclohexadienones was developed. Aliphatic aldehydes were oxidative decarbonylated into primary, secondary and tertiary alkyl radicals conveniently and allows for the subsequent cascade construction of dual $C(sp^3)$-$C(sp^3)$ and C=O bonds via radical addition, spirocyclization and oxidation sequence.

Keywords: decarbonylation; alkylation; spirocyclization; aldehyde; cinnamamide

1. Introduction

The azaspirocyclohexadienone ring represents an important structural motif in natural compounds and pharmaceuticals, such as Annosqualine, Erythratinone and compound **I** (acting as muscarinic antagonist, Figure 1) [1–4]. Thus, the development of efficient methods for the synthesis of these azaspirocyclohexadienones has attracted substantial attention. While traditional synthetic routes rely predominantly on the transition-metal-catalyzed intramolecular *ipso*-carbocyclization [5–9] and electrophilic *ipso*-cyclization [10–12], the radical cascade *ipso*-cyclization was expanding in recent years, to incorporate various functional groups into the azaspirocyclohexadienone framework [13–15]. Among them, the radical difunctionalization and *ipso*-cyclization of *N*-arylcinnamamides or *N*-arylpropiolamides has proven to be an straightforward pathway, where various carbon or heteroatom centered radicals added onto the α,β-unsaturated carbon-carbon multiple bond of substrates, followed by the intramolecular radical *ipso*-cyclization and dearomatization. In this context, simple ethers [16], alkanes [17,18], ketones [19], acetonitrile [20], acyl chloride [21], aldehydes [22], aryldiazonium salts [23], CF_3SO_2Na [24–26], arylsulfinic acids [27], sulfonylhydrazides [28], $AgSCF_3$ [29], thiophenols [30], disulfides [31], *N*-sulfanylsuccinimide [32], diselenide [33], *tert*-butyl nitrite [34], phosphonates [24,35,36] and silanes [37,38] have been demonstrated as the radical precursors for this spirocyclization (Scheme 1a), as independently reported by Li [16,17], Liang [25,29], and Zhu [18,20], etc. The addition of carbon radicals to *N*-arylcinnamamides and subsequent *ipso*-cyclization offers a direct synthesis of azaspiro-compounds with the concurrent construction of two C–C single bonds. However, the alkyl radical source for this spirocyclization of alkenes is still confined. When ethers, alkanes, ketones and acetonitrile acted as carbon radical precursors via the homolytic cleavage of sp^3 C-H bond, the functional groups (ether, ketone or cyano groups) would inevitably be imported into the products, and the generation of regio-isomers was also troublesome due to the possible existence several different sp^3 C-H bonds in these precursors, so readily available alkyl precursors that could introduce ordinary and simple linear/branched alkyl groups into azaspirocyclohexadienones are highly desirable, especially those could realize the convenient radical generation and be compatible with the primary, secondary and tertiary alkyls.

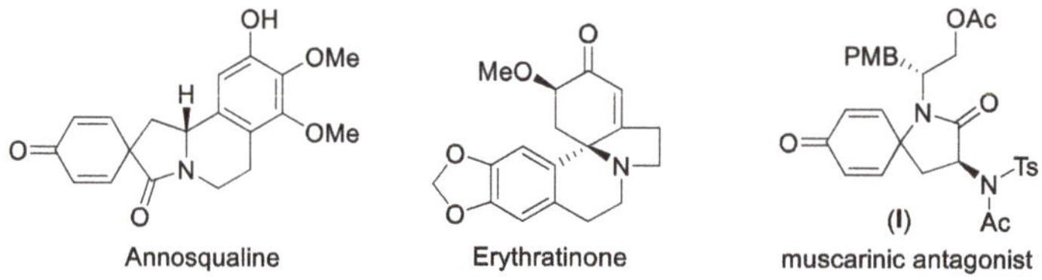

Figure 1. Representative azaspirocyclohexadienone-based bioactive molecules.

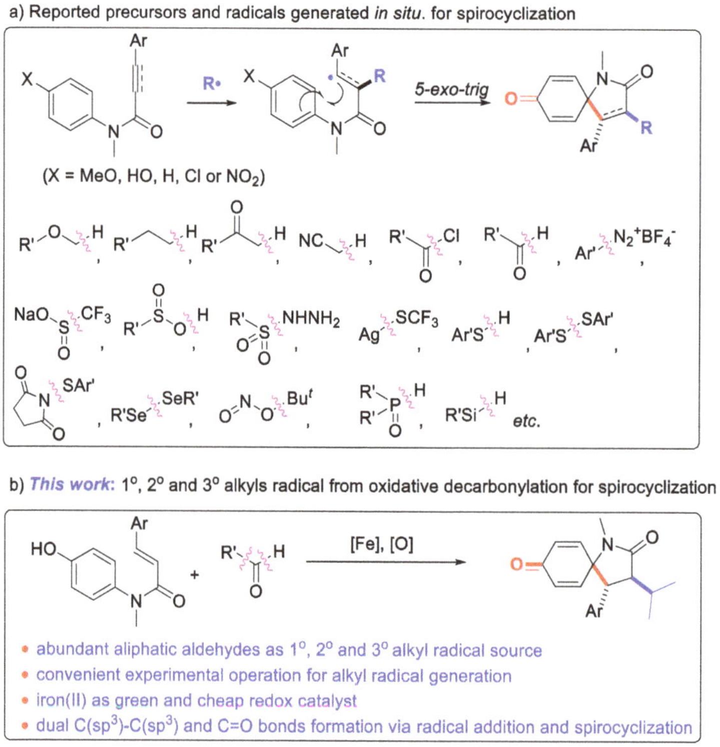

Scheme 1. Various radicals for the spirocyclization of *N*-aryl unsaturated amides.

On the other hand, aldehydes are cheap and readily available chemicals and have been directly used for decarbonylative couplings catalyzed by ruthenium or rhodium, as shown by the extensive studies of Li since 2009 [39–44]. In contrast, we are interested in the radical-type decarbonylative reactions of aldehydes in the absence of noble metals, with peroxides as radical initiator and oxidant [45–53]. The oxidative decarbonylative couplings of aldehydes with (hetero) arenes [45,46], styrenes [47–50], alkyne and electron-deficient alkenes [51–53] were successively developed by our group. These decarbonylative reactions were further updated by other groups, with dioxygen as the radical initiator and oxidant [54–56]. Similar radical type decarbonylative alkylations of C=C and C≡C bonds with aldehydes were also separately developed by Li [57,58], Li [59–62] and Yu [63,64].

The above studies have fully demonstrated that radical-type decarbonylation of aliphatic aldehydes was an economic and convenient way to obtain primary, secondary and tertiary alkyl radicals, thus we postulated that the merging the oxidative decarbonylation of aliphatic aldehydes into the radical difunctionalization of *N*-arylcinnamamides, would produce a benzyl radical and then facilitate the subsequent radical *ipso*-cyclization (Scheme 1b). Herein, we report a novel Fe-promoted oxidative decarbonylative alkylative spirocyclization of *N*-arylcinnamamides with aliphatic aldehydes to provide alkylated 1-azaspirocyclohexadienones.

2. Results

Based on the above speculation, *N*-(4-hydroxyphenyl)-*N*-methyl cinnamamide (**1a**) and isobutyraldehyde (**2a**) were chosen as the model substrates for this oxidative decarbonylative spirocyclization; with di-*tert*-butyl peroxide (DTBP) as the radical initiator and terminal oxidant, the desired alkylated 1-azaspirocyclohexadienone **3a** was isolated and characterized. Detailed optimization was performed focusing on different iron salt and its loading, the dosage of aldehyde and DTBP, the reaction temperature and reaction solvent, which revealed the combination of 2.5 mol% Fe(acac)$_2$/5 equiv aldehyde/3 equiv DTBP proved to be the most effective one, to afford the spirocyclization product **3a** in 67% isolated yield (Table 1, entry 1). For the reaction solvent, low polarity solvents including chlorobenzene (PhCl), trifluoromethylbenzene (PhCF$_3$) and much more polar acetonitrile, ethyl acetate (EA) all turned out to be compatible, and among these solvents tested, ethyl acetate provided the best result (entries 12–14). We are glad that this radical cascade reaction favors ethyl acetate, which should be the greenest solvent among the ordinary organic solvents.

Table 1. Optimization of reaction conditions [a].

Entry	Change from "Standard Conditions"	3a (%)
1	none	67
2	2.5 mol% FeCl$_2$ instead of Fe(acac)$_2$	51
3	2.5 mol% Fe(acac)$_3$ instead of Fe(acac)$_2$	20
4	1 mol% Fe(acac)$_2$	61
5	5 mol% Fe(acac)$_2$	56
6	**2a** reduced to 4 equiv	52
7	**2a** reduced to 3 equiv	39
8	DTBP reduced to 2 equiv	51
9	115 °C instead of 122 °C	53
10	120 °C instead of 122 °C	65
11	124 °C instead of 122 °C	58
12	PhCl instead of EA as solvent	35
13	PhCF$_3$ instead of EA as solvent	45
14	CH$_3$CN instead of EA as solvent	32

[a] Standard conditions: *N*-aryl cinnamamide **1a** (0.1 mmol), aldehyde **2a** (5 equiv, 0.5 mmol), Fe(acac)$_2$ (2.5 mol%) in EA (0.5 mL, prepared solution) and DTBP (3 equiv, 0.3 mmol) were reacted at 122 °C (oil bath temperature) for 24 h under air atmosphere. Isolated yield.

With the optimized reaction conditions in hand, we first examined the generality of this alkylative spirocyclization with different aliphatic aldehydes (**2a–2l**, Table 2). Various α-mono-substituted aliphatic aldehydes including isobutyraldehyde (**2a**), 2-ethylbutanal (**2b**), 2-methyl-butanal (**2c**), 2-methylpentanal (**2d**), 2-ethylhexanal (**2e**), cyclamen aldehyde (**2f**), cyclohexane-carbaldehyde (**2g**) and cyclopantanecarbaldehyde (**2h**) provided the corresponding secondary carbon radicals for the cascade spirocyclization after the oxidative decarbonylation. While the α-di-substituted pivaldehyde (**2i**) provided a tertiary carbon radical, the α-unsubstituted aliphatic aldehyde 3-methylbutanal (**2j**), propionaldehyde (**2k**) and 2-phenylacetaldehyde (**2l**) would provide primary carbon radicals, similarly. Gratifyingly, all of these aliphatic aldehydes underwent this decarbonylative alkylative spirocyclization witH-*N*-arylcinnamamide (**1a**) to produce the targeted alkylated 1-azaspirocyclohexadienone (**3a–3l**) smoothly. Moreover, the introduced alkyl group and the aryl group exhibit a trans-configuration, determined by the adjacent coupling constant of the proton on C3 and C4 (J = 12.0 Hz), which agreed well with the literature reports [18]. Considering the readily availability of these aliphatic aldehydes, avoidance the possible carbon radical rearrangement (to provide regioisomers as the alkane C-H bonds homolytic cleavage products) and simple operation for the radical generation (overriding pre-functionalization and photoreaction devices), this decarbonylative alkylative spirocyclization demonstrated again that the aliphatic aldehydes were convenient primary, secondary and tertiary alkyl precursors.

Table 2. Scope of the aliphatic aldehydes [a].

2a	isobutyraldehyde	**2b**	2-ethylbutanal
2c	2-methylbutanal	**2d**	2-methylpentanal
2e	2-ethylhexanal	**2f**	cyclamen aldehyde
2g	cyclohexanecarbaldehyde	**2h**	cyclopantanecarbaldehyde
2i	pivaldehyde	**2j**	3-methylbutanal
2k	propionaldehyde	**2l**	2-phenylacetaldehyde

3a, 67% **3b**, 71% **3c**, 68%[b]

3d, 69%[b] **3e**, 61%[b] **3f**, 74%[b]

3g, 72% **3h**, 61% **3i**, 52%

3j, 64% **3k**, 56% **3l**, 42%

[a] Conditions: cinnamamide **1a** (0.1 mmol), aldehyde **2a–2l** (5 equiv, 0.5 mmol), Fe(acac)$_2$ (2.5 mol%) in EA (0.5 mL, prepared solution) and DTBP (3 equiv, 0.3 mmol) were reacted at 122 °C (oil bath temperature) for 24 h under air atmosphere. Isolated yield. [b] Dr. = 1:1 as determined by GC.

After investigating the scope of aliphatic aldehydes, we next tested the generality of this decarbonylative alkylative spirocyclization on different *N*-aryl cinnamamides **1b–1k** under the optimized conditions. The effect of substituents on the cinnamamide moiety is listed in Figure 2.

A

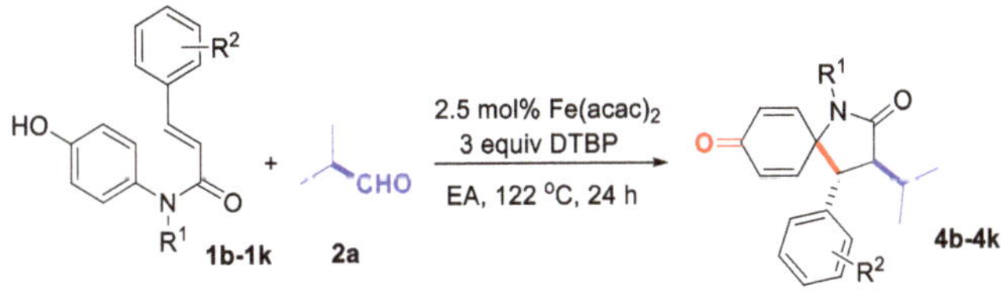

B

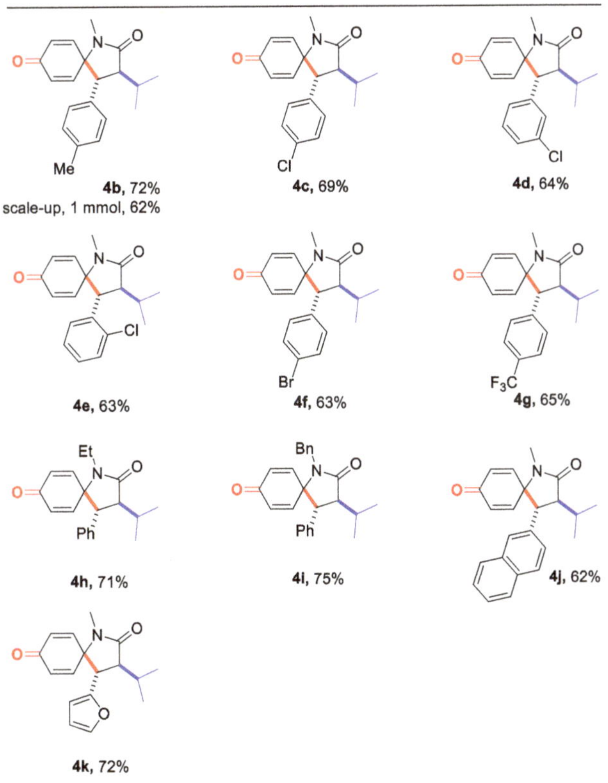

Figure 2. (**A**) Scope of the *N*-aryl cinnamamides. Conditions: cinnamamide **1b–1k** (0.1 mmol), aldehyde **2a** (5 equiv, 0.5 mmol), Fe(acac)$_2$ (2.5 mol%) in EA (0.5 mL, prepared solution) and DTBP (3 equiv, 0.3 mmol) were reacted at 122 °C (oil bath temperature) for 24 h under air atmosphere. Isolated yield. (**B**) **4b–4k**.

Various electron withdrawing or donating substituents were successfully incorporated into the cinnamamide unit of substrates **1b–1g**, such as methyl, halo and trifluoromethyl groups. Among them, the optimized reaction conditions were compatible with the cinnamamides with chloro groups substituted at the *para*, *meta* and *ortho* positions (compounds **1c–1e**), and similar yields were obtained, which revealed the substituents didn't cause obvious steric hindrance for this cascade reaction. For the substituent on the *N*-linkage, the ethyl and benzyl group-substituted cinnamamides **1h** and **1i** provided slightly better yields than the model substrate **1a**. To our delight, the 2-naphthalenyl and 2-furanyl units (**1j** and **1k**) could also be introduced onto the α,β-unsaturated C=C bond of amide substrates, and the cascade reaction provided the 1-azaspirocyclohexadienone **4j** and **4k** in moderate yields.

Several control experiments were carried out to understand this decarbonylative alkylative spirocyclization. First, the cascade reaction of cinnamamide **1a** and aliphatic aldehyde **2a** was inhibited in the presence of di-*tert*-butylhydroxytoluene (BHT); instead, the decarbonylated alkyl radical was captured as 2,6-di-*tert*-butyl-4-isopropyl-4-methylcyclohexa-2,5-dien-1-one (**5**), which confirmed the radical-type decarbonylation mechanism (Scheme 2a). Second, the control experiment using the *N*-(4-methoxyphenyl)-*N*-methyl cinnamamide (**1l**) to replace the *N*-(4-hydroxyphenyl)-*N*-methyl cinnamamide (**1a**) was conducted under the optimized conditions, however no desired *5-exo*-trig spirocyclization product (**3a**) could be detected; in contrast, the C3-alkylated 3,4-dihydroquinolin-2(1*H*)-one **6** was formed in 74% yield, via *6-endo*-trig cyclization pathway (Scheme 2b). The sharp difference on reactivity demonstrated the importance of the *para*-hydroxyl substituent for this spirocyclization, maybe due to its ability to stabilize the radical intermediate (**B**, Scheme 3) obtained from the *5-exo*-trig spirocyclization and accelerating the subsequent cyclohexadienone formation.

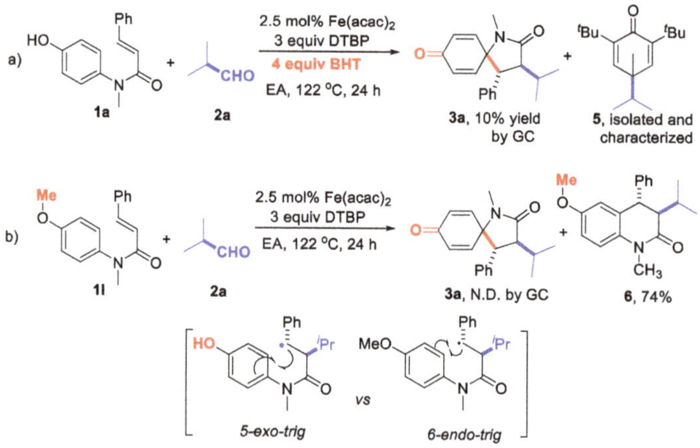

Scheme 2. Control experiments.

Based on the mechanistic experiments and the previous studies [46–53], a possible reaction pathway is depicted in Scheme 3, with the reaction of *N*-(4-hydroxyphenyl)-*N*-methyl cinnamamide (**1a**) and isobutyraldehyde (**2a**) as an example.

First, promoted by the iron-catalyst, the homolytic cleavage of DTBP at elevated temperature forms *tert*-butoxy radical. Subsequent intermolecular hydrogen atom abstraction of the aldehyde (**2a**), spontaneous decarbonylation and insertion into the C=C bond of the cinnamamide (**1a**) affords a metastable benzyl radical **A**, which then adds onto the *ispo*-carbon to give the spiro-cyclohexadienyl radical **B**. The radical **B** is preferably oxidized by Fe^{3+} and deprotonated by a *tert*-butoxide anion to afford the 1-azaspirocyclohexadienone (**3a**).

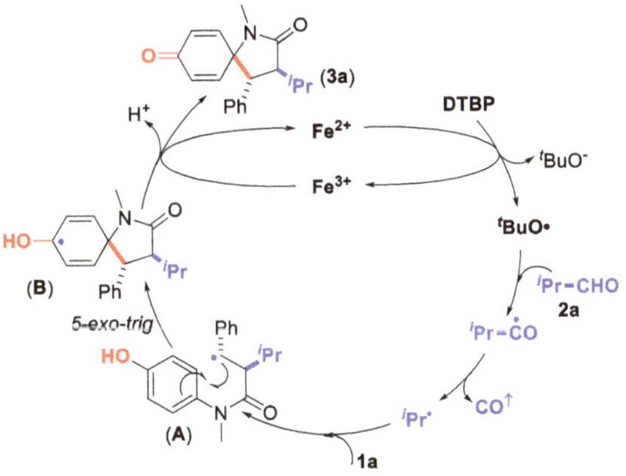

Scheme 3. Proposed mechanism.

3. Experimental

3.1. General Information

Unless otherwise noted, all commercially available compounds were used as provided without further purification. The substrates (various *N*-aryl cinnamamides) were synthesized from cinnamic acid and *para*-anisidine according to literature reports [18]. Dry solvents (toluene, ethyl acetate, dichloroethane, acetonitrile, chlorobenzene, fluorobenzene) were used as commercially available. Thin-layer chromatography (TLC) was performed using silica gel 60 F254 precoated plates (0.25 mm) or Sorbent Silica Gel 60 F254 plates (E. Merck). The developed chromatography was analyzed by UV lamp (254 nm). Unless other noted, high-resolution mass spectra (HRMS) were obtained from a JMS-700 instrument (ESI; JEOL). Melting points are uncorrected. Nuclear magnetic resonance (NMR) spectra were recorded on an Avance 400 spectrometer (Bruker) at ambient temperature. Chemical shifts for ^{1}H-NMR spectra are reported in parts per million (ppm) from tetramethylsilane with the solvent resonance as the internal standard (chloroform: δ 7.26 ppm). Chemical shifts for ^{13}C-NMR spectra are reported in parts per million (ppm) from tetramethylsilane with the solvent as the internal standard (CDCl$_3$: δ 77.16 ppm). Data are reported as following: chemical shift, multiplicity (s = singlet, d = doublet, dd = doublet of doublets, t = triplet, q = quartet, m = multiplet, br = broad signal), coupling constant (Hz), and integration.

3.2. General Experimental Procedures

An oven-dried microwave reaction vessel was charged with FeCl$_2$ (2.5 mol%) in EA (0.5 mL, pre-prepared solution), *N*-(4-hydroxyphenyl)-*N*-methylcinnamamide (**1a**, 0.1 mmol, 1.0 equiv), isobutyraldehyde (**2a**, 0.5 mmol, 5.0 equiv) and DTBP (0.3 mmol, 3.0 equiv). The vessel was sealed and heated at 122 °C (oil bath temperature) for 24 h. Afterwards the resulting mixture was cooled to room temperature, the solvent was removed in vacuo. The residue was purified by column chromatography on silica gel with a mixture of ethyl acetate/petroleum ether (1:3) as eluent to give the product **3a**.

3.3. Spectra Data of Products **3a–3l**, **4b–4k**, **5**, **6** (see "Supplementary Materials" for details)

3-Isopropyl-1-methyl-4-phenyl-1-azaspiro [4.5]deca-6,9-diene-2,8-dione (**3a**). The title compound was prepared according to the general procedure described above by the reaction between *N*-(4-hydroxyphenyl)-*N*-methylcinnamamide (**1a**) with isobutyraldehyde (**2a**), and purified by flash column chromatography as yellow oil (19.8 mg, 67%). ^{1}H-NMR (400 MHz, CDCl$_3$) δ 7.27–7.24 (m, 3H), 7.10 (dd, *J* = 7.6, 2.4 Hz, 2H), 6.78 (dd, *J* = 10.0, 3.2 Hz, 1H), 6.55 (dd, *J* = 10.2, 3.0 Hz, 1H), 6.39 (dd, *J* = 10.2, 2.0 Hz, 1H), 6.00 (dd, *J* = 10.2, 2.0 Hz, 1H), 3.43 (d, *J* = 12.0 Hz, 1H), 3.14 (dd, *J* = 11.8, 3.6 Hz, 1H),

2.73 (s, 3H), 2.38–2.34 (m, 1H), 1.01 (d, *J* = 6.8 Hz, 3H), 0.83 (d, *J* = 7.2 Hz, 3H). ^{13}C-NMR (100 MHz, CDCl$_3$) δ 184.46, 175.22, 149.40, 147.05, 135.04, 132.33, 131.51, 128.63, 128.21, 64.95, 50.84, 49.05, 28.07, 27.14, 20.14, 18.78. IR (cm^{-1}): 3032, 2965, 2932, 2875, 1672, 1630, 1606, 1498, 1446, 1454, 1418, 1393, 1374, 1260, 1141, 1119, 1065, 991, 865, 794, 724, 700. HRMS: calcd. for C$_{19}$H$_{21}$NO$_2$ Na$^+$ [M + Na]$^+$: 318.1465; Found: 318.1442.

1-Methyl-3-(pentan-3-yl)-4-phenyl-1-azaspiro[4 .5]deca-6,9-diene-2,8-dione (**3b**). The title compound was prepared according to the general procedure described above by the reaction between *N*-(4-hydroxyphenyl)-*N*-methylcinnamamide (**1a**) with 2-ethylbutanal (**2b**), and purified by flash column chromatography as yellow oil (23.0 mg, 71%). ^{1}H-NMR (400 MHz, CDCl$_3$) δ 7.26–7.24 (m, 3H), 7.09 (dd, *J* = 7.6, 2.8 Hz, 2H), 6.77 (dd, *J* = 10.0, 3.2 Hz, 1H), 6.60 (dd, *J* = 10.4, 3.2 Hz, 1H), 6.37 (dd, *J* = 10.0, 2.0 Hz, 1H), 6.00 (dd, *J* = 10.0, 2.0 Hz, 1H), 3.45 (d, *J* = 11.8 Hz, 1H), 3.31 (dd, *J* = 11.8, 2.6 Hz, 1H), 2.74 (s, 3H), 1.78–1.74 (m, 2H), 1.53–1.46 (m, 1H), 1.39–1.32 (m, 2H), 0.96 (t, *J* = 7.4 Hz, 3H), 0.80 (t, *J* = 7.3 Hz, 3H). ^{13}C-NMR (100 MHz, CDCl$_3$) δ 184.46, 175.15, 149.45, 146.94, 134.72, 132.36, 131.58, 128.68, 128.59, 128.28, 65.02, 51.48, 48.30, 34.68, 27.12, 26.47, 16.47, 12.44. IR (cm^{-1}): 3032, 2961, 2931, 2875, 1692, 1672, 1630, 1454, 1419, 1392, 1376, 1260, 1173, 1141, 1119, 1066, 992, 865, 723,700, 662, 563. HRMS: calcd. for C$_{21}$H$_{25}$NO$_2$ Na$^+$ [M + Na]$^+$: 346.1778; Found: 346.1753.

3-(sec-Butyl)-1-methyl-4-phenyl-1-azaspiro[4.5]deca-6,9-diene-2,8-dione (**3c**). The title compound was prepared according to the general procedure described above by the reaction between *N*-(4-hydroxyphenyl)-*N*-methylcinnamamide (**1a**) with 2-methylbutanal (**2c**), and purified by flash column chromatography as yellow oil (21.0 mg, 68%). ^{1}H-NMR (400 MHz, CDCl$_3$) δ 7.27–7.24 (m, 3H), 7.09 (d, *J* = 9.6, 2.4 Hz, 2H), 6.78 (dd, *J* = 10.0, 2.8 Hz, 1H), 6.54 (dd, *J* = 10.0, 3.2 Hz, 1H), 6.39 (dd, *J* = 10.4, 2.0 Hz, 1H), 6.00 (dd, *J* = 10.0, 2.0 Hz, 1H), 3.45 (d, *J* = 12.0 Hz, 1H), 3.20 (dd, *J* = 12.0, 2.8 Hz, 1H), 2.73 (s, 3H), 1.93–1.85 (m, 1H), 1.60–1.56 (m, 1H), 1.45–1.37 (m, 1H), 0.93 (t, *J* = 7.4 Hz, 3H), 0.85 (d, *J* = 7.0 Hz, 3H). ^{13}C-NMR (100 MHz, CDCl$_3$) δ 184.46, 175.15, 149.45, 146.94, 134.72, 132.36, 131.58, 128.68, 128.59, 128.28, 65.02, 51.48, 48.30, 34.68, 27.12, 26.47, 16.47, 12.44. IR (cm^{-1}): 3059, 3032, 2962, 2932, 2875, 1672, 1630, 1498, 1454, 1419, 1392, 1375, 1260, 1172, 1143, 1065, 990, 865, 767, 700, 621. HRMS: calcd. for C$_{20}$H$_{23}$NO$_2$ Na$^+$ [M + Na]$^+$: 332.1621; Found: 332.1605.

1-Methyl-3-(pentan-2-yl)-4-phenyl-1-azaspiro[4.5]deca-6,9-diene-2,8-dione (**3d**). The title compound was prepared according to the general procedure described above by the reaction between *N*-(4-hydroxyphenyl)-*N*-methylcinnamamide (**1a**) with 2-methylpentanal (**2d**), and purified by flash column chromatography as yellow oil (22.3 mg, 69%). ^{1}H-NMR (400 MHz, CDCl$_3$) δ 7.26–7.23 (m, 3H), 7.09 (dd, *J* = 7.6, 2.4 Hz, 2H), 6.76 (dd, *J* = 10.0, 2.8 Hz, 1H), 6.57 (dd, *J* = 10.0, 2.8 Hz, 1H), 6.38 (dd, *J* = 10.0, 2.0 Hz, 1H), 6.00 (dd, *J* = 10.0, 2.0 Hz, 1H), 3.44 (d, *J* = 10.8 Hz, 1H), 3.21 (dd, *J* = 11.6, 3.2 Hz, 1H), 2.74 (s, 3H), 2.30–2.23 (m, 1H), 1.37–1.23 (m, 2H), 1.16–1.01 (m, 2H), 0.96 (d, *J* = 6.9 Hz, 3H), 0.70 (t, *J* = 7.4 Hz, 3H). ^{13}C-NMR (100 MHz, CDCl$_3$) δ 184.48, 175.64, 149.44, 147.26, 146.96, 135.09, 132.34, 132.24, 131.58, 131.47, 128.66, 128.58, 128.33, 128.29, 128.23, 65.05, 50.43, 48.48, 48.25, 36.50, 35.77, 32.70, 27.24, 20.94, 20.67, 15.97, 14.31, 13.98. IR (cm^{-1}): 3059, 3032, 2958, 2928, 2872, 1672, 1630, 1498, 1454, 1421, 1377, 1261, 1172, 1119, 1067, 990, 865, 794, 724, 700, 621. HRMS: calcd. for C$_{21}$H$_{25}$NO$_2$ Na$^+$ [M + Na]$^+$: 346.1778; Found: 346.1752.

3-(Heptan-3-yl)-1-methyl-4-phenyl-1-azaspiro[4.5]deca-6,9-diene-2,8-dione (**3e**) [18]. The title compound was prepared according to the general procedure described above by the reaction between *N*-(4-hydroxyphenyl)-*N*-methylcinnamamide (**1a**) with 2-ethylhexanal (**2e**), and purified by flash column chromatography as yellow oil (21.4 mg, 61%). ^{1}H-NMR (400 MHz, CDCl$_3$) δ 7.25–7.22 (m, 3H), 7.09 (dd, *J* = 7.6, 2.4 Hz, 2H), 6.77 (dt, *J* = 10.0, 2.8 Hz, 1H), 6.60 (dt, *J* = 10.2, 3.2 Hz, 1H), 6.37 (dd, *J* = 10.0, 2.0 Hz, 1H), 6.00 (dt, *J* = 10.2, 2.2 Hz, 1H), 3.44 (dd, *J* = 11.8, 3.2 Hz, 1H), 3.30 (dd, *J* = 12.0, 3.0 Hz, 1H), 2.74 (s, 3H), 1.84–1.79 (m, 1H), 1.53–1.37 (m, 1H), 1.37–1.26 (m, 4H), 1.23–1.11 (m, 3H), 0.95 (t, *J* = 8.6 Hz, 3H), 0.81–0.72 (m, 3H). ^{13}C-NMR (100 MHz, CDCl$_3$) δ 184.48, 175.79, 149.53, 147.23, 134.93,

132.23, 131.54, 128.61, 128.44, 128.29, 65.08, 51.27, 46.04, 45.96, 40.21, 40.10, 31.12, 30.60, 30.31, 30.01, 27.22, 24.47, 24.24, 23.10, 22.80, 14.24, 14.02, 12.69, 12.24. IR (cm^{-1}): 3060, 3032, 2958, 2929, 2872, 1691, 1672, 1630, 1499, 1455, 1393, 1376, 1259, 1173, 1119, 1066, 991, 865, 766,722, 700, 662.

(**3f**) *3-(1-(4-isopropylphenyl)propan-2-yl)-1-methyl-4-phenyl-1-azaspiro[4.5]deca-6,9-diene-2,8-dione.* The title compound was prepared according to the general procedure described above by the reaction between *N*-(4-hydroxyphenyl)-*N*-methylcinnamamide (**1a**) with 3-(4-isopropylphenyl)-2-methylpropanal (**2f**), and purified by flash column chromatography as yellow oil (30.6 mg, 74%). ^{1}H-NMR (400 MHz, CDCl$_3$) δ 7.24–7.23 (m, 1H), 7.17–7.13 (m, 2.5H), 7.20–7.07 (m, 2.5H), 7.00 (dd, *J* = 6.0, 2.0 Hz, 1H), 6.89 (dd, *J* = 6.0, 1.6 Hz, 1H), 6.80–6.73 (m, 1.5H), 6.65 (dd, *J* = 10.0, 3.2 Hz, 0.5H), 6.53 (dd, *J* = 11.4, 3.2 Hz, 0.5H), 6.43–6.32 (m, 1.5H), 5.96 (td, *J* = 10.1, 2.0 Hz, 1H), 3.44–3.38 (m, 1H), 3.22–3.14 (m, 1H), 3.08–3.02 (m, 1H), 2.91–2.84 (m, 1H), 2.72 (d, *J* = 8 Hz, 3H), 2.67 (t, *J* = 6.8 Hz, 0.5H), 2.49 (q, *J* = 6.8 Hz, 0.5H), 2.28 (d, 5.6 Hz, 0.5H), 2.06–2.00(m, 1H), 1.28–1.21 (m, 6H), 0.99 (dd, *J* = 6.8, 4.4 Hz, 3H). ^{13}C-NMR (100 MHz, CDCl$_3$) δ 174.77, 149.42, 149.36, 147.09, 146.84, 146.76, 146.58, 137.97, 137.48, 134.68, 133.85, 132.44, 132.18, 131.53, 131.45, 129.49, 129.06, 128.61, 128.58, 128.54, 128.30, 128.20, 128.00, 126.41, 126.33, 64.95, 64.89, 51.61, 50.78, 47.81, 45.16, 40.13, 40.02, 34.93, 34.56, 33.86, 33.79, 27.19, 26.98, 24.33, 24.19, 16.74, 15.80. IR (cm^{-1}): 3049, 3030, 2960, 2928, 2873, 1689, 1631, 1499, 1454, 1392, 1378, 1265, 1172, 1114, 1058, 990, 864, 735, 700, 570. HRMS: calcd. for C$_{28}$H$_{31}$NO$_2$ Na$^+$ [M + Na]$^+$: 436.2247; Found: 436.2217.

3-Cyclohexyl-1-methyl-4-phenyl-1-azaspiro[4.5]deca-6,9-diene-2,8-dione (**3g**) [18]. The title compound was prepared according to the general procedure described above by the reaction between *N*-(4-hydroxyphenyl)-*N*-methylcinnamamide (**1a**) with cyclohexanecarbaldehyde (**2g**), and purified by flash column chromatography as yellow oil (24.1 mg, 72%). ^{1}H-NMR (400 MHz, CDCl$_3$) δ 7.27–7.24(m, 3H), 7.09 (dd, *J* = 7.6, 2.4 Hz, 2H), 6.76 (dd, *J* = 10.0, 3.2 Hz, 1H), 6.54 (dd, *J* = 10.2, 3.0 Hz, 1H), 6.39 (dd, *J* = 10.4, 2.0 Hz, 1H), 5.98 (dd, *J* = 10.2, 2.0 Hz, 1H), 3.48 (d, *J* = 12.0 Hz, 1H), 3.11 (dd, *J* = 11.8, 3.6 Hz, 1H), 2.73 (s, 3H), 2.01–1.94 (m, 1H), 1.73 (d, *J* = 13.2, 1H), 1.63–1.56 (m, 3H), 1.38–1.13 (m, 4H), 1.08–0.99 (m, 1H), 0.87–0.81 (m, 1H). ^{13}C-NMR (100 MHz, CDCl$_3$) δ 184.50, 175.30, 149.45, 147.09, 135.06, 132.33, 131.45, 128.63, 128.22, 128.19, 65.00, 51.02, 48.81, 38.27, 30.98, 29.14, 27.19, 26.71, 26.56, 26.23. IR (cm^{-1}): 3057, 3032, 2925, 2852, 1690, 1672, 1499, 1450, 1419, 1393, 1377, 1260, 1172, 1134, 1096, 1069, 991, 864, 796, 732, 657, 569.

3-Cyclopentyl-1-methyl-4-phenyl-1-azaspiro[4.5]deca-6,9-diene-2,8-dione (**3h**) [18]. The title compound was prepared according to the general procedure described above by the reaction between *N*-(4-hydroxyphenyl)-*N*-methylcinnamamide (**1a**) with cyclopentanecarbaldehyde (**2h**), and purified by flash column chromatography as yellow oil (19.6 mg, 61%). ^{1}H-NMR (400 MHz, CDCl$_3$) δ 7.27–7.24 (m, 3H), 7.09 (dd, *J* = 7.6, 2.4 Hz, 2H), 6.77 (dd, *J* = 10.2, 3.0 Hz, 1H), 6.54 (dd, *J* = 10.2, 3.1 Hz, 1H), 6.39 (dd, *J* = 10.2, 2.0 Hz, 1H), 5.99 (dd, *J* = 10.2, 2.0 Hz, 1H), 3.38 (d, *J* = 11.6 Hz, 1H), 3.21 (dd, *J* = 11.8, 6.2 Hz, 1H), 2.73 (s, 3H), 2.24–2.18 (m, 1H), 2.05–1.96 (m, 1H), 1.87–1.76 (m, 3H), 1.50–1.38 (m, 3H), 1.25–1.21 (m, 1H). ^{13}C-NMR (100 MHz, CDCl$_3$) δ 184.50, 175.30, 149.45, 147.09, 135.06, 132.33, 131.45, 128.63, 128.22, 128.19, 65.06, 53.45, 46.83, 41.20, 29.85, 29.56, 27.16, 25.18, 24.99. IR (cm^{-1}): 3059, 3031, 2923, 2869, 1730, 1692, 1671, 1630, 1453, 1442, 1393, 1375, 1260, 1172, 1075, 991, 865, 794, 732,700, 645, 569.

3-(tert-Butyl)-1-methyl-4-phenyl-1-azaspiro[4.5]deca-6,9-diene-2,8-dione (**3i**). The title compound was prepared according to the general procedure described above by the reaction between *N*-(4-hydroxyphenyl)-*N*-methylcinnamamide (**1a**) with pivalaldehyde (**2i**), and purified by flash column chromatography as yellow oil (16.7 mg, 52%). ^{1}H-NMR (400 MHz, CDCl$_3$) δ 7.27 (s, 3H), 7.21 (s, 2H), 6.77 (dd, *J* = 10.0, 3.2 Hz, 1H), 6.46 (dd, *J* = 10.2, 3.2 Hz, 1H), 6.38 (dd, *J* = 10.0, 2.0 Hz, 1H), 5.92 (dd, *J* = 10.2, 2.0 Hz, 1H), 3.40 (d, *J* = 11.2 Hz, 1H), 3.01 (d, *J* = 11.6 Hz, 1H), 2.70 (s, 3H), 1.00 (s, 9H). ^{13}C-NMR (100 MHz, CDCl$_3$) δ 184.48, 175.09, 149.54, 147.98, 136.60, 132.11, 131.04, 128.02, 64.35, 52.50, 51.33, 33.87, 28.08, 27.22. IR (cm^{-1}): 3031, 2958, 2870, 1688, 1672, 1630, 1605, 1468, 1420, 1392, 1370,

1260, 1244, 1171, 1119, 1093, 864, 794, 720, 735,700, 610, 563. HRMS: calcd. for $C_{20}H_{23}NO_2$ Na^+ [M + Na]$^+$: 332.1621; Found: 332.1597.

3-Isobutyl-1-methyl-4-phenyl-1-azaspiro[4.5]deca-6,9-diene-2,8-dione (**3j**). The title compound was prepared according to the general procedure described above by the reaction between *N*-(4-hydroxyphenyl)-*N*-methylcinnamamide (**1a**) with 3-methylbutanal (**2j**), and purified by flash column chromatography as yellow oil (19.8 mg, 64%). ^{1}H-NMR (400 MHz, CDCl$_3$) δ 7.27–7.24 (m, 3H), 7.09 (dd, *J* = 7.6, 2.4 Hz, 2H), 6.76 (dd, *J* = 10.0, 3.2 Hz, 1H), 6.59 (dd, *J* = 10.4, 3.0 Hz, 1H), 6.39 (dd, *J* = 10.0, 2.0 Hz, 1H), 6.01 (dd, *J* = 10.2, 2.0 Hz, 1H), 3.25 (d, *J* = 11.6 Hz, 1H), 3.15 – 3.09 (m, 1H), 2.74 (s, 3H), 1.89–1.83 (m, 1H), 1.74–1.67 (m, 1H), 1.38–1.31 (m, 1H), 0.86 (d, *J* = 7.2 Hz, 3H), 0.80 (d, *J* = 6.4 Hz, 3H). ^{13}C-NMR (100 MHz, CDCl$_3$) δ 184.45, 176.38, 149.36, 146.81, 134.34, 132.40, 131.59, 128.66, 128.36, 128.25, 65.23, 56.35, 41.66, 41.10, 27.23, 25.20, 22.93, 22.37. IR (cm^{-1}): 3056, 3033, 2956, 2927, 2869, 1692, 1672, 1630, 1467, 1454, 1393, 1376, 1262, 1172, 1137, 1080, 1060, 866, 790, 721, 700. HRMS: calcd. for $C_{20}H_{23}NO_2$ Na^+ [M + Na]$^+$: 332.1621; Found: 332.1597.

3-Ethyl-1-methyl-4-phenyl-1-azaspiro[4.5]deca-6,9-diene-2,8-dione (**3k**). The title compound was prepared according to the general procedure described above by the reaction between *N*-(4-hydroxyphenyl)-*N*-methylcinnamamide (**1a**) with propionaldehyde (**2k**), and purified by flash column chromatography as yellow oil (16.1 mg, 56%). ^{1}H-NMR (400 MHz, CDCl$_3$) δ 7.27–7.24 (m, 3H), 7.10 (dd, *J* = 8.0, 2.4 Hz 2H), 6.79 (dd, *J* = 10.0, 3.2 Hz, 1H), 6.57 (dd, *J* = 10.2, 3.2 Hz, 1H), 6.41 (dd, *J* = 10.0, 2.0 Hz, 1H), 6.02 (dd, *J* = 10.2, 2.0 Hz, 1H), 3.34 (d, *J* = 12.0 Hz, 1H), 3.13–3.06 (m, 1H), 2.75 (s, 3H), 1.96–1.86 (m, 1H), 1.77–1.68 (m, 1H), 0.89 (t, *J* = 7.4 Hz, 3H). ^{13}C-NMR (100 MHz, CDCl$_3$) δ 184.46, 175.72, 149.32, 146.57, 134.25, 132.53, 131.70, 128.70, 128.34, 128.17, 65.14, 54.39, 44.64, 27.17, 23.01, 11.18. IR (cm^{-1}): 3056, 3032, 2965, 2931, 2877, 1692, 1672, 1629, 1454, 1420, 1394, 1376, 1262, 1174, 1125, 1060, 988, 865, 719, 699, 659. HRMS: calcd. for $C_{18}H_{19}NO_2$ Na^+ [M + Na]$^+$: 304.1308; Found: 304.1286.

3-Benzyl-1-methyl-4-phenyl-1-azaspiro[4.5]deca-6,9-diene-2,8-dione (**3l**). The title compound was prepared according to the general procedure described above by the reaction between *N*-(4-hydroxyphenyl)-*N*-methylcinnamamide (**1a**) with 2-phenylacetaldehyde (**2l**), and purified by flash column chromatography as yellow oil (14.4 mg, 42%). ^{1}H-NMR (400 MHz, CDCl$_3$) δ 7.24–7.16 (m, 6H), 7.07–7.04 (m, 2H), 6.99–6.97 (m, 2H), 6.56 (dd, *J* = 10.2, 3.0 Hz, 1H), 6.47 (dd, *J* = 10.0, 3.0 Hz, 1H), 6.30 (dd, *J* = 10.2, 2.0 Hz, 1H), 6.00 (dd, *J* = 10.2, 2.0 Hz, 1H), 3.42 (dt, *J* = 12.2, 5.4 Hz, 1H), 3.30–3.20 (m, 2H), 2.93 (dd, *J* = 13.6, 5.8 Hz, 1H), 2.71 (s, 3H). ^{13}C-NMR (100 MHz, CDCl$_3$) δ 184.41, 174.77, 149.24, 146.36, 137.36, 133.43, 132.39, 131.76, 129.98, 128.69, 128.47, 128.28, 128.27, 126.71, 65.01, 52.74, 45.16, 34.06, 27.23. IR (cm^{-1}): 3060, 3029, 2922, 2853, 1693, 1672, 1630, 1496, 1453, 1393, 1377, 1261, 1172, 1130, 1092, 1075, 865, 788, 721, 699. HRMS: calcd. for $C_{23}H_{21}NO_2$ Na^+ [M + Na]$^+$: 366.1465; Found: 366.1443.

3-Isopropyl-1-methyl-4-phenyl-1-azaspiro[4.5]deca-6,9-diene-2,8-dione (**4b**). The title compound was prepared according to the general procedure described above by the reaction between *N*-(4-hydroxyphenyl)-*N*-methyl-3-(p-tolyl)acrylamide (**1b**) with isobutyraldehyde (**2a**), and purified by flash column chromatography as yellow oil (22.2 mg, 72%). ^{1}H-NMR (400 MHz, CDCl$_3$) δ 7.05 (d, *J* = 7.6 Hz, 2H), 6.53 (d, *J* = 7.6 Hz, 2H), 6.77 (dd, *J* = 10.0, 3.2 Hz, 1H), 6.56 (dd, *J* = 10.2, 3.0 Hz, 1H), 6.38 (dd, *J* = 10.0, 2.0 Hz, 1H), 6.01 (dd, *J* = 10.2, 1.8 Hz, 1H), 3.40 (d, *J* = 11.6 Hz, 1H), 3.11 (dd, *J* = 12.0, 3.6 Hz, 1H), 2.73 (s, 3H), 2.36–2.31 (m, 1H), 2.28 (s, 3H), 1.01 (d, *J* = 6.8 Hz, 3H), 0.83 (d, *J* = 7.2 Hz, 3H). ^{13}C-NMR (100 MHz, CDCl$_3$) δ 184.60, 175.34, 149.56, 147.20, 137.93, 132.28, 131.95, 131.50, 129.33, 128.09, 65.05, 50.58, 49.13, 28.07, 27.14, 21.16, 20.13, 18.83. IR (cm^{-1}): 3046, 3030, 2960, 2929, 2876, 1690, 1672, 1629, 1516, 1465, 1419, 1393, 1375, 1263, 1065, 992, 864, 736, 639, 575. HRMS: calcd. for $C_{20}H_{23}NO_2$ Na^+ [M + Na]$^+$: 332.1621; Found: 332.1602.

4-(4-Chlorophenyl)-3-isopropyl-1-methyl-1-azaspiro[4.5]deca-6,9-diene-2,8-dione (**4c**). The title compound was prepared according to the general procedure described above by the reaction between

3-(4-chlorophenyl)-*N*-(4-hydroxyphenyl)-*N*-methylacrylamide (**1c**) with isobutyraldehyde (**2a**), and purified by flash column chromatography as yellow oil (22.7 mg, 69%). ^{1}H-NMR (400 MHz, CDCl$_3$) δ 7.24 (dd, *J* = 6.4, 2.0 Hz, 2H), 7.05 (dd *J* = 6.4, 2.0 Hz, 2H), 6.76 (dd, *J* = 10.2, 3.0 Hz, 1H), 6.54 (dd, *J* = 10.2, 3.2 Hz, 1H), 6.40 (dd, *J* = 10.0, 2.0 Hz, 1H), 6.05 (dd, *J* = 10.2, 2.0 Hz, 1H), 3.40 (d, *J* = 12.0 Hz, 1H), 3.08 (dd, *J* = 12.0, 3.6 Hz, 1H), 2.73 (s, 3H), 2.38–2.30 (m, 1H), 1.00 (d, *J* = 6.8 Hz, 3H), 0.82 (d, *J* = 6.8 Hz, 3H). ^{13}C-NMR (100 MHz, CDCl$_3$) δ 184.23, 174.90, 149.09, 146.66, 134.10, 133.70, 132.53, 131.86, 129.53, 128.91, 64.83, 50.23, 49.23, 28.06, 27.19, 20.29, 18.68. IR (cm^{-1}): 3050, 2961, 2931, 2874, 1692, 1672, 1630, 1494, 1466, 1417, 1392, 1373, 1259, 1173, 1121, 1092, 1014, 866, 830, 736. HRMS: calcd. for C$_{19}$H$_{20}$ClNO$_2$ Na$^+$ [M + Na]$^+$: 352.1075; Found: 352.1059.

4-(3-Chlorophenyl)-3-isopropyl-1-methyl-1-azaspiro[4.5]deca-6,9-diene-2,8-dione (**4d**). The title compound was prepared according to the general procedure described above by the reaction between 3-(3-chlorophenyl)-*N*-(4-hydroxyphenyl)-*N*-methylacrylamide (**1d**) with isobutyraldehyde (**2a**), and purified by flash column chromatography as yellow oil (21.1 mg, 64%). ^{1}H-NMR (400 MHz, CDCl$_3$) δ 7.25–7.20 (m, 2H), 7.10 (s 1H), 7.00 (dt, *J* = 7.0, 1.8 Hz, 1H), 6.76 (dd, *J* = 10.2, 3.2 Hz, 1H), 6.56 (dd, *J* = 10.2, 3.0 Hz, 1H), 6.42 (dd, *J* = 10.0, 2.0 Hz, 1H), 6.06 (dd, *J* = 10.2, 2.0 Hz, 1H), 3.40 (d, *J* = 12.0 Hz, 1H), 3.08 (dd, *J* = 12.0, 3.6 Hz, 1H), 2.73 (s, 3H), 2.38–2.30 (m, 1H), 1.01 (d, *J* = 6.8 Hz, 3H), 0.83 (d, *J* = 6.8 Hz, 3H). ^{13}C-NMR (100 MHz, CDCl$_3$) δ 184.22, 174.78, 148.99, 146.52, 137.35, 134.58, 132.62, 131.87, 129.96, 128.55, 128.35, 126.53, 64.76, 50.52, 49.20, 28.09, 27.17, 20.23, 18.75. IR (cm^{-1}): 3056, 2961, 2931, 2874, 1691, 1672, 1630, 1468, 1422, 1392, 1375, 1261, 1173, 1119, 1082, 880, 852, 736, 690, 623. HRMS: calcd. for C$_{19}$H$_{20}$ClNO$_2$ Na$^+$ [M + Na]$^+$: 352.1075; Found: 352.1053.

4-(2-Chlorophenyl)-3-isopopyl-1-methyl-1-azaspiro[4.5]deca-6,9-diene-2,8-dione (**4e**). The title compound was prepared according to the general procedure described above by the reaction between 3-(2-chlorophenyl)-*N*-(4-hydroxyphenyl)-*N*-methylacrylamide (**1e**) with isobutyraldehyde (**2a**), and purified by flash column chromatography as yellow oil (21.0 mg, 63%). ^{1}H-NMR (400 MHz, CDCl$_3$) δ 7.34–7.24 (m, 3H), 7.22–7.16 (m, 2H), 6.88 (dd, *J* = 10.2, 3.0 Hz, 1H), 6.68 (dd, *J* = 10.2, 3.2 Hz, 1H), 6.33 (dd, *J* = 10.2, 2.0 Hz, 1H), 6.12 (dd, *J* = 10.2, 2.0 Hz, 1H), 4.25 (d, *J* = 11.8 Hz, 1H), 3.04 (dd, *J* = 11.8, 3.6 Hz, 1H), 2.74 (s, 3H), 2.39–2.31 (m, 1H), 1.00 (d, *J* = 6.8 Hz, 3H), 0.76 (d, *J* = 7.0 Hz, 3H). ^{13}C-NMR (100 MHz, CDCl$_3$) δ 184.41, 174.88, 149.76, 146.58, 133.44, 131.83, 131.81, 130.45, 135.01, 129.37, 129.22, 126.68, 64.64, 51.18, 46.06, 28.23, 26.94, 20.50, 18.43. IR (cm^{-1}): 3059, 2960, 2931, 2873, 1692, 1672, 1631, 1468, 1422, 1392, 1374, 1260, 1174, 1116, 1065, 1037, 864, 750, 736, 698. HRMS: calcd. for C$_{19}$H$_{20}$ClNO$_2$ Na$^+$ [M + Na]$^+$: 352.1075; Found: 352.1054.

4-(4-Bromophenyl)-3-isopropyl-1-methyl-1-azaspiro[4.5]deca-6,9-diene-2,8-dione (**4f**). The title compound was prepared according to the general procedure described above by the reaction between 3-(4-bromophenyl)-*N*-(4-hydroxyphenyl)-*N*-methylacrylamide (**1f**) with isobutyraldehyde (**2a**), and purified by flash column chromatography as yellow oil (23.5 mg, 63%). ^{1}H-NMR (400 MHz, CDCl$_3$) δ 7.41–7.38 (m, 2H), 6.99 (dd, *J* = 6.4, 2.0 Hz, 2H), 6.75 (dd, *J* = 10.0, 3.2 Hz, 1H), 6.53 (dd, *J* = 10.2, 3.0 Hz, 1H), 6.40 (dd, *J* = 10.2, 2.0 Hz, 1H), 6.05 (dd, *J* = 10.2, 2.0 Hz, 1H), 3.39 (d, *J* = 12.0 Hz, 1H), 3.07 (dd, *J* = 12.0, 3.6 Hz, 1H), 2.73 (s, 3H), 2.38–2.30 (m, 1H), 0.99 (d, *J* = 7.0 Hz, 3H), 0.82 (d, *J* = 7.0 Hz, 3H). ^{13}C-NMR (100 MHz, CDCl$_3$) δ 184.22, 174.83, 149.09, 146.60, 134.26, 132.56, 131.90, 131.88, 129.86, 122.24, 64.73, 50.31, 49.19, 28.09, 27.19, 20.31, 18.70. IR (cm^{-1}): 3048, 2960, 2929, 2873, 1693, 1671, 1630, 1491, 1466, 1392, 1374, 1260, 1173, 1120, 1010, 866, 827, 755, 629, 571. HRMS: calcd. for C$_{19}$H$_{20}$BrNO$_2$ Na$^+$[M + Na]$^+$: 396.0570; Found: 396.0550.

3-Isopropyl-1-methyl-4-(4-(trifluoromethyl)phenyl)-1-azaspiro[4.5]deca-6,9-diene-2,8-dione (**4g**). The title compound was prepared according to the general procedure described above by the reaction between *N*-(4-hydroxyphenyl)-*N*-methyl-3-(4-(trifluoromethyl)phenyl)acrylamide (**1g**) with isobutyraldehyde (**2a**), and purified by flash column chromatography as yellow oil (23.6 mg, 65%). ^{1}H-NMR (400 MHz, CDCl$_3$) δ 7.54 (d, *J* = 7.8 Hz, 2H), 7.25 (t, *J* = 7.4 Hz, 2H), 6.79 (dd, *J* = 10.2, 3.2 Hz, 1H), 6.55 (dd, *J* =

10.2, 3.2 Hz, 1H), 6.42 (dd, *J* = 10.0, 2.0 Hz, 1H), 6.04 (dd, *J* = 10.2, 2.0 Hz, 1H), 3.49 (d, *J* = 11.8 Hz, 1H), 3.15 (dd, *J* = 12.0, 3.8 Hz, 1H), 2.74 (s, 3H), 2.39–2.31 (m, 1H), 1.01 (d, *J* = 7.0 Hz, 3H), 0.82 (d, *J* = 7.0 Hz, 3H). ^{13}C-NMR (100 MHz, CDCl$_3$) δ 184.05, 174.67, 148.90, 146.28, 139.43, 132.68, 131.97, 128.69, 125.73, 125.69, 64.68, 50.57, 49.24, 28.13, 27.17, 20.30, 18.72. ^{19}F-NMR (376 MHz, CDCl3) δ 62.65 (s, 1F). IR (cm^{-1}): 3050, 2962, 2933, 2876, 1692, 1672, 1631, 1468, 1422, 1392, 1375, 1327, 1166, 1124, 1069, 1017, 869, 853, 737, 602. HRMS: calcd. for C$_{20}$H$_{20}$F$_3$NO$_2$ Na$^+$ [M + Na]$^+$: 386.1338; Found: 386.1319.

(1-Ethyl-3-isopropyl-4-phenyl-1-azaspiro[4.5]deca-6,9-diene-2,8-dione (**4h**). The title compound was prepared according to the general procedure described above by the reaction between *N*-ethyl-*N*-(4-hydroxyphenyl)cinnamamide (**1h**) with isobutyraldehyde (**2a**), and purified by flash column chromatography as yellow oil (22.0 mg, 71%). ^{1}H-NMR (400 MHz, CDCl$_3$) δ 7.27–7.23 (m, 3H), 7.09 (dd, *J* = 8.0, 2.0 Hz, 2H), 6.82 (dd, *J* = 10.1, 3.1 Hz, 1H), 6.62 (dd, *J* = 10.2, 3.0 Hz, 1H), 6.37 (dd, *J* = 10.1, 2.0 Hz, 1H), 5.95 (dd, *J* = 10.2, 2.0 Hz, 1H), 3.42 (d, *J* = 12.0 Hz, 1H), 3.36–3.27 (m, 1H), 3.14 (dd, *J* = 12.0, 3.6 Hz, 1H), 3.10–3.01 (m, 1H), 2.38–2.30 (m, 1H), 1.13 (t, *J* = 7.2 Hz, 3H), 1.01 (d, *J* = 7.0 Hz, 3H), 0.82 (d, *J* = 7.2 Hz, 3H). ^{13}C-NMR (100 MHz, CDCl$_3$) δ 184.71, 175.02, 149.52, 148.09, 135.02, 131.86, 130.60, 128.61, 128.28, 128.23, 65.24, 51.25, 49.04, 36.86, 28.11, 20.17, 18.71, 15.29. IR (cm^{-1}): 3057, 3033, 2962, 2934, 2874, 1678, 1629, 1498, 1454, 1402, 1376, 1310, 1262, 1140, 1125, 1064, 940, 866, 724, 700. HRMS: calcd. for C$_{20}$H$_{23}$NO$_2$ Na$^+$ [M + Na]$^+$: 332.1621; Found: 332.1597.

1-Benzyl-3-isopropyl-4-phenyl-1-azaspiro[4.5]deca-6,9-diene-2,8-dione (**4i**). The title compound was prepared according to the general procedure described above by the reaction between *N*-benzyl-*N*-(4-hydroxyphenyl)cinnamamide (**1i**) with isobutyraldehyde (**2a**), and purified by flash column chromatography as yellow oil (27.8 mg, 75%). ^{1}H-NMR (400 MHz, CDCl$_3$) δ 7.27–7.23 (m, 3H), 7.22–7.18 (m, 5H), 7.04 (dd, *J* = 6.8, 3.0 Hz, 2H), 6.54 (dd, *J* = 10.4, 3.2 Hz, 1H), 6.47 (dd, *J* = 10.2, 3.0 Hz, 1H), 6.18 (dd, *J* = 10.0, 2.0 Hz, 1H), 5.81 (dd, *J* = 10.2, 2.0 Hz, 1H), 4.66 (d, *J* = 15.0 Hz, 1H), 4.07 (d, *J* = 14.8 Hz, 1H), 3.42 (d, *J* = 12.0 Hz, 1H), 3.22 (dd, *J* = 12.0, 3.8 Hz, 1H), 2.43–2.35 (m, 1H), 1.07 (d, *J* = 6.8 Hz, 3H), 0.85 (d, *J* = 6.8 Hz, 3H). ^{13}C-NMR (100 MHz, CDCl$_3$) δ 184.70, 175.31, 149.35, 147.30, 137.95, 134.66, 131.35, 130.71, 128.64, 128.60, 128.53, 128.30, 128.24, 127.78, 65.26, 51.27, 48.88, 45.36, 28.14, 20.11, 18.86. IR (cm^{-1}): 3061, 3031, 2961, 2930, 2873, 1670, 1629, 1496, 1454, 1399, 1264, 1179, 1065, 1011, 958, 930, 863, 792, 734, 700. HRMS: calcd. for C$_{25}$H$_{25}$NO$_2$ Na$^+$ [M + Na]$^+$: 394.1778; Found: 394.1761.

3-Isopropyl-1-methyl-4-(naphthalen-2-yl)-1-azaspiro[4.5]deca-6,9-diene-2,8-dione (**4j**). The title compound was prepared according to the general procedure described above by the reaction between *N*-(4-hydroxyphenyl)-*N*-methyl-3-(naphthalen-2-yl)acrylamide (**1j**) with isobutyraldehyde (**2a**), and purified by flash column chromatography as white solid (21.7 mg, 62%). m.p. 84.4–86.8 °C. ^{1}H-NMR (400 MHz, CDCl$_3$) δ 7.80–7.73 (m, 3H), 7.58–7.54 (m, 1H), 7.50–7.45 (m, 2H), 7.22 (dd, *J* = 8.4, 2.0 Hz, 1H), 6.84 (dd, *J* = 10.2, 3.2 Hz, 1H), 6.63 (dd, *J* = 10.2, 3.0 Hz, 1H), 6.41 (dd, *J* = 10.2, 2.0 Hz, 1H), 5.94 (dd, *J* = 10.2, 2.0 Hz, 1H), 3.61 (d, *J* = 11.8 Hz, 1H), 3.27 (dd, *J* = 11.8, 3.8 Hz, 1H), 2.76 (s, 3H), 2.42–2.34 (m, 1H), 1.04 (d, *J* = 6.8 Hz, 3H), 0.83 (d, *J* = 7.0 Hz, 3H). ^{13}C-NMR (100 MHz, CDCl$_3$) δ 184.38, 175.20, 149.49, 146.96, 133.11, 132.76, 132.40, 131.58, 133.02, 128.46, 127.88, 127.80, 127.46, 126.63, 126.46, 125.83, 65.04, 50.97, 49.34, 28.21, 27.15, 20.25, 18.80. IR (cm^{-1}): 3052, 2960, 2931, 2874, 1691, 1672, 1629, 1466, 1418, 1392, 1374, 1261, 1173, 1141, 1064, 992, 854, 822, 735, 607. HRMS: calcd. for C$_{23}$H$_{23}$NO$_2$ Na$^+$ [M + Na]$^+$: 368.1621; Found: 368.1605.

4-(Furan-2-yl)-3-isopropyl-1-methyl-1-azaspiro[4.5]deca-6,9-diene-2,8-dione (**4k**). The title compound was prepared according to the general procedure described above by the reaction between 3-(furan-2-yl)-*N*-(4-hydroxyphenyl)-*N*-methylacrylamide (**1k**) with isobutyraldehyde (**2a**), and purified by flash column chromatography as yellow oil (20.5 mg, 72%). ^{1}H-NMR (400 MHz, CDCl$_3$) δ 7.29–7.28 (m, 1H), 6.70 (dd, *J* = 10.0, 3.1 Hz, 1H), 6.63 (dd, *J* = 10.2, 3.0 Hz, 1H), 6.43 (dd, *J* = 10.0, 2.0 Hz, 1H), 6.24 (dd, *J* = 3.2, 2.0 Hz, 1H), 6.10–6.07 (m, 2H), 3.48 (d, *J* = 11.8 Hz, 1H), 3.17 (dd, *J* = 12.0, 4.0 Hz, 1H), 2.72 (s, 3H), 2.42–2.34 (m, 1H), 0.98 (d, *J* = 6.8 Hz, 3H), 0.86 (d, *J* = 7.2 Hz, 3H). ^{13}C-NMR (100 MHz, CDCl$_3$)

δ 184.51, 174.69, 149.14, 149.04, 146.70, 142.45, 132.12, 131.24, 110.67, 108.59, 64.25, 48.72, 44.06, 27.76, 27.04, 19.70, 18.14. IR (cm^{-1}): 3049, 2962, 2933, 2876, 1697, 1674, 1631, 1505, 1420, 1390, 1373, 1261, 1173, 1149, 1118, 1065, 1021, 859, 736, 599. HRMS: calcd. for $C_{17}H_{19}NO_3$ Na$^+$ [M + Na]$^+$: 308.1257; Found: 308.1248.

2,6-di-tert-Butyl-4-isopropyl-4-methylcyclohexa-2,5-dienone (**5**) [52]. The title compound was prepared from mechanistic experiment. ^{1}H-NMR (400 MHz, CDCl$_3$) δ 6.44 (s, 2H), 1.80–174 (m, 1H), 1.24 (s, 18H), 1.17 (s, 3H), 0.84 (d, *J* = 6.8 Hz, 6H). ^{13}C-NMR (100 MHz, CDCl$_3$) δ 186.90, 146.95, 145.80, 42.43, 37.55, 34.89, 29.70, 24.72, 18.09. IR (cm^{-1}): 3001, 2961, 2876, 1658, 1643, 1460, 1374, 1267, 1061, 867, 751.

3-Isopropyl-6-methoxy-1-methyl-4-phenyl-3,4-dihydroquinolin-2(1H)-one (**6**).   The title compound was prepared according to the general procedure described above by the reaction between *N*-(4-methoxyphenyl)-*N*-methylcinnamamide with isobutyraldehyde (**2a**), and purified by flash column chromatography as colorless oil (22.9 mg, 74%). ^{1}H-NMR (400 MHz, CDCl3) δ 7.24–7.13 (m, 3H), 7.00–6.97 (m, 3H), 6.85 (dd, J = 8.8, 3.2 Hz, 1H), 6.75 (d, J = 2.8 Hz, 1H), 4.13 (s, 1H), 3.78 (s, 3H), 3.33 (s, 3H), 2.57 (dd, J = 9.2, 2.0 Hz, 1H), 1.70–1.63 (m, 1H), 1.04 (d, J = 6.6 Hz, 3H), 0.98 (d, J = 6.8 Hz, 3H). ^{13}C-NMR (100 MHz, CDCl3) δ 170.31, 155.69, 141.99, 133.86, 128.82, 128.23, 127.19, 126.84, 115.88, 115.52, 112.61, 56.64, 55.56, 45.37, 29.73, 28.55, 21.19, 21.05. IR (cm^{-1}): 3060, 3025, 2960, 2933, 2872, 2835, 1731, 1666, 1590, 1503, 1469, 1432, 1387, 1341, 1249, 1034, 910, 810, 699, 621. HRMS: calcd. for $C_{20}H_{23}NO_2$ Na+ [M + Na]+: 332.1621; Found: 332.1610.

4. Conclusions

We have developed a convenient Fe-catalyzed decarbonylative alkylative spirocyclization of *N*-aryl cinnamamide with aliphatic aldehydes to provide 1-azaspirocyclohexadienones. Readily available aliphatic aldehydes were decarbonylated into primary, secondary and tertiary alkyl radicals readily for the cascade construction of dual C(sp^3)-C(sp^3) and C=O bonds. The application of cheap and readily available aliphatic aldehydes as alkyl source, convenient experimental operation for the alkyl radical generation, green ferrous catalyst and reaction solvent, C=C bond difunctionalization strategy and versatile synthetic utilities of the 1-azaspirocyclohexadienone, would render this decarbonylative alkylative spirocyclization attractive for organic synthesis and medicinal chemistry.

Supplementary Materials: The following are available online, experimental details, characterisation of products and copies of NMR spectrums.

Author Contributions: X.P. and R.-X.L. contributed equally to this work on data collection, data analysis, data interpretation; data collection, X.-Y.X.; conception and supervision, L.Y.; All X.P., R.-X.L., X.-Y.X. and L.Y. have read and agreed to the published version of the manuscript.

Funding: This work was supported by the National Natural Science Foundation of China (21772168), Opening Fund of Beijing National Laboratory for Molecular Sciences, Key Foundation of Education Bureau of Hunan Province (17A208), Hunan 2011 Collaborative Innovation Centre of Chemical Engineering & Technology with Environmental Benignity and Effective Resource Utilization, the project of innovation team of the ministry of education (IRT_17R90).

Conflicts of Interest: The authors declare no conflict of interest.

References

1.   Kotha, S.; Deb, A.C.; Lahiri, K.; Manivannan, E. Selected Synthetic Strategies to Spirocyclics. *Synthesis* **2009**, *2009*, 165–193. [CrossRef]

2.   Hu, J.-F.; Fan, H.; Xiong, J.; Wu, S.-B. Discorhabdins and Pyrroloiminoquinone-Related Alkaloids. *Chem. Rev.* **2011**, *111*, 5465–5491. [CrossRef]

3.   Sannigrahi, M. Stereocontrolled synthesis of spirocyclics. *Tetrahedron* **1999**, *55*, 9007–9071. [CrossRef]

4.   Rosenberg, S.; Leino, R. Synthesis of Spirocyclic Ethers. *Synthesis* **2009**, *2009*, 2651–2673.

5.   Nemoto, T.; Zhao, Z.-D.; Yokosaka, T.; Suzuki, Y.; Wu, R.; Hamada, Y. Palladium-Catalyzed Intramolecular *ipso*-Friedel-Crafts Alkylation of Phenols and Indoles: Rearomatization-Assisted Oxidative Addition. *Angew. Chem. Int. Ed.* **2013**, *52*, 2217–2220. [CrossRef]

6.   Wu, Q.-F.; Liu, W.-B.; Zhuo, C.-X.; Rong, Z.-Q.; Ye, K.-Y.; You, S.-L. Iridium-Catalyzed Intramolecular Asymmetric Allylic Dearomatization of Phenol. *Angew. Chem. Int. Ed.* **2011**, *50*, 4455–4458. [CrossRef] [PubMed]

7.   Matsuura, B.S.; Condie, A.G.; Buff, R.C.; Karahalis, G.J.; Stephenson, C.R.J. Intercepting Wacker Intermediates with Arenes: C-H Functionalization and Dearomatization. *Org. Lett.* **2011**, *13*, 6320–6323. [CrossRef] [PubMed]

8.   Chiba, S.; Zhang, L.; Lee, J.-Y. Copper-Catalyzed Synthesis of Azaspirocyclohexadienones from α-Azido-*N*-arylamides under an Oxygen Atmosphere. *J. Am. Chem. Soc.* **2010**, *132*, 7266–7267. [CrossRef]

9.   Pigge, F.C.; Dhanya, R.; Hoefgen, E.R. Morita-Baylis-Hillman Cyclizations of Arene-Ruthenium-Functionalized Acrylamides. *Angew. Chem. Int. Ed.* **2007**, *46*, 2887–2890. [CrossRef]

10. Tang, B.-X.; Tang, D.-J.; Tang, S.; Yu, Q.-F.; Zhang, Y.-H.; Liang, Y.; Zhong, P.; Li, J.-H. Intramolecular *ipso*-Halocyclization of 4-(*p*-Unsubstituted-aryl)-1-alkynes Leading to Spiro[4,5]trienones: Scope, Application, and Mechanistic Investigations. *J. Org. Chem.* **2012**, *77*, 2837–2849. [CrossRef]

11. Zhang, X.-X.; Larock, R.C. Synthesis of Spiro[4,5]trienones by Intramolecular *ipso*-Halocyclization of 4-(*p*-Methoxyaryl)-1-alkynes. *J. Am. Chem. Soc.* **2005**, *127*, 12230–12231. [CrossRef] [PubMed]

12. Tang, B.-X.; Tang, D.-J.; Tang, S.; Yu, Q.-F.; Zhang, Y.-H.; Liang, Y.; Zhong, P.; Li, J.-H. Selective Synthesis of Spiro[4,5]trienyl Acetates via an Intramolecular Electrophilic *ipso*-Iodocyclization Process. *Org. Lett.* **2008**, *10*, 1063–1066. [CrossRef] [PubMed]

13. Song, R.-J.; Xie, Y.-X. Recent Advances in Oxidative *ipso*-Annulation of *N*-Arylpropiolamides. *Chin. J. Chem.* **2017**, *35*, 280–288. [CrossRef]

14. Yang, X.-H.; Song, R.-J.; Xie, Y.-X.; Li, J.-H. Iron Catalyzed Oxidative Coupling, Addition, and Functionalization. *ChemCatChem.* **2016**, *8*, 2429–2445. [CrossRef]

15. Yi, H.; Zhang, G.-T.; Wang, H.-M.; Huang, Z.-Y.; Wang, J.; Singh, A.K.; Lei, A.-W. Recent Advances in Radical C-H Activation/Radical Cross-Coupling. *Chem. Rev.* **2017**, *17*, 9016–9085. [CrossRef] [PubMed]

16. Wei, W.-T.; Song, R.-J.; Ouyang, X.-H.; Li, Y.; Li, H.-B.; Li, J.-H. Copper-catalyzed oxidative *ipso*-carboalkylation of activated alkynes with ethers leading to 3-etherified azaspiro[4.5]trienones. *Org. Chem. Front.* **2014**, *1*, 484–489. [CrossRef]

17. Ouyang, X.-H.; Song, R.-J.; Liu, B.; Li, J.-H. Synthesis of 3-Alkyl Spiro[4,5]trienones by Copper-Catalyzed Oxidative *ipso*-Annulation of Activated Alkynes with Unactivated Alkanes. *Chem. Commun.* **2016**, *52*, 2573–2576. [CrossRef]

18. Zhang, H.-L.; Gu, Z.-X.; Xu, P.; Hu, H.-W.; Cheng, Y.-X.; Zhu, C.-J. Metal-free tandem oxidative C(sp^3)-H bond functionalization of alkanes and dearomatization of *N*-phenyl-cinnam-amides: Access to alkylated 1-azaspiro[4.5]decanes. *Chem. Commun.* **2016**, *52*, 477–480. [CrossRef]

19. Wang, C.-S.; Roisnel, T.; Dixneuf, P.H.; Soulé, J.F. Access to 3-(2-Oxoalkyl)-azaspiro[4.5]trienones via AcidTriggered Oxidative Cascade Reaction through Alkenyl Peroxide Radical Intermediate. *Adv. Synth. Catal.* **2019**, *361*, 445–450. [CrossRef]

20. Zhang, H.-L.; Zhu, C.-J. Iron-Catalyzed Cascade Cyanoalkylation/Radical Dearomatization of *N*-phenylcinnamamides: Access to Cyanoalkylated 1-Azaspiro[4.5]decanes. *Org. Chem. Front.* **2017**, *4*, 1272–1275. [CrossRef]

21. Liu, Y.; Wang, Q.-L.; Zhou, C.-S.; Xiong, B.-Q.; Zhang, P.-L.; Yang, C.-A.; Tang, K.-W. Visible-Light-Mediated *Ipso*-Carboacylation of Alkynes: Synthesis of 3-Acylspiro[4,5]trienones from *N*-(*p*-Methoxyaryl)propiolamides and Acyl Chlorides. *J. Org. Chem.* **2018**, *83*, 2210–2218. [CrossRef]

22. Ouyang, X.-H.; Song, R.-J.; Li, Y.; Liu, B.; Li, J.-H. Metal-Free Oxidative *Ipso*-Carboacylation of Alkynes: Synthesis of 3-Acylspiro[4,5]trienones from *N*-Arylpropiolamides and Aldehydes. *J. Org. Chem.* **2014**, *79*, 4582–4589. [CrossRef] [PubMed]

23. Yuan, L.; Jiang, S.-M.; Li, Z.-Z.; Zhu, Y.; Yu, J.; Li, L.; Li, M.-Z.; Tang, S.; Sheng, R.-R. Photocatalyzed cascade Meerwein addition cyclization of *N*-benzylacrylamides toward azaspirocycles. *Org. Biomol. Chem.* **2018**, *16*, 2406–2410. [CrossRef] [PubMed]

24. Wu, J.; Ma, D.-M.; Tang, G.; Zhao, Y.-F. Copper-Catalyzed Phosphonylation/Trifluoromethylation of *N-p-*NO$_2$-Benzoylacrylamides Coupled with Dearomatization and Denitration. *Org. Lett.* **2019**, *21*, 7674–7678. [CrossRef] [PubMed]

25. Hua, H.-L.; He, Y.-T.; Qiu, Y.-F.; Li, Y.-X.; Song, B.; Gao, P.; Song, X.-R.; Guo, D.-H.; Liu, X.-Y.; Liang, Y.-M. Copper-Catalyzed Difunctionalization of Activated Alkynes by Radical Oxidation-Tandem Cyclization/Dearomatization to Synthesize 3-Trifluoromethyl Spiro[4.5]trienones. *Chem. Eur. J.* **2015**, *21*, 1468–1473. [CrossRef] [PubMed]

26. Han, G.-F.; Wang, Q.; Liu, Y.-X.; Wang, Q.-M. Copper-Mediated α-Trifluoromethylation of *N*-Phenylcinnamamides Coupled with Dearomatization: Access to Trifluoromethylated 1-Azaspiro [4.5]decanes. *Org. Lett.* **2014**, *16*, 5914–5917. [CrossRef] [PubMed]

27. Wei, W.; Cui, H.-H.; Yang, D.-S.; Yue, H.-L.; He, C.-L.; Zhang, Y.-L., Wang, H. Visible-light enabled spirocyclization of alkynes leading to 3-sulfonyl and 3- sulfenyl azaspiro[4,5]trienones. *Green Chem.* **2017**, *19*, 5608–5613. [CrossRef]

28. Wen, J.-W.; Wei, W.; Xue, S.-N.; Yang, D.-S.; Lou, Y.; Gao, C.-Y.; Wang, H. Metal-Free Oxidative Spirocyclization of Alkynes with Sulfonylhydrazides Leading to 3-Sulfonated Azaspiro[4,5]trienones. *J. Org. Chem.* **2015**, *80*, 4966–4972. [CrossRef] [PubMed]

29. Jin, D.-P.; Gao, P.; Chen, D.-Q.; Chen, S.; Wang, J.; Liu, X.-Y.; Liang, Y.-M. AgSCF$_3$-Mediated Oxidative Trifluoromethythiolation of Alkynes with Dearomatization to Synthesize SCF$_3$-Substituted Spiro[4,5]trienones. *Org. Lett.* **2016**, *18*, 3486–3489. [CrossRef] [PubMed]

30. Cui, H.-H.; Wei, W.; Yang, D.-S.; Zhang, J.-M.; Xu, Z.-H.; Wen, J.-W.; Wang, H. Silver-catalyzed direct spirocyclization of alkynes with thiophenols: A simple and facile approach to 3-thioazaspiro[4,5]trienones. *RSC Adv.* **2015**, *5*, 84657–84661. [CrossRef]

31. Qian, P.-C.; Liu, Y.; Song, R.-J.; Xiang, J.-N.; Li, J.-H. Copper-Catalyzed Oxidative *ipso*-Cyclization of *N*-(*p*-Methoxyaryl)propiolamides with Disulfides and Water Leading to 3-(Arylthio)-1-azaspiro[4.5]deca-3,6,9-triene-2,8-diones. *Synlett* **2015**, *26*, 1213–1216. [CrossRef]

32. Gao, W.-C.; Liu, T.; Cheng, Y.-F.; Chang, H.-H.; Li, X.; Zhou, R.; Wei, W.-L.; Qiao, Y. AlCl3-Catalyzed Intramolecular Cyclization of *N*-Arylpropynamides witH-*N*-Sulfanylsuccinimides: Divergent Synthesis of 3-Sulfenyl Quinolin-2-ones and Azaspiro[4,5]trienones. *J. Org. Chem.* **2017**, *82*, 13459–13467. [CrossRef] [PubMed]

33. Sahoo, H.; Mandal, A.; Dana, S.; Baidya, M. Visible Light-Induced Synthetic Approach for Selenylative Spirocyclization of *N*-Aryl Alkynamides with Molecular Oxygen as Oxidant. *Adv. Synth. Catal.* **2018**, *360*, 1099–1130. [CrossRef]

34. Yang, X.-H.; Ouyang, X.-H.; Wei, W.-T.; Song, R.-J.; Li, J.-H. Nitrative Spirocyclization Mediated by TEMPO: Synthesis of Nitrated Spirocycles from *N*-Arylpropiolamides, tert-Butyl Nitrite and Water. *Adv. Synth. Catal.* **2015**, *357*, 1161–1166. [CrossRef]

35. Wang, L.-J.; Wang, A.-Q.; Xia, Y.; Wu, X.-X.; Liu, X.-Y.; Liang, Y.-M. Silver-catalyzed carbon–phosphorus functionalization of *N*-(*p*-methoxyaryl)propiolamides coupled with dearomatization: Access to phosphorylated aza-decenones. *Chem. Commun.* **2014**, *50*, 13998–14001. [CrossRef]

36. Zhang, Y.; Zhang, J.-H.; Hu, B.-Y.; Ji, M.-M.; Ye, S.-Y.; Zhu, G.-G. Synthesis of Difluoromethylated and Phosphorated Spiro[5.5]trienones via Dearomative Spirocyclization of Biaryl Ynones. *Org. Lett.* **2018**, *20*, 2988–2992. [CrossRef]

37. Gao, P.; Zhang, W.-W.; Zhang, Z.-C. Copper-Catalyzed Oxidative *ipso*-Annulation of Activated Alkynes with Silanes: An Approach to 3-Silyl Azaspiro[4,5] trienones. *Org. Lett.* **2016**, *18*, 5820–5823. [CrossRef]

38. Wu, L.-J.; Tan, F.-L.; Li, M.; Song, R.-J.; Li, J.-H. Fe-Catalyzed Oxidative Spirocyclization of *N*-Arylpropiolamides with Silanes and TBHP Involving the Formation of C-Si Bond. *Org. Chem. Front.* **2017**, *4*, 350–353. [CrossRef]

39. Guo, X.; Wang, J.; Li, C.-J. An Olefination via Ruthenium-Catalyzed Decarbonylative Addition of Aldehydes to Terminal Alkynes. *J. Am. Chem. Soc.* **2009**, *131*, 15092–15093. [CrossRef]

40. Shuai, Q.; Yang, L.; Guo, X.; Basle, O.; Li, C.-J. Rhodium-Catalyzed Oxidative C–H Arylation of 2-Arylpyridine Derivatives via Decarbonylation of Aromatic Aldehydes. *J. Am. Chem. Soc.* **2010**, *132*, 12212–12213. [CrossRef]

41. Guo, X.; Wang, J.; Li, C.-J. Ru-Catalyzed Decarbonylative Addition of Aliphatic Aldehydes to Terminal Alkynes. *Org. Lett.* **2010**, *12*, 3176–3178. [CrossRef] [PubMed]

42. Yang, L.; Correia, C.A.; Guo, X.; Li, C.-J. A novel catalytic decarbonylative Heck-type reaction and conjugate addition of aldehydes to unsaturated carbonyl compounds. *Tetrahedron Lett.* **2010**, *51*, 5486–5489. [CrossRef]

43. Yang, L.; Guo, X.; Li, C.-J. The First Decarbonylative Coupling of Aldehydes and Norbornenes Catalyzed by Rhodium. *Adv. Synth. Catal.* **2010**, *352*, 2899–2904. [CrossRef]

44. Yang, L.; Zeng, T.; Shuai, Q.; Guo, X.; Li, C.-J. Phosphine ligand triggered oxidative decarbonylative homocoupling of aromatic aldehydes: Selectively generating biaryls and diarylketones. *Chem. Commun.* **2011**, *47*, 2161–2163. [CrossRef] [PubMed]

45. Tang, R.-J.; He, Q.; Yang, L. Metal-free oxidative decarbonylative coupling of aromatic aldehydes with arenes: Direct access to biaryls. *Chem. Commun.* **2015**, *51*, 5925–5928. [CrossRef] [PubMed]

46. Tang, R.-J.; Kang, L.; Yang, L. Metal-Free Oxidative Decarbonylative Coupling of Aliphatic Aldehydes with Azaarenes: Successful Minisci-Type Alkylation of Various Heterocycles. *Adv. Synth. Catal.* **2015**, *357*, 2055–2060. [CrossRef]

47. Wu, C.-S.; Li, R.; Wang, Q.-Q.; Yang, L. Fe-Catalyzed decarbonylative alkylation–peroxidation of alkenes with aliphatic aldehydes and hydroperoxide under mild conditions. *Green Chem.* **2019**, *21*, 269–274. [CrossRef]

48. Peng, Y.; Jiang, Y.-Y.; Du, X.-J.; Ma, D.-Y.; Yang, L. Co-Catalyzed decarbonylative alkylative esterification of styrenes with aliphatic aldehydes and hypervalent iodine(III) reagents. *Org. Chem. Front.* **2019**, *6*, 3065–3070. [CrossRef]

49. Li, Y.-X.; Wang, Q.-Q.; Yang, L. Metal-free decarbonylative alkylation-aminoxidation of styrene derivatives with aliphatic aldehydes and *N*-hydroxyphthalimide. *Org. Biomol. Chem.* **2017**, *15*, 1338–1342. [CrossRef]

50. Li, W.-Y.; Wang, Q.-Q.; Yang, L. Metal-free decarbonylative alkylation–aminoxidation of styrene derivatives with aliphatic aldehydes and *N*-hydroxyphthalimide. *Org. Biomol. Chem.* **2017**, *15*, 9987–9991. [CrossRef]

51. Yang, L.; Lu, W.; Zhou, W.; Zhang, F. Metal-free cascade oxidative decarbonylative alkylarylation of acrylamides with aliphatic aldehydes: A convenient approach to oxindoles via dual C(sp^2)-H bond functionalization. *Green Chem.* **2016**, *18*, 2941–2945. [CrossRef]

52. Gao, R.-X.; Luan, X.-Q.; Xie, Z.-Y.; Yang, L.; Pei, Y. Fe-Catalyzed decarbonylative cascade reaction of *N*-aryl cinnamamides with aliphatic aldehydes to construct 3,4-dihydroquinolin-2(1H)-ones. *Org. Biomol. Chem.* **2019**, *17*, 5262–5268. [CrossRef] [PubMed]

53. Wu, C.-S.; Liu, R.-X.; Ma, D.-Y.; Luo, C.-P.; Yang, L. Four-Component Radical Dual Difunctionalization (RDD) of Two Different Alkenes with Aldehydes and tert-Butyl Hydroperoxide (TBHP): An Easy Access to β,δ-Functionalized Ketones. *Org. Lett.* **2019**, *21*, 6117–6121. [CrossRef]

54. Paul, S.; Guin, J. Dioxygen-Mediated Decarbonylative C-H Alkylation of Heteroaromatic Bases with Aldehydes. *Chem. Eur. J.* **2015**, *21*, 17618–17622. [CrossRef] [PubMed]

55. Biswas, P.; Paul, S.; Guin, J. Aerobic Radical-Cascade Alkylation/Cyclization of α,β-Unsaturated Amides: An Efficient Approach to Quaternary Oxindoles. *Angew. Chem. Int. Ed.* **2016**, *55*, 7756–7760. [CrossRef] [PubMed]

56. Kumar, A.; Shah, B.A. Synthesis of Biaryls via Benzylic C-C Bond Cleavage of Styrenes and Benzyl Alcohols. *Org. Lett.* **2015**, *17*, 5232–5235. [CrossRef]

57. Zong, Z.; Wang, W.; Bai, X.; Xi, H.; Li, Z.-P. Manganese-Catalyzed Alkyl-Heck-Type Reaction via Oxidative Decarbonylation of Aldehydes. *Asian J. Org. Chem.* **2015**, *4*, 622–625.

58. Lv, L.; Bai, X.; Yan, X.; Li, Z.-P. Iron-catalyzed decarbonylation initiated [2+2+m] annulation of benzene-linked 1,n-enynes with aliphatic aldehydes. *Org. Chem. Front.* **2016**, *3*, 1509–1513. [CrossRef]

59. Ouyang, X.-H.; Song, R.-J.; Liu, B.; Li, J.-H. Metal-Free Oxidative Decarbonylative Hydroalkylation of Alkynes with Secondary and Tertiary Alkyl Aldehydes. *Adv. Synth. Catal.* **2016**, *358*, 1903–1909. [CrossRef]

60. Li, Y.; Pan, J.-H.; Hu, M.; Liu, B.; Song, R.-J.; Li, J.-H. Intermolecular oxidative decarbonylative [2 + 2 + 2] carbocyclization of *N*-(2-ethynylaryl)acrylamides with tertiary and secondary alkyl aldehydes involving C(sp$_3$)-H functionalization. *Chem. Sci.* **2016**, *7*, 7050–7054. [CrossRef]

61. Zou, H.-X.; Li, Y.; Yang, H.-X.; Xiang, J.; Li, J.-H. Metal-Free Oxidative Decarbonylative [3 + 2] Annulation of Terminal Alkynes with Tertiary Alkyl Aldehydes toward Cyclopentenes. *J. Org. Chem.* **2018**, *83*, 8581–8588. [CrossRef] [PubMed]

62. Li, Y.; Li, J.-H. Decarbonylative Formation of Homoallyl Radical Capable of Annulation witH-N-Arylpropiolamides via Aldehyde Auto-oxidation. *Org. Lett.* **2018**, *20*, 5323–5326. [CrossRef] [PubMed]

63. Pan, C.; Chen, Y.; Song, J.; Li, L.; Yu, J.-T. Metal-Free Cascade Oxidative Decarbonylative Alkylation/Arylation of Alkynoates with Alphatic Aldehydes. *J. Org. Chem.* **2016**, *81*, 12065–12069. [CrossRef] [PubMed]
64. Pan, C.; Huang, B.; Hu, W.; Feng, X.; Yu, J.-T. Metal-Free Radical Oxidative Annulation of Ynones with Alkanes To Access Indenones. *J. Org. Chem.* **2016**, *81*, 2087–2093. [CrossRef] [PubMed]

Sample Availability: Samples of the compounds are available from the authors.

 *molecules* 

Article

Solvent-Free Iron(III) Chloride-Catalyzed Direct Amidation of Esters

Blessing D. Mkhonazi [1]**, Malibongwe Shandu** [1]**, Ronewa Tshinavhe** [1]**, Sandile B. Simelane** [2] **and Paseka T. Moshapo** [1],*****

[1] Research Centre in Synthesis and Catalysis, Department of Chemical Science, University of Johannesburg, P.O. Box 524, Auckland Park 2006, South Africa; 201585722@student.uj.ac.za (B.D.M.); 201305188@student.uj.ac.za (M.S.); 201424658@student.uj.ac.za (R.T.)

[2] Department of Chemistry, University of Eswatini, Private Bag 4, Kwaluseni M201, Eswatini; bssimelane@uniswa.sz

* Correspondence: pasekam@uj.ac.za; Tel.: +27-11-559-2372

Academic Editor: Hans-Joachim Knölker
Received: 10 January 2020; Accepted: 21 February 2020; Published: 26 February 2020

Abstract: Amide functional groups are prominent in a broad range of organic compounds with diverse beneficial applications. In this work, we report the synthesis of these functional groups via an iron(iii) chloride-catalyzed direct amidation of esters. The reactions are conducted under solvent-free conditions and found to be compatible with a range of amine and ester substrates generating the desired amides in short reaction times and good to excellent yields at a catalyst loading of 15 mol%.

Keywords: amidation; iron(III) chloride; amides; esters; solvent-free

1. Introduction

Amide functional groups are present in a plethora of natural and organic molecules with beneficial applications in the pharmaceutical, agrochemical, and textile industries. Examples include proteins; medicinal drugs such as apixaban, lenalidomide, and enzalutamide; as well as pesticides such as Boscalid® [1–5]. In 2018, amide functional groups were present in 68% of new small molecule drugs approved by the FDA [6]. In general, amides are synthesized by a reaction between carboxylic acids and amines using activating reagents such as HATU, PPh₃, and DCC in excess amounts [7]. Alternatively, carboxylic acids can be converted to their acid chloride or anhydride derivatives that readily react with amines to form the desired amide bonds. These reactions are often not catalytic and generate large quantities of undesired co-products. The synthesis of amides using greener and more atom economic reaction conditions is therefore an attractive approach to curb this problem [8,9]. Alternative amidation methodologies have been reported and some of these methods include the use of boron-based catalysts for efficient coupling of carboxylic acids at mild reaction temperatures; however, these reactions generate boronic acid side products [10–12]. Transition metal catalysts have also been used to couple amines and azides with a variety of substrates such as carboxylic acids, aldehydes, alcohols, alkynes and aryl halides [13–15]. Esters are key features in several organic synthesis substrates and have also been reported as useful substrates for amide coupling under acidic or basic conditions [16–20]. Some of the recent reports in direct ester amidation reactions include the use of lanthanum trifluoromethanesulfonate (Ln(Otf)₃) in low catalyst loading as reported by Ohshima and co-workers [21]. Hu and co-workers reported the use of a nickel catalyst for reductive amidation of esters using nitroarenes [22]. Similar reductive coupling reactions using chromium catalysts have also been recently reported by Zeng and co-workers [23]. Nickel has also been employed by Newman and co-workers for direct amidation of methyl esters at high reaction temperatures [24]. Therefore, there is a growing interest in the use of esters as substrates for direct amidation reactions. This interest provides an opportunity for the investigation of more catalysts as well as different amino substrates that can be used to promote ester amidation [25,26]. In this study, we report the direct amidation of esters using relatively cheap and readily available $FeCl_3$ catalyst [27–29] under solvent-free reaction conditions.

2. Results and Discussion

To establish the optimal reaction conditions for amidation of esters, readily available ethyl acetate and benzyl amine were selected as the ester and amine of choice, respectively (Scheme 1). Iron catalysts, namely, $FeCl_3$, $FeCl_2$, and $FeBr_3$, were evaluated as suitable promoters for the direct amidation of 1a and 2a under solvent-free conditions at 50 °C. These catalysts were also compared to other readily available catalysts such as $AlCl_3$ and $BiCl_3$ and as can be seen in Figure 1, $FeCl_3$ was found to be the most active catalyst as it afforded a higher yield of the desired product **3a**. A control experiment conducted at 50 °C in the absence of a catalyst did not form any product after 24 h. Thus, $FeCl_3$ was selected as the preferred catalyst for further reaction optimization and evaluation.

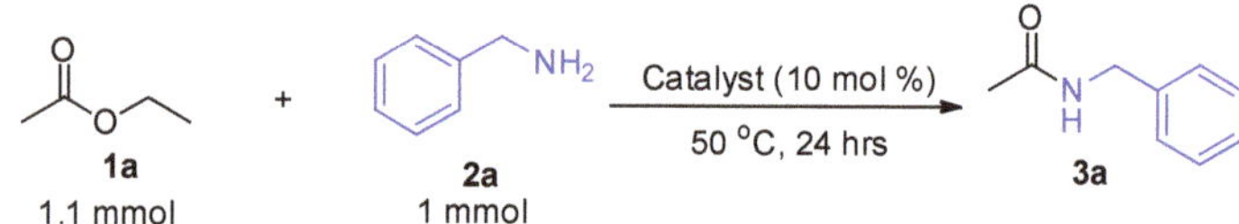

Scheme 1. Catalyst = $AlCl_3$, $FeCl_3$, $FeCl_2$, $FeBr_3$ or $BiCl_3$.

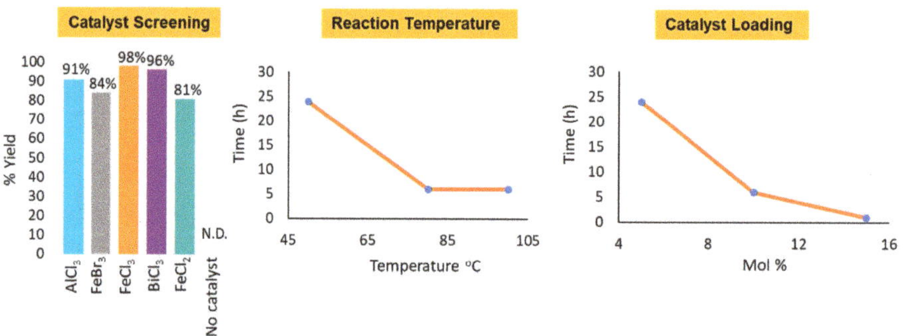

Figure 1. Reaction optimization. N.D. = Not Detected.

We then investigated the optimal reaction temperature and $FeCl_3$ catalyst loading (Figure 1). Eighty degrees Celsius was the most favorable reaction temperature as the reaction went to completion within 5 h at 10 mol% catalyst loading. Increasing the reaction temperature to 100 °C did not effect a notable change in the rate of the reaction. Furthermore, increasing the catalyst loading to 15 mol% improved the rate of the reaction as TLC indicated completion of the reaction within 90 min. As a result, 15 mol% catalyst loading and 80 °C were selected as our optimal reaction conditions. In anticipation of reaction conditions that may require solvents (i.e., cases where both ester and amine substrates are solids or reactions that solidify upon mixing of substrates), we also investigated suitable solvents for $FeCl_3$ catalyzed direct amidation of esters. We randomly selected four common organic solvents for investigation and the reactions between ester **1a** and amine **2a** (Scheme 1) were left to proceed for 90 min upon which TLC analysis indicated reaction completion for reactions conducted in acetonitrile and 1,2 dichloroethane (Table 1). Starting amine **2a** was observed in the tetrahydrofuran and toluene reactions after 90 min. Therefore, acetonitrile was found to be the most favorable solvent as it is relatively greener in comparison to 1,2 dichloroethane and it also afforded the product in high yield. The steric effect of alkoxy groups around the ester carbonyl carbon was also investigated (Scheme 2). An increase in steric bulk (isopropoxy) and chain length of the alkoxy group (n-butoxy) resulted in a decrease in ester reactivity and the reactions required longer reaction times to reach completion. However, there was no significant difference between esters with methoxy and ethoxy groups.

Table 1. Solvent evaluation.

Entry	Solvent	Product Yield
1	Toluene	84% [a]
2	Acetonitrile	96%
3	Tetrahydrofuran	85% [a]
4	1,2 Dichloroethane	94%
5	No Solvent	99%

Reaction conditions: ester **1a** (1.1 mmol), amine **2a** (1 mmol), 15 mol% FeCl$_3$, 80 °C. [a] Reaction did not reach completion within 90 min.

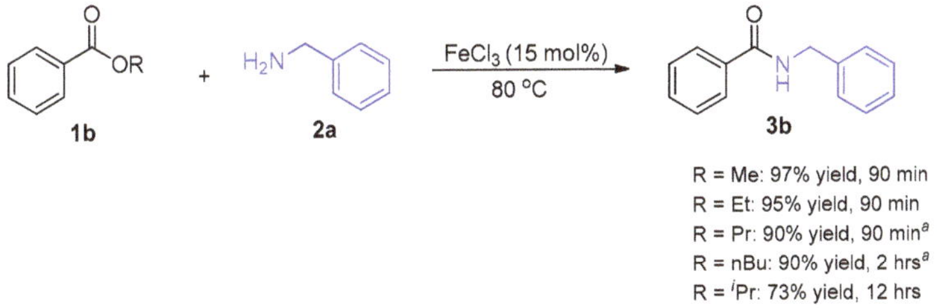

R = Me: 97% yield, 90 min
R = Et: 95% yield, 90 min
R = Pr: 90% yield, 90 min[a]
R = nBu: 90% yield, 2 hrs[a]
R = ^{i}Pr: 73% yield, 12 hrs

Scheme 2. [a] Reaction mixture solidified and 0.5 mL CH$_3$CN was added.

Having established the optimal reaction conditions, we then investigated the catalyst's reaction scope by varying both ester 1 and amine 2 substrates according to Scheme 3. Reactions were monitored by TLC. Aqueous work-up and flash column chromatography afforded the various amide products shown in Scheme 3. Aryl esters containing electron withdrawing groups were more active towards amidation than those with electron donating groups (product **3d** vs. **3e**). Both primary and secondary amines reacted to form amide products 3 in good yields. No reaction was observed between aryl ester **1b** (R = Et) and aniline. Interestingly, amidation of 2-pyridinecarboxylates with aryl amines was successful and afforded 2-pyridinecarboxamide products (**3f** to **3o**) in high yields. The formation of these pyridinecarboxamide products can be attributed to the possible coordination of the carbonyl oxygen and pyridine nitrogen to iron, thus, forming a stabilized intermediate complex that can undergo the observed amidation.

To test the proposed nitrogen and oxygen iron coordination hypothesis, 3-pyridinecarboxylate was reacted with aniline and as expected, the reaction failed to afford the desired product. Reactions with 2-pyridinecarboxamide were successful with aryl amines possessing both electron donating and electron withdrawing groups as well as alkyl amines. No product formation was observed when 4-nitroaniline was used as a substrate and the starting materials were recovered. Phenyloxy esters were also found to be suitable substrates for direct amidation and afforded the desired amides in good yields ranging from 70% to 89% (**3p** to **3s**).

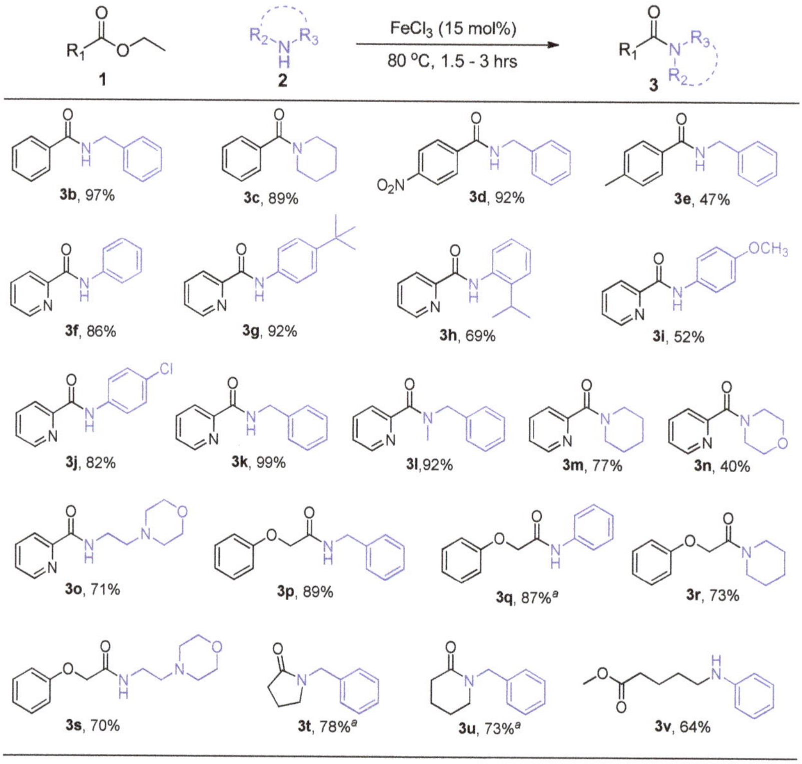

Scheme 3. General reaction scheme for FeCl$_3$ substrate scope evaluation. [a] Reaction mixture solidified and 0.5 mL of CH$_3$CN was added.

We further investigated the direct amidation of esters possessing alkyl bromide functional groups according to Scheme 4. Amidation of 4-bromobutanoate **4a** with benzyl amine **2a** afforded lactam **3t**. We then envisaged that increasing the alkyl chain length in Scheme 4 would enable us to synthesize lactams with varying ring sizes. 5-bromobutanoate **4b** afforded lactam **3u** in 73%; however, the reaction could not be replicated with 6-bromobutanoate as it formed several side products. Amidation using aniline afforded the substitution product **3v** which failed to undergo an intramolecular cyclization to form the desired lactam product.

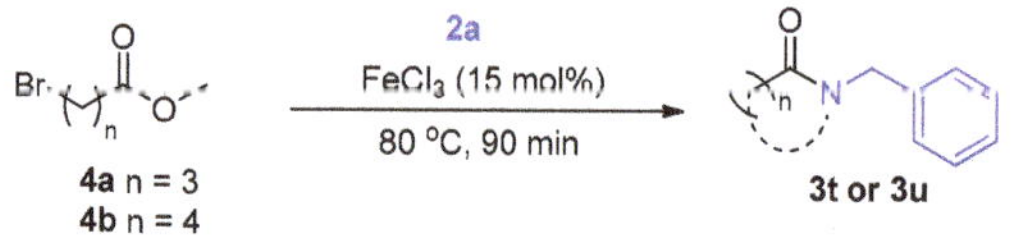

Scheme 4. Lactam synthesis.

Finally, and having succeeded in the amidation of 2-pyridinecarboxylates, we then utilized the developed methodology in the synthesis of pyrimidine carboxamides **7**. Although their mode of action is unknown, these molecules are worth investigating due to their reported potent anti-tubercular activities against clinical *Mycobacterium tuberculosis* (*Mtb*) strains [30]. Direct amidation of the corresponding ester 6 synthesized from commercially available dichloropyrimidine **5** afforded the amide **7** in good isolated yields (Scheme 5).

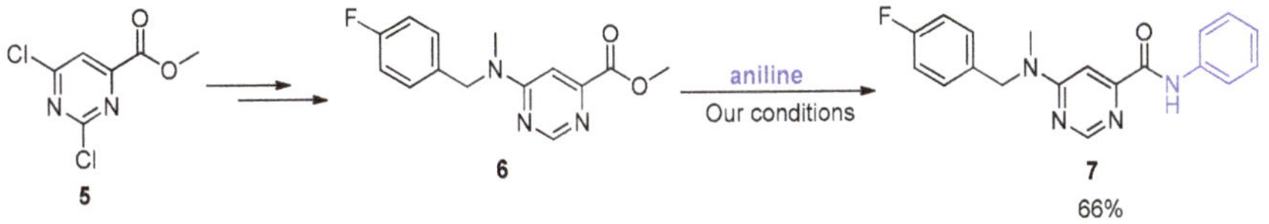

Scheme 5. Application of the developed protocol in the synthesis of anti-TB pyrimidine carboxamides.

In conclusion, we have successfully demonstrated a solvent-free direct amidation of esters using $FeCl_3$ as a Lewis acid catalyst at 80 °C. Amidation was effective with both primary and secondary amines and most of the reactions were complete in short reaction times (1.5–12 h) and good yields (47–99%). Linear and branched alkoxy groups on the corresponding esters could be displaced with ease to afford the desired amides. We have also demonstrated the synthesis of lactams with varying ring sizes from reactions between primary alkyl amines and 4-bromo or 5-bromoalkyl esters. The successful amidation of 2-pyridine carboxylates is currently being used in the synthesis of a library of novel aminopyrimidine carboxamides that will be screened for anti-TB activity.

3. Materials and Methods

All the solvents used were freshly distilled and dried by appropriate techniques. All reagents were purchased from Sigma Aldrich $FeCl_3$ reagent grade 97% was used as the catalyst. All reactions were monitored by thin layer chromatography (TLC) on aluminum-backed Merck silica gel 60 F254 plates using an ascending technique. The plates were visualized under UV light at 254 nm. Gravity column chromatography was done on Merck silica gel 60 (70–230 mesh). All proton nuclear magnetic resonance (^{1}H NMR) spectra were recorded using deuterated chloroform solutions on a Bruker Ultrashield (400 or 500 MHz) spectrometer. Carbon-13 nuclear magnetic resonance (^{13}C NMR) spectra were recorded on the same instruments at 100 MHz. All chemical shifts are reported in ppm. The chemical structures of the synthesized compounds were confirmed by comparison of their NMR data to literature reported data. Optimization reactions, ^{1}H and ^{13}C NMR data of all the compounds are available online as a Supplementary Materials.

Typical Experimental Procedure for FeCl₃-Catalyzed Direct Amidation of Esters. An oven-dried pressure tube equipped with a magnetic stirrer was evacuated with nitrogen. To this was added of ester 1a (492 uL, 5.04 mmol), followed by amine 2a (0.5 mL, 4.58 mmol), and finally $FeCl_3$ (111 mg, 0.684 mmol). The mixture was then sealed and stirred at 80 °C (0.5 mL of CH_3CN was added if the reaction mixture solidified). The reaction was monitored by TLC until completion upon which it was diluted with EtOAc and washed once with saturated $NaHCO_3$ and once with distilled H_2O. The combined aqueous layers were extracted once with ethyl acetate. The combined organic layers were then dried over $MgSO_4$, filtered and solvents removed under reduced pressure. The crude product was purified by silica gel flash column chromatography using a combination of hexane and ethyl acetate (3:2).

N-Benzylacetamide (3a). Yield: 99%; ^{1}H NMR (400 MHz, $CDCl_3$) δ 7.30–7.15 (m, 5H), 6.44 (br, 1H), 4.34 (d, *J* = 5.6 Hz, 2H), 1.94, (s, 3H); ^{13}C NMR (100 MHz, $CDCl_3$) δ 170.0, 138.2, 128.5, 127.6, 127.3, 43.5, 23.0. (The obtained NMR data agreed with the literature data for this compound.) [31]

N-Benzylbenzamide (3b). Yield: 97%; ^{1}H NMR (400 MHz, $CDCl_3$) δ 7.36 (d, *J* = 7.2 Hz, 2H), 7.55–7.20 (m, 8H), 6.49 (s, 1H), 4.62 (d, *J* = 5.6 Hz, 2H); ^{13}C NMR (100 MHz, $CDCl_3$) δ 167.3, 138.2, 134.4, 131.5, 128.8, 128.6, 127.9, 127.6, 126.9, 44.1. (The obtained NMR data agreed with the literature data for this compound.) [32]

Phenyl(piperidin-1-yl)methanone (3c). Yield: 89%; ^{1}H NMR (400 MHz, $CDCl_3$) δ7.36 (br, 5H), 3.68 (br, 2H), 3.30 (br, 2H), 1.70–1.40 (m, 6H). (The obtained NMR data agreed with the literature data for this compound.) [33]

N-(4-Nitrobenzyl)benzamide (3d). Yield: 92%; ^{1}H NMR (500 MHz, CDCl$_3$) δ 8.20 (d, *J* = 8.5 Hz, 2H), 7.91 (d, *J* = 8.5 Hz, 2H), 7.40–7.23 (m, 5H), 6.83 (br, 1H), 4.60 (d, *J* = 5.5 Hz, 2H); ^{13}C NMR (125 MHz, CDCl$_3$) δ 165.4, 149.6, 139.9, 137.5, 128.9, 128.2, 127.9, 127.8, 123.7, 44.4. (The obtained NMR data agreed with the literature data for this compound.) [32]

N-(4-Methylbenzyl)benzamide (3e). Yield: 47%; ^{1}H NMR (500 MHz, CDCl$_3$) δ 7.67 (d, *J* = 7.5 Hz, 2H), 7.40–7.18 (m, 7H), 6.51 (br, 1H), 4.60 (d, *J* = 5.0 Hz, 2H), 2.37 (s, 3H); ^{13}C NMR (125 MHz, CDCl$_3$) δ 167.3, 141.9, 138.4, 131.6, 129.2, 128.7, 127.9, 127.5, 127.0, 44.1, 21.4. (The obtained NMR data agreed with the literature data for this compound.) [32]

N-Phenylpicolinamide (3f). Yield: 86%; ^{1}H NMR (500 MHz, CDCl$_3$) δ 10.1 (br, 1H), 8.59 (d, *J* = 3.2 Hz, 1H), 8.28 (d, *J* = 6.4 Hz, 1H), 7.92–7.83 (m, 1H), 7.77 (d, *J* = 6.0 Hz, 1H), 7.50–7.31 (m, 3H), 7.13 (t. *J* = 5.8 Hz, 1H); ^{13}C NMR (100 MHz, CDCl$_3$) δ 161.9, 149.8, 147.9, 137.7, 137.6, 129.0, 126.3, 124.2, 122.3, 119.6. (The obtained NMR data agreed with the literature data for this compound.) [34]

N-(4-(tert-Butyl)phenyl)picolinamide (3g). Yield: 92%; ^{1}H NMR (500 MHz, CDCl$_3$) δ 9.97 (br, 1H), 8.57 (d, *J* = 4.0 Hz, 1H), 8.28 (d, *J* = 8.0 Hz, 1H), 7.84 (t, *J* = 7.8 Hz, 1H), 7.70 (d, *J* = 8.5 Hz, 2H), 7.48–7.36 (m, 3H), 1.32 (s, 9H); ^{13}C NMR (125 MHz, CDCl$_3$) δ 161.8, 149.9, 147.8, 147.1, 137.5, 135.1, 126.2, 125.9, 125.8, 122.2, 119.4, 34.6, 31.3. (The obtained NMR data agreed with the literature data for this compound.) [34]

(N-(2-Isopropylphenyl)picolinamide (3h). Yield: 69%; ^{1}H NMR (500 MHz, CDCl$_3$) δ 10.2 (br, 1H), 8.62 (d, *J* = 4.0 Hz, 1H), 8.32 (d, *J* = 8.0 Hz, 1H), 8.22 (d, *J* = 8.0 Hz, 1H), 7.88 (t, *J* = 8.0 Hz, 1H), 7.51–7.12 (m, 4H), 3.34–3.20 (m, 1H), 1.34 (d, *J* = 7.0 Hz, 6H); ^{13}C NMR (100 MHz, CDCl$_3$) δ 162.1, 150.3, 148.1, 139,0, 137.5, 134.5, 126.5, 126.4, 125.6, 125.2, 122.4, 118.9, 28.2, 22.9. (The obtained NMR data agreed with the literature data for this compound.) [34]

N-(4-Methoxyphenyl)picolinamide (3i). Yield: 52%; ^{1}H NMR (500 MHz, CDCl$_3$) δ 9.89 (br, 1H), 8.57 (d, *J* = 4.0 Hz, 1H), 7.86 (t, *J* = 8.0 Hz, 1H), 7.78–7.62 (m, 2H), 7.48–7.40 (m, 1H), 6.90 (d, *J* = 8.5 Hz, 2H), 3.78 (s, 3H); ^{13}C NMR (125 MHz, CDCl$_3$) δ 161.7, 156.4, 150.0, 147.9, 137.6, 131.0, 126.2, 122.2, 121.2, 114.2, 55.4. (The obtained NMR data agreed with the literature data for this compound.) [34]

N-(4-Chlorophenyl)picolinamide (3j). Yield: 82%; ^{1}H NMR (500 MHz, CDCl$_3$) δ 9.98 (br, 1H), 8.52 (d, *J* = 3.5 Hz, 1H), 8.20 (d, *J* = 7.5 Hz, 1H), 7.82 (t, *J* = 7.5 Hz, 1H), 7.68 (d, *J* = 8.5 Hz, 2H), 7.40 (t, *J* = 5.6 Hz, 1H), 7.27 (d, *J* = 8.5 Hz, 2H); ^{13}C NMR (100 MHz, CDCl$_3$) δ 161.8, 149.4, 147.8, 137.5, 136.3, 129.0, 128.9, 126.4, 122.2, 120.8. (The obtained NMR data agreed with the literature data for this compound.) [34]

N-Benzylpicolinamide (3k). Yield: 99%; ^{1}H NMR (500 MHz, CDCl$_3$) δ 8.52–8.38 (m, 2H), 8.20 (d, *J* = 8.0 Hz, 1H), 7.77 (t, *J* = 8.0 Hz, 1H), 7.45–7.20 (m, 6H), 4.63 (d, *J* = 6.0 Hz, 2H); ^{13}C NMR (100 MHz, CDCl$_3$) δ 164.0, 149.7, 147.8, 138.1, 137.1, 128.4, 127.6, 127.2, 125.9,122.1, 43.2. (The obtained NMR data agreed with the literature data for this compound.) [34]

N-Benzyl-N-methylpicolinamide (3l). Yield: 92%; mixture of rotamers, ^{1}H NMR (400 MHz, CDCl$_3$) δ 8.52 (dd, *J* = 4.4 amd 14.0 Hz, 2H), 7.80 –7.58 (m, 4H), 7.35–7.10 (m, 12H), 4.72 (s, 2H), 4.62 (s. 2H), 2.97 (s, 3H), 2.91 (s, 3H); ^{13}C NMR (100 MHz, CDCl$_3$) δ 169.1, 168.8, 154.4, 154.3, 148.1, 136.8, 136.6, 128.5, 128.4, 128.0, 127.4, 127.3, 124.2, 123.5, 123.4, 54.4, 51.0, 36.3, 33.1. (The obtained NMR data agreed with the literature data for this compound.) [34]

Piperidin-1-yl(pyridin-2-yl)methanone (3m). Yield: 77%; ^{1}H NMR (500 MHz, CDCl$_3$) δ 8.51 (d, *J* = 4.0 Hz, 1H), 7.70 (t, *J* = 7.8 Hz, 1H), 7.49 (d, *J* = 7.5 Hz, 1H), 7.30–7.15 (m, 1H), 3.81–3.65 (m, 2H), 3.31–3.48 (m, 2H), 1.72–1.43 (m, 6H); ^{13}C NMR (100 MHz, CDCl$_3$) δ 167.5, 154.7, 148.3, 136.8, 124.0, 123.1, 48.1, 43.2, 26.4, 25.4, 24.4. (The obtained NMR data agreed with the literature data for this compound.) [35]

Morpholino(pyridin-2-yl)methanone (3n). Yield: 40%; ^{1}H NMR (500 MHz, CDCl$_3$) δ 8.50 (d, *J* = 4.0 Hz, 1H), 7.23 (t, *J* = 7.8 Hz, 1H), 7.60 (d, *J* = 7.5 Hz, 1H), 7.27 (t, *J* = 6.0 Hz, 1H), 3.88–3.50 (m, 8H); ^{13}C NMR (125 MHz, CDCl$_3$) δ 167.3, 153.6, 148.1, 137.0, 124.5, 124.0, 66.9, 66.6, 47.6, 42.7. (The obtained NMR data agreed with the literature data for this compound.) [36]

N-(2-morpholinoethyl)picolinamide (3o). Yield: 71%; ^{1}H NMR (500 MHz, CDCl$_3$) δ 8.46 (d, *J* = 4.0 Hz, 1H), 8.25 (br, 1H), 8.08 (d, *J* = 8.0 Hz, 1H), 7.73 (t, *J* = 8.0 Hz, 1H), 7.31 (t, *J* = 6.0 Hz, 1H), 3.70–3.41 (m, 6H), 2.61–2.31 (m, 6H); ^{13}C NMR (100 MHz, CDCl$_3$) δ 164.2, 149.9, 147.9, 127.0, 125.8, 122.0, 66.8, 57.2, 53.3, 35.8. (The obtained NMR data agreed with the literature data for this compound.) [37]

N-Benzyl-2-phenoxyacetamide (3p). Yield: 89%; ^{1}H NMR (500 MHz, CDCl$_3$) δ 7.48–7.21 (m, 7H), 7.10–6.87 (m, 4H); ^{13}C NMR (100 MHz, CDCl$_3$) δ 168.1, 157.1, 137.7, 129.7, 128.6, 127.6, 127.5, 122.1, 114.6, 67.3, 42.9. (The obtained NMR data agreed with the literature data for this compound.) [38]

2-Phenoxy-N-phenylacetamide (3q). Yield: 87%; ^{1}H NMR (500 MHz, CDCl$_3$) δ 8.32 (bs, 1H), 7.59 (d, *J* = 8.0 Hz, 2H), 7.42–7.29 (m, 4H), 7.25–6.92 (m, 4H); ^{13}C NMR (100 MHz, CDCl$_3$) δ 166.2, 157.0, 136.8, 129.8, 129.0, 124.7, 122.3, 120.1, 114.8, 67.6. (The obtained NMR data agreed with the literature data for this compound.) [39]

2-Phenoxy-1-(piperidin-1-yl)ethanone (3r). Yield: 70%; ^{1}H NMR (500 MHz, CDCl$_3$) δ 7.38–7.20 (m, 2H), 7.12–6.87 (m, 3H), 4.63 (s, 2H), 3.60–3.42 (m, 4H), 1.69–1.48 (m, 6H); ^{13}C NMR (100 MHz, CDCl$_3$) δ 166.0, 157.9, 129.4, 121.3, 114.5, 67.6, 46.3, 43.0, 26.3, 25.4, 24.3.

N-(2-Morpholinoethyl)-2-phenoxyacetamide (3s). Yield: 70%; ^{1}H NMR (500 MHz, CDCl$_3$) δ 7.37–7.25 (m, 2H), 7.09 (br, 1H), 7.09–6.70 (m, 3H), 4.45 (s, 2H), 3.65–3.47 (m, 4H), 3.44–3.35 (m, 2H), 2.52–2.27 (m, 6H); ^{13}C NMR (100 MHz, CDCl$_3$) δ 168.0, 157.2, 129.6, 121.9, 67.2, 66.8, 56.5, 53.0, 35.1.

1-Benzylpyrrolidin-2-one (3t). Yield: 78%; ^{1}H NMR (400 MHz, CDCl$_3$) δ 7.24–7.71 (m, 5H), 4.35 (s, 2H), 3.15 (t, *J* = 6.8 Hz, 3H), 2.33 (t, *J* = 8.0 Hz, 2H), 1.91–1.83 (m, 2H); ^{13}C NMR (100 MHz, CDCl$_3$) δ 174.1, 135.9, 127.9, 127.3, 126.7, 45.8, 45.7, 30.1, 16.9. (The obtained NMR data agreed with the literature data for this compound.) [39]

1-Benzylpiperidin-2-one (3u). Yield: 73%; ^{1}H NMR (500 MHz, CDCl$_3$) δ 7.35–7.18 (m, 5H), 4.53 (s, 2H), 3.13 (t, *J* = 5.0 Hz, 2H), 2.59–2.31 (m, 2H), 1.81–1.67 (m, 4H); ^{13}C NMR (125 MHz, CDCl$_3$) δ 169.7, 137.11, 128.4, 127.8, 127.1, 49.9, 47.1, 32.2, 23.0, 21.2. (The obtained NMR data agreed with the literature data for this compound.) [39]

Methyl 5-(phenylamino)pentanoate (3v). Yield: 64%; ^{1}H NMR (500 MHz, CDCl$_3$) δ 7.16 (t, *J* = 7.8 Hz, 2H), 6.68 (t, *J* = 7.0 Hz, 1H), 6.59 (d, *J* = 7.5 Hz, 2H), 3.66 (s, 3H), 3.27 (t, *J* = 6.8 Hz, 1H), 3.12 (t, *J* = 7.0 Hz, 1H), 2.45–2.30 (m, 2H), 1.82–1.59 (m, 4H); ^{13}C NMR (100 MHz, CDCl$_3$) δ 173.7, 148.1, 129.2, 129.1, 117.2, 112.7, 51.4, 43.4, 33.7, 33.6, 28.8, 26.5, 22.4. (The obtained NMR data agreed with the literature data for this compound.) [40]

6-((4-Fluorobenzyl)(methyl)amino)-N-phenylpyrimidine-4-carboxamide (7). Yield: 66%; Clear oil; ^{1}H NMR (400 MHz, CDCl$_3$) δ 9.94 (s, 1H), 8.62 (s, 1H), 7.74 (d, *J* = 8.0 Hz, 2H), 7.31–7.43 (m, 3H), 7.10–7.28 (m, 3H), 6.90–7.08 (m, 2H), 4.86 (br, 2H), 3.11 (br, 3H). ^{13}C NMR (100 MHz, CDCl$_3$) δ 163.2, 161.4, 157.1, 155.2, 137.4, 129.1, 124.6, 119.7, 115.8, 115.6, 100.3, 33.40. ^{19}F NMR (376 MHz, CDCl$_3$) δ -114.8 (The obtained NMR data agreed with the literature data for this compound.) [30]

Supplementary Materials: The following are available online. Materials and methods, results of optimization conditions as well as NMR spectra of products.

Author Contributions: Conceptualization, P.T.M.; methodology, B.D.M., M.S. and R.T.; formal analysis, P.T.M., S.B.S., B.D.M., M.S. and R.T.; writing—original draft preparation, P.T.M.; writing—review and editing, P.T.M., S.B.S. All authors have read and agreed to the published version of the manuscript.

Funding: This research was funded by the National Research Foundation of South Africa, Grant number 118082 and the article processing charge (APC) was funded by Prof. H.-J. Knölker.

Acknowledgments: We thank the University of Johannesburg and Chemical Science Department for providing the research infrastructure. We also express gratitude to our NMR technician Mutshinyalo Nwamadi for assisting us in acquiring the NMR data.

Conflicts of Interest: The authors declare no conflict of interest.

References

1. Humphrey, J.M.; Chamberlin, A.R. Chemical Synthesis of Natural Product Peptides: Coupling Methods for the Incorporation of Noncoded Amino Acids into Peptides. *Chem. Rev.* **1997**, *97*, 2243–2266. [CrossRef]
2. Agrawal, R.; Jain, P.; Dikshit, S.N. Apixaban: A New Player in the Anticoagulant Class. *Curr. Drug Targets* **2012**, *13*, 863–875. [CrossRef]
3. Ponomaryov, Y.; Krasikova, V.; Lebedev, A.; Chernyak, D.; Varacheva, L.; Chernobroviy, A. Scalable and green process for the synthesis of anticancer drug lenalidomide. *Chem. Heterocycl. Comp.* **2015**, *51*, 133–138. [CrossRef]
4. Zhou, A.-N.; Li, B.; Ruan, L.; Wang, Y.; Duan, G.; Li, J. An improved and practical route for the synthesis of enzalutamide and potential impurities study. *Chin. Chem. Lett.* **2017**, *28*, 426–430. [CrossRef]
5. Volovych, I.; Neumann, M.; Schmidt, M.; Buchner, G.; Yang, J.-Y.; Wölk, J.; Sottmann, T.; Strey, R.; Schomäcker, R.; Schwarze, M. A novel process concept for the three step Boscalid®synthesis. *RSC Adv.* **2016**, *6*, 58279–58287. [CrossRef]
6. Jarvis, L.M. The new drugs of 2018. *C&EN* **2019**, *97*, 33–37.
7. Dunetz, J.R.; Magano, J.; Weisenburger, G.A. Large-Scale Applications of Amide Coupling Reagents for the Synthesis of Pharmaceuticals. *Org. Process Res. Dev.* **2016**, *20*, 140–177. [CrossRef]
8. Constable, D.J.C.; Dunn, P.J.; Hayler, J.D.; Humphrey, G.R.; Leazer, Jr.J.L.; Linderman, R.J.; Lorenz, K.; Manley, J.; Pearlman, B.A.; Wells, A.; et al. Key green chemistry research areas—a perspective from pharmaceutical manufacturers. *Green Chem.* **2007**, *9*, 411–420. [CrossRef]
9. Pattabiraman, V.R.; Bode, J.W. Rethinking amide bond synthesis. *Nature* **2011**, *480*, 471–479. [CrossRef]
10. Sabatini, M.T.; Boulton, L.T.; Sheppard, T.D. Borate esters: Simple catalysts for the sustainable synthesis of complex amides. *Sci. Adv.* **2017**, *3*, e1701028. [CrossRef]
11. Moore, J.A. An assessment of boric acid and borax using the IEHR evaluative process for assessing human developmental and reproductive toxicity of agents. *Reprod. Toxicol.* **1997**, *11*, 123–160. [CrossRef]
12. Sabatini, M.T.; Boulton, L.T.; Sneddon, H.F.; Sheppard, T.D. A green chemistry perspective on catalytic amide bond formation. *Nat. Catal.* **2019**, *2*, 10–17. [CrossRef]
13. Allen, C.L.; Williams, J.M.J. Metal-catalysed approaches to amide bond formation. *Chem. Soc. Rev.* **2011**, *40*, 3405–3415. [CrossRef]
14. de Figueiredo, R.M.; Suppo, J.-S.; Campagne, J.-M. Nonclassical routes for amide bond formation. *Chem. Rev.* **2016**, *116*, 19–12029. [CrossRef]
15. Ojeda-Porras, A.; Gamba-Sánchez, D. Recent developments in amide synthesis using nonactivated starting materials. *J. Org. Chem.* **2016**, *81*, 11548–11555. [CrossRef] [PubMed]
16. Ranu, B.C.; Dutta, P. A simple and convenient procedure for the conversion of esters to secondary amides. *Synth. Commun.* **2003**, *33*, 297–301. [CrossRef]
17. Price, K.E.; Larriveé-Aboussafy, C.; Lillie, B.M.; McLaughlin, R.W.; Mustakis, J.; Hettenbach, K.W.; Hawkins, J.M.; Vaidyanathan, R. Mild and efficient DBU-catalyzed amidation of ayanoacetates. *Org. Lett.* **2009**, *11*, 2003–2006. [CrossRef] [PubMed]
18. Kim, B.R.; Lee, H.-G.; Kang, S.-B.; Sung, G.H.; Kim, J.-J.; Park, J.K.; Lee, S.-G.; Yoon, Y.-J. *tert*-Butoxide-Assisted Amidation of Esters under Green Conditions. *Synthesis* **2012**, *44*, 42–50.
19. Ohshima, T.; Hayashi, Y.; Agura, K.; Fujii, Y.; Yoshiyama, A.; Mashima, K. Sodium methoxide: A simple but highly efficient catalyst for the direct amidation of esters. *Chem. Commun.* **2012**, *48*, 5434–5436. [CrossRef]
20. Rzhevskiy, S.A.; Ageshina, A.A.; Chesnokov, G.A.; Gribanov, P.S.; Topchiy, M.A.; Nechaev, M.S.; Asachenko, A.F. Solvent- and transition metal-free amide synthesis from phenyl esters and aryl amines. *RSC Adv.* **2019**, *9*, 1536–1540. [CrossRef]
21. Morimoto, H.; Fujiwara, R.; Shimizu, Y.; Morisaki, K.; Ohshima, T. Lanthanum(III) triflate catalyzed direct amidation of esters. *Org. Lett.* **2014**, *16*, 2018–2021. [CrossRef] [PubMed]
22. Cheung, C.W.; Ploeger, M.L.; Hu, X. Direct amidation of esters with nitroarenes. *Nat. Commun.* **2017**, *8*, 14878. [CrossRef] [PubMed]
23. Ling, L.; Chen, C.; Luo, M.; Zeng, X. Chromium-catalyzed activation of acyl C−O Bonds with magnesium for amidation of esters with nitroarenes. *Org Lett.* **2019**, *21*, 1912–1916. [CrossRef] [PubMed]
24. Ben Halima, T.; Masson-Makdissi, J.; Newman, S.G. Nickel catalysed amide bond formation from methyl esters. *Angew. Chem. Int. Ed.* **2018**, *57*, 12925–12929. [CrossRef] [PubMed]

25. Li, G.; Szostak, M. Highly selective transition-metal-free transamidation of amides and amidation of esters at room temperature. *Nat Commun.* **2018**, *9*, 4165. [CrossRef]

26. Caldwell, N.; Campbell, P.S.; Jamieson, C.; Potjewyd, F.; Simpson, I.; Watson, A.J.B. Amidation of esters with amino alcohols using organobase catalysis. *J. Org. Chem.* **2014**, *79*, 9347–9354. [CrossRef]

27. Correa, A.; Mancheño, O.G.; Bolm, C. Iron-catalysed carbon–heteroatom and heteroatom–heteroatom bond forming processes. *Chem. Soc. Rev.* **2008**, *37*, 1108–1117. [CrossRef]

28. Sun, C.-L.; Li, B.-J.; Shi, Z.-J. Direct C−H Transformation via Iron Catalysis. *Chem. Rev.* **2011**, *111*, 1293–1314. [CrossRef]

29. Bauer, I.; Knölker, H.-J. Iron Catalysis in Organic Synthesis. *Chem. Rev.* **2015**, *115*, 3170–3387. [CrossRef]

30. Wilson, C.R.; Gessner, R.K.; Moosa, A.; Seldon, R.; Warner, D.F.; Mizrahi, V.; de Melo, C.S.; Simelane, S.B.; Nchinda, A.; Abay, E.; et al. Novel antitubercular 6-dialkylaminopyrimidine carboxamides from phenotypic whole-cell high throughput screening of a softfocus library: Structure−activity relationship and target identification studies. *J. Med. Chem.* **2017**, *60*, 10118–10134. [CrossRef]

31. Taylor, J.E.; Jones, M.D.; Williams, J.M.J.; Bull, S.D. N-Acyl DBN tetraphenylborate salts as N-acylating agents. *J. Org. Chem.* **2012**, *77*, 2808–2818. [CrossRef] [PubMed]

32. Sawant, D.N.; Bagal, D.B.; Ogawa, S.; Selvam, K.; Saito, S. Diboron-Catalyzed Dehydrative Amidation of Aromatic Carboxylic Acids with Amines. *Org. Lett.* **2018**, *20*, 4397–4400. [CrossRef] [PubMed]

33. Wu, W.; Zhang, Z.; Liebeskind, L.S. In Situ Carboxyl Activation Using a Silatropic Switch: A New Approach to Amide and Peptide Constructions. *J. Am. Chem. Soc.* **2011**, *133*, 14256–14259. [CrossRef]

34. Zhu, M.; Fujita, K.-I.; Yamaguchi, R. Aerobic Oxidative Amidation of Aromatic and Cinnamic Aldehydes with Secondary Amines by CuI/2-Pyridonate Catalytic System. *J. Org. Chem.* **2012**, *77*, 9102–9109. [CrossRef] [PubMed]

35. Li, Y.; Chen, H.; Liu, J.; Wan, X.; Xu, Q. Clean synthesis of primary to tertiary carboxamides by CsOH-catalyzed aminolysis of nitriles in water. *Green Chem.* **2016**, *18*, 4865–4870. [CrossRef]

36. Kiselyov, A.S. Reaction of *N*-fluoropyridinium fluoride with isonitriles and diazo compounds: A one-pot synthesis of (pyridin-2-yl)-1H-1,2,3-triazoles. *Tetrahedron Lett.* **2006**, *47*, 2631–2634. [CrossRef]

37. Lundberg, H.; Tinnis, F.; Adolfsson, H. Direct Amide Coupling of Non-activated Carboxylic Acids and Amines Catalysed by Zirconium(IV) Chloride. *Chem. Eur. J.* **2012**, *18*, 3822–3826. [CrossRef]

38. Nagarajan, S.; Ran, P.; Shanmugavelan, P.; Sathishkumar, M.; Ponnuswamy, A.; Suk, N.K.; Gnana, K. The catalytic activity of titania nanostructures in the synthesis of amides under solvent-free conditions. *New J. Chem.* **2012**, *36*, 1312–1319. [CrossRef]

39. Kim, K.; Hong, S.H. Iridium-Catalyzed Single-Step N-Substituted Lactam Synthesis from Lactones and Amines. *J. Org. Chem.* **2015**, *80*, 4152–4156. [CrossRef]

40. Yuan, M.-L.; Xie, J.-H.; Zhou, Q.-L. Boron Lewis Acid Promoted Ruthenium-Catalyzed Hydrogenation of Amides: An Efficient Approach to Secondary Amines. *ChemCatChem* **2016**, *8*, 3036–3040. [CrossRef]

Sample Availability: Samples of all the compounds **3a–3v** and **7** are available from the authors.

 *molecules*

Article

Enantioselective Iron/Bisquinolyldiamine Ligand-Catalyzed Oxidative Coupling Reaction of 2-Naphthols

Lin-Yang Wu, Muhammad Usman and Wen-Bo Liu *

Sauvage Center for Molecular Sciences; Engineering Research Center of Organosilicon Compounds & Materials, Ministry of Education; College of Chemistry and Molecular Sciences; Wuhan University, Wuhan 430072, Hubei, China; 2016282030199@whu.edu.cn (L.-Y.W.); whitestar399@hotmail.com (M.U.)
* Correspondence: wenboliu@whu.edu.cn

Academic Editors: Hans-Joachim Knölker and Jean-Luc Renaud
Received: 15 January 2020; Accepted: 12 February 2020; Published: 14 February 2020

Abstract: An iron-catalyzed asymmetric oxidative homo-coupling of 2-naphthols for the synthesis of 1,1'-Bi-2-naphthol (BINOL) derivatives is reported. The coupling reaction provides enantioenriched BINOLs in good yields (up to 99%) and moderate enantioselectivities (up to 81:19 er) using an iron-complex generated in situ from $Fe(ClO_4)_2$ and a bisquinolyldiamine ligand $[(1R,2R)-N^1,N^2$-di(quinolin-8-yl)cyclohexane-1,2-diamine, **L1**]. A number of ligands (**L2–L8**) and the analogs of **L1**, with various substituents and chiral backbones, were synthesized and examined in the oxidative coupling reactions.

Keywords: iron catalysis; asymmetric catalysis; nitrogen ligand; oxidative coupling; BINOL synthesis

---

1. Introduction

Axially chiral compounds (atropisomers) have aroused much attention from organic chemists due to their prevalence in natural products, bioactive molecules, functional materials, and their wide applications in asymmetric transformations [1]. Many elegant methods have been established for the asymmetric synthesis of axially chiral compounds, both employing transition-metal catalysts [2] and organocatalysts [1]. In particular, 1,1'-Bi-2-naphthol (BINOL) is one of the most useful structural motifs and ligand substructures in asymmetric catalysis [3–9]. Since the pioneering report by the Noyori group utilizing enantioenriched BINOL as the ligand in asymmetric catalysis [10], numerous BINOL-derived ligands/catalysts (i.e., BINAP [11], BINAM [12], chiral phosphoric acid [13], chiral phosphoramidite [14], and BINSA [15]; Figure 1) have been designed and synthesized. The emergence of such a library of ligands/catalysts has brought marvelous contributions to the synthetic community, with tremendously efficient asymmetric transformations such as reductive coupling [16], allylation [17], ene-type [18], and Aldol [19] reactions and axial chirality assembly [20].

In the past few decades, enormous efforts have been devoted to the enantioselective assembly of BINOL scaffolds. Transition-metal-catalyzed asymmetric oxidative coupling of 2-naphthols have shown its power in the synthesis of BINOLs (Figure 2a). Efficient vanadium-catalyzed [21] protocols were reported by the Uang [22], Chen [23], Gong [24–26], and Sasai [27–29] groups, independently. Many research groups, including Nakajima, Kozlowski, and others, have successfully developed a series of copper-catalyzed coupling reactions of 2-naphthols [30–44]. Recently, a notable work by the Tu group [45] established a Cu/SPDO (spirocyclic pyrrolidine oxazoline) complex-catalyzed cross-coupling reaction to synthesize 3,3'-disubstituted BINOLs. Nevertheless, iron-catalyzed coupling strategies in this area have not been explored so far. Only a handful of remarkable iron catalysts, namely, an iron-salen complex reported by Katsuki and co-workers [46,47], a chiral diphosphine oxide–iron(II) complex developed by the Ishihara group [48], and an iron-chiral phosphoric acid (CPA) catalyst introduced by the Pappo group [49,50], have been disclosed to date (Figure 2a).

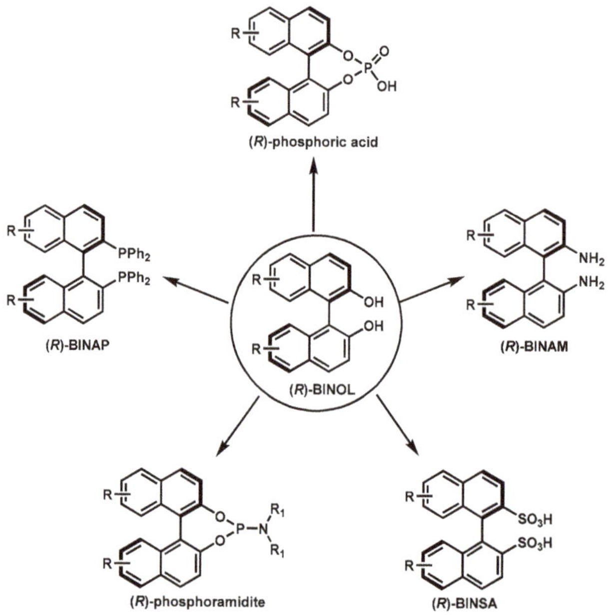

Figure 1. Representative 1,1′-Bi-2-naphthol (BINOL)-derived ligands/catalysts.

Recently, we have developed an iron-catalyzed direct amination of aliphatic C-H bonds [51,52], and it was interesting to find that the catalysts used were simply generated by in situ mixing of an iron salt and an aminopyridine ligand. Inspired by these results, we envisaged that introducing a chiral aminopyridine-type ligand might impart chirality to the products. Our attention was drawn to *N,N′*-dimethyl-*N,N′*-bis(8-quinolyl)-cyclohexanediamine (BQCN), developed by the Che group and successfully applied in iron-catalyzed asymmetric *cis*-dihydroxylation of alkenes [53] and, most recently, in Friedel-Crafts reactions [54]. Since our previous studies revealed that free secondary amine ligand presented a good reactivity [51], we were wondering whether the *N*-unprotected bis(8-quinolyl)-cyclohexanediamine ligand (bisquinolyldiamine, **L1**), which is synthesized straightforwardly from 8-haloquinoline and diamine, is capable of controlling the selectivities in the iron-catalyzed oxidative coupling of 2-naphthols. Herein, we report the studies toward the synthesis of the amino ligands and their applications in the synthesis of optically active BINOL derivatives via an oxidative homo-coupling reaction of 2-naphthols under mild conditions (Figure 2b).

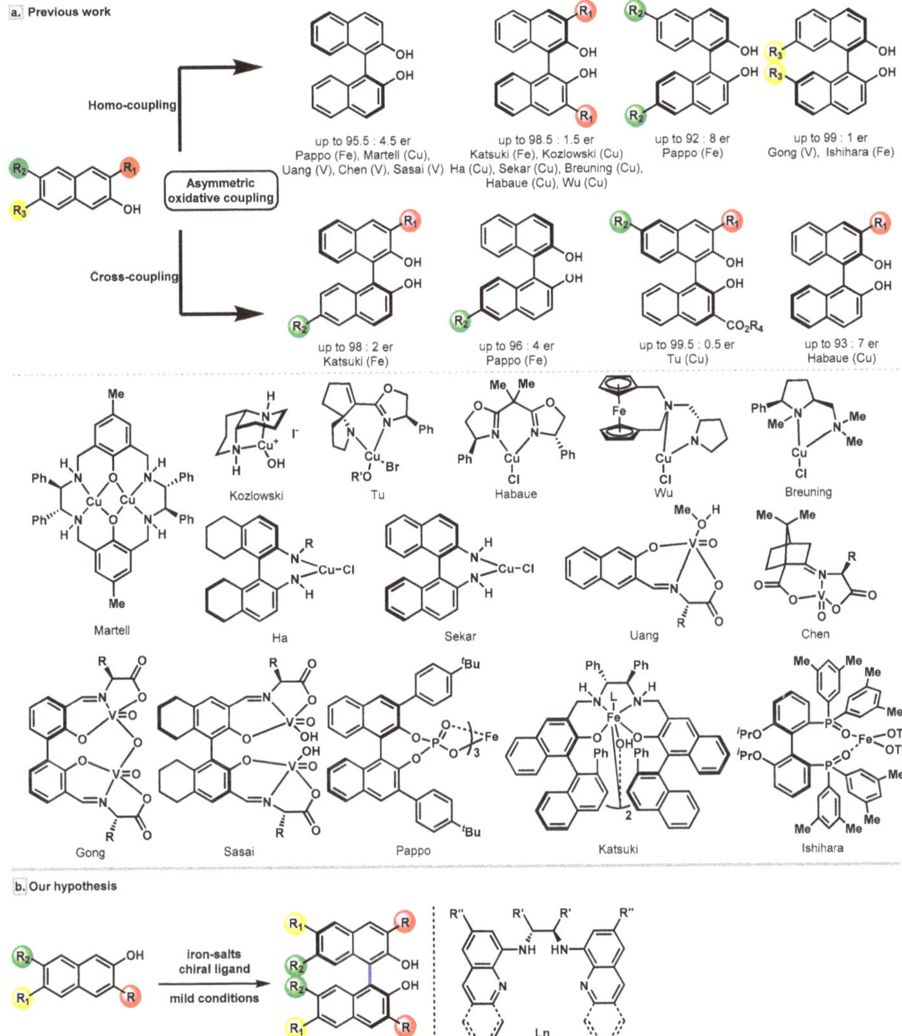

Figure 2. Transition metal-catalyzed coupling for the enantioselective synthesis of BINOLs. (**a**) The previous examples of homo- or cross-coupling of 2-naphthols. (**b**) Our proposed asymmetric oxidative homo-coupling of 2-naphthols using iron/bisquinolyldiamine catalyst.

2. Results and Discussion

2.1. Synthesis of Bisquinolyldiamine Ligands

The Buchwald-Hartwig C–N coupling reaction [55,56] was used for the synthesis of a variety of bisquinolyldiamine ligands following the literature procedure [53]. As shown in Figure 3, with the catalyst derived from 5 mol% Pd$_2$(dba)$_3$ and 10 mol% *rac*-BINAP, eight ligands were synthesized in moderate to good yields. The chiral backbones of these ligands include (1*R*,2*R*)-cyclohexane-1,2-diamine (**L1**), (*R*)-[1,1′-binaphthalene]-2,2′-diamine (**L6**), (12*R*)-9,10-dihydro-9,10-ethanoanthracene-11,12-diamine (**L7**), and (1*R*,2*R*)-1,2-diphenylethane-1,2-diamine (**L8**). Ligands with a variety of electronically differentiated substituents at the C6 position of the quinoline moiety (**L2–L4**) and an acridine-derived ligand (**L5**) were also prepared to probe the electronic and steric effects of the ligands on the reaction.

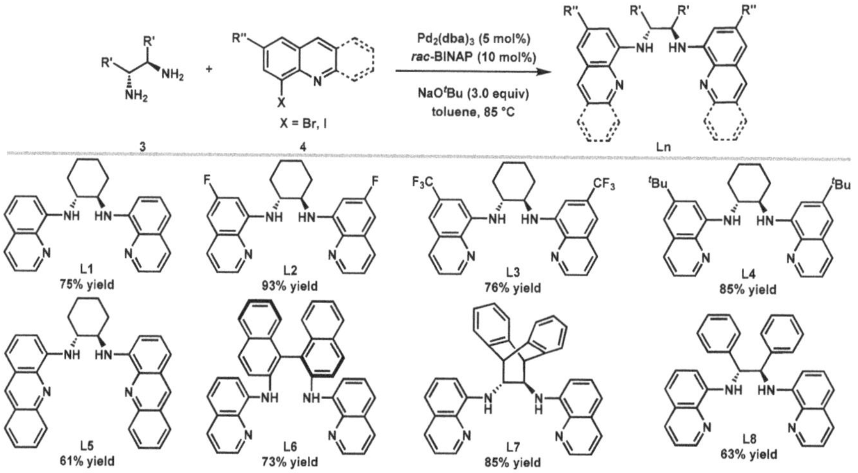

Figure 3. Ligands synthesis.

2.2. Reaction Investigation

To test our hypothesis, we selected 2-naphthol (**1a**) as the model substrate for optimization of reaction conditions (Table 1). All reactions were carried out with the catalyst generated in situ by stirring $Fe(ClO_4)_2$ and **L1** for 30 min before the addition of the substrate, and under atmospheric dioxygen. Various solvents (methanol, dichloroethane, chloroform, toluene, and chlorobenzene) were screened first (entries 1–5) and resulted in moderate conversions and enantioselectivities, except toluene and methanol. The reaction in chlorobenzene gave a better balance between reactivity and selectivity (53% conv. and 79:21 er, entry 5). The addition of 4Å MS led to a slightly higher enantioselectivity with partial conversion in a much shorter reaction time (entries 5 vs. 6). We were delighted to find that increasing the temperature from 30 °C to 50 °C delivered 39% conversion with comparable enantioselectivity (entry 7). Further increase in temperature (i.e., 70 °C and 90 °C) resulted in lower enantioselectivities (entries 8 and 9). Then, the reaction with different iron salts were investigated, including $Fe(ClO_4)_3$, $Fe(OAc)_2$, $Fe(OTf)_2$, $Fe(acac)_2$, and $FeCl_2$, but failed to provide better results (entries 10–14). The ratio of iron precursor versus **L1** was also examined (entries 14–19). Surprisingly, increasing $Fe(ClO_4)_2$-loading from 5 mol% to 10 mol% improves the efficiency without affecting the enantioselectivity and delivered the coupling product **2a** in 84% isolated yield with 80:20 er (entries 15 and 16). Although further increasing $Fe(ClO_4)_2$ to 12.5 mol% improved the yield, a slightly diminished er was also observed (entry 17). Finally, reactions with the catalysts derived from the diamine ligand (**L2–L8**) were inspected. Electron-withdrawing groups-substituted ligands (**L2** and **L3**) showed excellent reactivities but with low enantioselectivities (entries 20 and 21). In contrast, ligand **L4** with an electron-donating substituent delivered the product in reduced yield, albeit with good er (entry 22). Upon further screening, the ligands (**L5–L8**) bearing different chiral backbones, lower yield, and er were obtained with **L5**, while ligands **L6–L8** failed to give any product (entries 23–26).

Table 1. Conditions Screening [a]

Entry	Fe (x mol%)	Ln (y mol%)	Additives	Solvent	T °C	t/h	conv. (%) [b]	er (%) [c]
1	Fe(ClO$_4$)$_2$ (5.0)	L1 (5.0)	-	MeOH	30	28.5		n.d. [f]
2	Fe(ClO$_4$)$_2$ (5.0)	L1 (5.0)	-	DCE	30	28.5	57 [d]	73:27
3	Fe(ClO$_4$)$_2$ (5.0)	L1 (5.0)	-	CHCl$_3$	30	28.5	68 [d]	77:23
4	Fe(ClO$_4$)$_2$ (5.0)	L1 (5.0)	-	toluene	30	28.5	22 [d]	80:20
5	Fe(ClO$_4$)$_2$ (5.0)	L1 (5.0)	-	PhCl	30	28.5	53 [d]	79:21
6	Fe(ClO$_4$)$_2$ (5.0)	L1 (5.0)	MS 4Å	PhCl	30	5.0	12 [d]	80:20
7	Fe(ClO$_4$)$_2$ (5.0)	L1 (5.0)	MS 4Å	PhCl	50	5.0	39 [d]	78:22
8	Fe(ClO$_4$)$_2$ (5.0)	L1 (5.0)	MS 4Å	PhCl	70	5.0	38 [d]	73:27
9	Fe(ClO$_4$)$_2$ (5.0)	L1 (5.0)	MS 4Å	PhCl	90	5.0	28 [d]	70:30
10	Fe(ClO$_4$)$_3$ (5.0)	L1 (5.0)	MS 4Å	PhCl	50	5.0	76	76:24
11	Fe(OAc)$_2$ (5.0)	L1 (5.0)	MS 4Å	PhCl	50	5.0	16	n.d. [f]
12	Fe(OTf)$_2$ (5.0)	L1 (5.0)	MS 4Å	PhCl	50	5.0	19	n.d. [f]
13	Fe(acac)$_2$ (5.0)	L1 (5.0)	MS 4Å	PhCl	50	5.0	23	n.d. [f]
14	FeCl$_2$ (5.0)	L1 (5.0)	MS 4Å	PhCl	50	5.0	20	n.d. [f]
15	Fe(ClO$_4$)$_2$ (7.5)	L1 (5.0)	MS 4Å	PhCl	50	5.0	78	77:23
16	Fe(ClO$_4$)$_2$ (10.0)	L1 (5.0)	MS 4Å	PhCl	50	5.0	88(84 [e])	80:20
17	Fe(ClO$_4$)$_2$ (12.5)	L1 (5.0)	MS 4Å	PhCl	50	5.0	95	72:28
18	Fe(ClO$_4$)$_2$ (5.0)	L1 (10.0)	MS 4Å	PhCl	50	5.0	39	77:23
19	Fe(ClO$_4$)$_2$ (10.0)	L1 (10.0)	MS 4Å	PhCl	50	5.0	86	77:23
20	Fe(ClO$_4$)$_2$ (10.0)	L2 (5.0)	MS 4Å	PhCl	50	5.0	90	70:30
21	Fe(ClO$_4$)$_2$ (10.0)	L3 (5.0)	MS 4Å	PhCl	50	5.0	85	60:40
22	Fe(ClO$_4$)$_2$ (10.0)	L4 (5.0)	MS 4Å	PhCl	50	5.0	53 [e]	78:22
23	Fe(ClO$_4$)$_2$ (10.0)	L5 (5.0)	MS 4Å	PhCl	50	5.0	67 [e]	55:45
24	Fe(ClO$_4$)$_2$ (10.0)	L6 (5.0)	MS 4Å	PhCl	50	5.0	-	n.d. [f]
25	Fe(ClO$_4$)$_2$ (10.0)	L7 (5.0)	MS 4Å	PhCl	50	5.0	-	n.d. [f]
26	Fe(ClO$_4$)$_2$ (10.0)	L8 (5.0)	MS 4Å	PhCl	50	5.0	-	n.d. [f]

[a] All reactions carried out with **1a** (0.5 mmol), MS 4Å (150 mg) in 5 mL solvent under O$_2$ atmosphere (1 atm). [b] Conversions determined by GC using dodecane as an internal standard. [c] Determined by HPLC (Chiralpak AS-H). [d] Determined by ^{1}H-NMR analysis. [e] Isolated yield. [f] n.d.: not detected.

2.3. Substrates Scope

Next, we investigated the scope of the iron-catalyzed asymmetric oxidative coupling reaction, and the results were summarized in Scheme 1. A variety of substituted 2-naphthols with electronic and steric properties were examined. Substrate **1** containing functionalities at the C3-position, including OMe, OBn, *o*-tolyl, and Ph were successfully converted into the coupling products (**2b–2e**) in 56–88% yields with 56:44 to 81:19 er. The substrate bearing an electron-withdrawing Br substitution at the C3-position was also converted into the corresponding product (**2f**) in 51% yield with 79:21 er. The electronic effects of different functionalities are clearly demonstrated by the observation that electron-donating groups delivered higher yield (e.g., C3-OMe 88%, **2b** vs. C3-Br 51%, **2f**). However, C3-substituted substrates with carbonyl functionalities like CO$_2$H, COPh, and CO$_2$Bn failed to give any products (**2g–2i**). When C6-Br-substituted 2-naphthol was applied, the desired product (**2j**) was obtained in 82% yield with 79:21 er. A C6-phenyl-substituted 2-naphthol also resulted in 64% yield and 74:26 er (**2k**). Moreover, C7-substituted (BnO, ^{n}BuO, TBSO, and MeO) substrates were also effectively coupled and delivered the corresponding products (**2l–2o**) in 65–99% yields, albeit with dramatically diminished er.

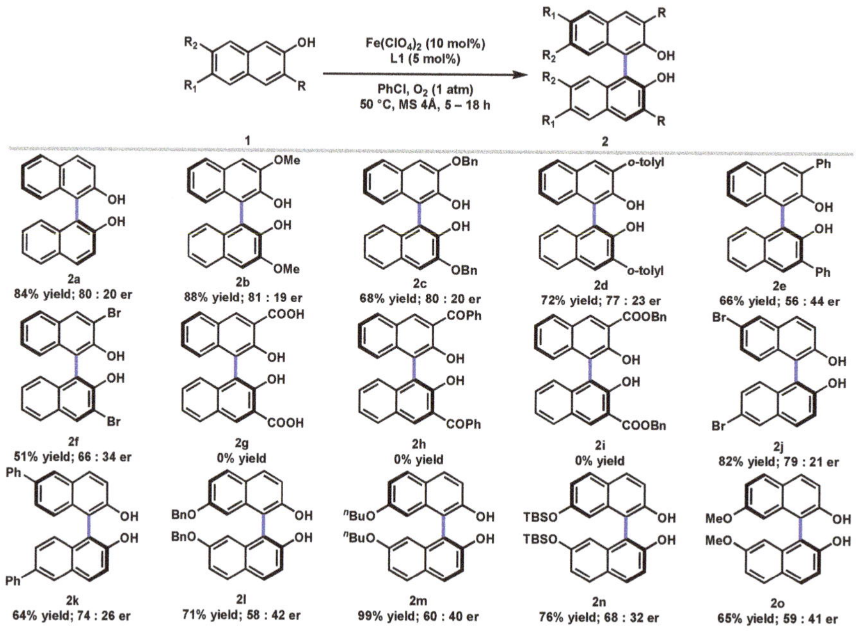

Scheme 1. Reaction conditions:

$$1 \xrightarrow[\substack{\text{PhCl, O}_2 \text{ (1 atm)} \\ 50\,°\text{C, MS 4Å, 5 – 18 h}}]{\substack{\text{Fe(ClO}_4)_2 \text{ (10 mol\%)} \\ \textbf{L1} \text{ (5 mol\%)}}} 2$$

Compound	Yield / er
2a	84% yield; 80 : 20 er
2b (OMe)	88% yield; 81 : 19 er
2c (OBn)	68% yield; 80 : 20 er
2d (o-tolyl)	72% yield; 77 : 23 er
2e (Ph)	66% yield; 56 : 44 er
2f (Br)	51% yield; 66 : 34 er
2g (COOH)	0% yield
2h (COPh)	0% yield
2i (COOBn)	0% yield
2j (Br)	82% yield; 79 : 21 er
2k (Ph)	64% yield; 74 : 26 er
2l (BnO)	71% yield; 58 : 42 er
2m (^{n}BuO)	99% yield; 60 : 40 er
2n (TBSO)	76% yield; 68 : 32 er
2o (MeO)	65% yield; 59 : 41 er

Scheme 1. Substrates Scope [a]. [a] Reactions conducted with Fe(ClO$_4$)$_2$ (10 mol%), **L1** (5 mol%), substrate **1** (0.5 mmol), and 4Å MS (150 mg) in PhCl (5 mL) under oxygen atmosphere (1 atm) at 50 °C. Percentage represented isolated yields. Er determined by HPLC.

3. Materials and Methods

3.1. General Information

Unless otherwise noted, all reagents were purchased commercially and used without further purification. Petroleum ether (PE) (60–90 °C), ethyl acetate (EA), and dichloromethane (DCM) were used for silica gel chromatography. MeCN, toluene, DMF, and THF were purchased commercially or were dried by passage through an activated alumina column under argon [57]. PhCl, CHCl$_3$, MeOH, and acetone were freshly distilled after drying over CaH$_2$. ^{1}H-NMR spectra were recorded at room temperature on a Bruker ADVANCE III 400 MHz spectrometer and were reported relative to residual Chloroform-*d* (δ 7.26 ppm) or DMSO-d6 (δ 2.50 ppm). ^{13}C-NMR spectra were recorded on a Bruker ADVANCE III 400 MHz spectrometer (100 MHz) and were reported relative to Chloroform-*d* (δ 77.16 ppm). ^{19}F-NMR spectra were recorded on a Bruker ADVANCE III 400 MHz spectrometer (376 MHz). Data for ^{1}H-NMR were reported as chemical shift (δ ppm) (multiplicity, coupling constant (Hz), and integration) using standard abbreviations for multiplicities: s = singlet, d = doublet, t = triplet, q = quartet, and m = multiplet. Data for ^{13}C-NMR and ^{19}F-NMR were reported in terms of chemical shifts (δ ppm). High-resolution mass spectra (HRMS) were obtained by using a Bruker Compact TOF mass spectrometer in electrospray ionization mode (ESI). Enantiomeric ratio (er) was determined by an Agilent 1260 Series HPLC utilizing DAICEL Chiralpak (AD-H, AS-H, or IC) or Chiralcel (OD-H) columns (4.6 mm × 250 mmL). Optical rotations were measured with a Perkin Elmer 343 polarimeter and were reported as: [α]$_D^T$ (concentration in g/100 mL, solvent). The NMR spectra of all new compounds and HPLC spectra of oxidative coupling products were provided in the Supplementary Materials.

3.2. Preparation of Ligands

General Procedure (Scheme 2): To an oven-dried Schlenk flask were added diamine **3** (1.0 equiv), $Pd_2(dba)_3$ (5 mol%), *rac*-BINAP (10 mol%), NaO^tBu (3.0 equiv), and toluene under Ar atmosphere. Then 8-haloqunoline **4** (2.2 equiv) was added directly. The flask was sealed, and the reaction was stirred at 85 °C until the complete consumption of the starting material **3**. The mixture was cooled to room temperature, filtered through a silica plug, and the plug was washed with EA. The combined filtrates were concentrated under reduced pressure, and the residue was purified by silica gel chromatography to give the desired product **Ln**.

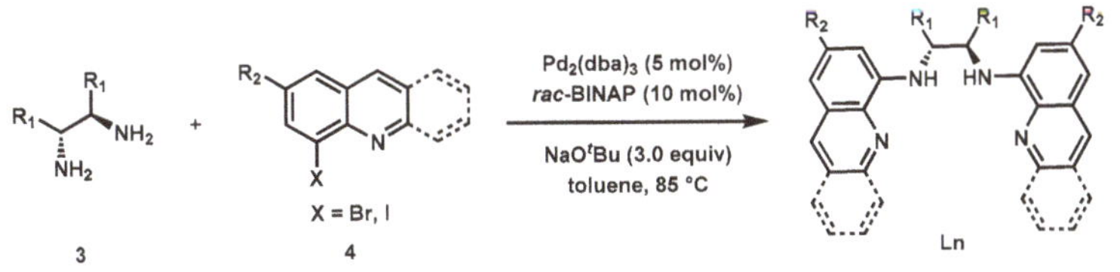

Scheme 2. Synthesis of Bisquinolyldiamine Ligands.

(1R,2R)-N^1,N^2-Di(quinolin-8-yl)cyclohexane-1,2-diamine (**L1**) [53]: Following the general procedure, the reaction was carried out with (1R,2R)-cyclohexane-1,2-diamine **3a** (0.36 g, 3.2 mmol, 1.0 equiv); $Pd_2(dba)_3$ (0.15 g, 0.16 mmol, 5 mol%); *rac*-BINAP (0.20 g, 0.32 mmol, 10 mol%); NaO^tBu (0.92 g, 9.6 mmol, 3.0 equiv); and 8-bromoqunoline **4a** (1.46 g, 7.0 mmol, 2.2 equiv) in 30 mL of toluene. The desired product was obtained (0.88 g, 75% yield) as a pale yellow solid after purification by silica gel chromatography (PE/EA = 30/1 to 10/1). ^{1}H-NMR (400 MHz, Chloroform-*d*) δ 8.57 (dd, *J* = 4.2, 1.7 Hz, 2H), 7.98 (dd, *J* = 8.3, 1.7 Hz, 2H), 7.36 (td, *J* = 8.0, 0.9 Hz, 2H), 7.30–7.25 (m, 2H), 6.98 (dd, *J* = 8.2, 1.3 Hz, 2H), 6.84 (dd, *J* = 7.7, 1.1 Hz, 2H), 6.43 (brs, 2H), 3.86–3.67 (m, 2H), 2.49–2.31 (m, 2H), 1.86 (td, *J* = 4.6, 4.1, 2.2 Hz, 2H), 1.67–1.48 (m, 4H).

(1R,2R)-N^1,N^2-Bis(6-fluoroquinolin-8-yl)cyclohexane-1,2-diamine (**L2**): Following the general procedure, the reaction was carried out with (1R,2R)-cyclohexane-1,2-diamine **3a** (81.6 mg, 0.7 mmol, 1.0 equiv); $Pd_2(dba)_3$ (32.8 mg, 0.04 mmol, 5 mol%); *rac*-BINAP (45.1 mg, 0.07 mmol, 10 mol%); NaO^tBu (206.9 mg, 2.2 mmol, 3.0 equiv); and 8-bromo-6-fluoroquinoline **4b** (356.7 mg, 1.6 mmol, 2.2 equiv) in 2 mL of toluene. The desired product was obtained (262.3 mg, 93% yield) as a pale yellow solid after purification by silica gel chromatography (PE/EA = 5/1). ^{1}H-NMR (400 MHz, Chloroform-*d*) δ 8.49 (dd, *J* = 4.2, 1.5 Hz, 2H), 7.87 (dd, *J* = 8.3, 1.2 Hz, 2H), 7.34–7.21 (m, 2H), 6.68–6.49 (m, 4H), 6.46 (dd, *J* = 9.3, 2.5 Hz, 2H), 3.76–3.56 (m, 2H), 2.45–2.28 (m, 2H), 1.97–1.79 (m, 2H), 1.66–1.47 (m, 4H) (Figure S2); ^{13}C-NMR (100 MHz, Chloroform-*d*) δ 162.4 (d, J_{C-F} = 243.3 Hz), 146.3 (d, J_{C-F} = 13.5 Hz), 145.6 (d, J_{C-F} = 2.4 Hz), 135.6, 135.4 (d, J_{C-F} = 5.8 Hz), 129.3 (d, J_{C-F} = 12.8 Hz), 122.2, 96.1 (d, J_{C-F} = 22.8 Hz), 95.1 (d, J_{C-F} = 30.6 Hz), 56.6, 31.7, 24.5 (Figure S3); ^{19}F NMR (376 MHz, Chloroform-*d*) δ −110.9 (Figure S4); HRMS (ESI$^+$) calcd for $C_{24}H_{23}F_2N_4$ [M + H]$^+$: 405.1885, found 405.1880; $[\alpha]_D^{24}$ = −315.6 (c = 0.2, CHCl3); M. p. 182–186 °C.

(1R,2R)-N^1,N^2-Bis(6-(trifluoromethyl)quinolin-8-yl)cyclohexane-1,2-diamine (**L3**): Following the general procedure, the reaction was carried out with (1R,2R)-cyclohexane-1,2-diamine **3a** (0.31 g, 2.7 mmol, 1.0 equiv); $Pd_2(dba)_3$ (0.13 g, 0.14 mmol, 5 mol%); *rac*-BINAP (0.17 g, 0.28 mmol, 10 mol%); NaO^tBu (0.79 g, 8.2 mmol, 3.0 equiv); and 8-bromo-6-trifluoromethylquinoline **4c** (1.57 g, 5.7 mmol, 2.1 equiv) in 25 mL of toluene. The desired product was obtained (1.04 g, 76% yield) as a yellow green solid after purification by silica gel chromatography (PE/DCM = 10/1 to 5/1 to 1/1). ^{1}H-NMR (400 MHz, Chloroform-*d*) δ 8.61 (dd, *J* = 4.2, 1.5 Hz, 2H), 7.95 (d, *J* = 8.1 Hz, 2H), 7.32 (dd, *J* = 8.2, 4.2 Hz, 2H), 7.14–7.03 (m, 2H), 6.95–6.83 (m, 2H), 6.55 (s, 2H), 3.87–3.62 (m, 2H), 2.45–2.24 (m, 2H), 2.03–1.85 (m,

2H), 1.70–1.48 (m, 4H) (Figure S5); ^{13}C-NMR (100 MHz, Chloroform-*d*) δ 148.4, 144.9, 138.8, 136.9, 129.4 (q, J_{C-F} = 31.7 Hz), 127.5, 124.5 (q, J_{C-F} = 272.7 Hz), 122.3, 110.6 (q, J_{C-F} = 4.7 Hz), 100.0, 57.3, 32.3, 24.8 (Figure S6); ^{19}F NMR (376 MHz, Chloroform-*d*) δ –62.8 (Figure S7); HRMS (ESI$^+$) calcd for $C_{26}H_{23}F_6N_4$ [M + H]$^+$: 505.1821, found 505.1817; $[\alpha]_D^{24}$ = –329.1 (*c* = 1.0, CHCl$_3$); M. p. 120–124 °C.

(1R,2R)-N^1,N^2-Bis(6-(tert-butyl)quinolin-8-yl)cyclohexane-1,2-diamine (**L4**): Following the general procedure, the reaction was carried out with (1*R*,2*R*)-cyclohexane-1,2-diamine **3a** (0.68 g, 6.0 mmol, 1.0 equiv); Pd$_2$(dba)$_3$ (0.28 g, 0.3 mmol, 5 mol%); *rac*-BINAP (0.37 g, 0.6 mmol, 10 mol%); NaO^{t}Bu (1.73 g, 18 mmol, 3.0 equiv); and 8-bromo-6-(*tert*-butyl)quinoline **4d** (3.46 g, 13.1 mmol, 2.2 equiv) in 35 mL of toluene. The desired product was obtained (2.44 g, 85% yield) as a yellow solid after purification by silica gel chromatography (PE/DCM = 10/1 to PE/EA = 5/1). ^{1}H-NMR (400 MHz, Chloroform-*d*) δ 8.55 (dd, *J* = 4.3, 1.7 Hz, 2H), 7.99 (d, *J* = 7.5 Hz, 2H), 7.31–7.25 (m, 2H), 7.02–6.89 (m, 4H), 3.87–3.81 (m, 2H), 2.38 (d, *J* = 12.2 Hz, 2H), 1.91–1.83 (m, 2H), 1.71–1.50 (m, 4H), 1.36 (s, 18H) (Figure S8); ^{13}C-NMR (100 MHz, Chloroform-*d*) δ 150.6, 146.2, 143.5, 137.4, 136.0, 128.5, 121.3, 109.4, 104.3, 55.5, 35.2, 31.4, 30.7, 24.0 (Figure S9); HRMS (ESI$^+$) calcd for $C_{32}H_{41}N_4$ [M + H]$^+$: 481.3326, found 481.3323; $[\alpha]_D^{24}$ = –39.2 (*c* = 1.0, CHCl$_3$); M. p. 172–174 °C.

(1R,2R)-N^1,N^2-Di(acridin-4-yl)cyclohexane-1,2-diamine (**L5**): Following the general procedure, the reaction was carried out with (1*R*,2*R*)-cyclohexane-1,2-diamine **3a** (0.19 g, 1.6 mmol, 1.0 equiv); Pd$_2$(dba)$_3$ (0.08 g, 0.08 mmol, 5 mol%); *rac*-BINAP (0.10 g, 0.16 mmol, 10 mol%); NaO^{t}Bu (0.47 g, 4.9 mmol, 3.0 equiv); and 4-iodoacridine **4e** (1.07 g, 3.5 mmol, 2.2 equiv) in 30 mL of toluene. The desired product was obtained (0.46 g, 61% yield) as a yellow solid after purification by silica gel chromatography (PE/DCM = 2/1 to PE/DCM = 1/1 to DCM). ^{1}H-NMR (400 MHz, DMSO-*d$_6$*) δ 8.78 (s, 2H), 8.01 (d, *J* = 8.3 Hz, 2H), 7.91 (d, *J* = 8.7 Hz, 2H), 7.67 (t, *J* = 8.0 Hz, 2H), 7.49 (t, *J* = 8.0 Hz, 2H), 7.43 (t, *J* = 7.9 Hz, 2H), 7.18 (d, *J* = 8.4 Hz, 2H), 6.97 (d, *J* = 7.5 Hz, 2H), 6.75 (d, *J* = 7.0 Hz, 2H), 3.98–3.90 (m, 2H), 2.37–2.30 (m, 2H), 1.86–1.80 (m, 2H), 1.62–1.55 (m, 4H) (Figure S10); ^{13}C-NMR (100 MHz, Chloroform-*d*) δ 146.5, 144.2, 140.7, 135.1, 129.7, 128.9, 127.8, 127.3, 127.2, 127.0, 125.5, 113.8, 103.1, 56.6, 31.6, 24.4 (Figure S11); HRMS (ESI$^+$) calcd for $C_{32}H_{29}N_4$ [M + H]$^+$: 469.2387, found 469.2371; $[\alpha]_D^{24}$ = –678.0 (*c* = 0.5, CHCl$_3$); M. p. 198–202 °C.

(R)-N^2,N$^{2'}$-Di(quinolin-8-yl)-[1,1'-binaphthalene]-2,2'-diamine (**L6**) [53]: Following the general procedure, the reaction was carried out with (*R*)-[1,1'-binaphthalene]-2,2'-diamine **3b** (141.9 mg, 0.5 mmol, 1.0 equiv); Pd$_2$(dba)$_3$ (23.2 mg, 0.025 mmol, 5 mol%); *rac*-BINAP (31.4 mg, 0.05 mmol, 10 mol%); NaO^{t}Bu (148.2 mg, 1.5 mmol, 3.0 equiv); and 8-bromoqunoline **4a** (224.0 mg, 1.1 mmol, 2.2 equiv) in 10 mL of toluene. The desired product was obtained (196.6 mg, 73% yield) as a yellow solid after purification by recrystallization from EA. ^{1}H-NMR (400 MHz, Chloroform-*d*) δ 8.42 (d, *J* = 3.0 Hz, 2H), 7.99–7.95 (m, 4H), 7.94–7.86 (m, 6H), 7.37 (dt, *J* = 8.0, 4.0 Hz, 2H), 7.30–7.19 (m, 8H), 6.94–6.89 (m, 4H).

(12R)-N^{11},N^{12}-Di(quinolin-8-yl)-9,10-dihydro-9,10-ethanoanthracene-11,12-diamine (**L7**) [53]: Following the general procedure, the reaction carried out with (12*R*)-9,10-dihydro-9,10-ethanoanthracene-11,12-diamine **3c** (20 mg, 0.08 mmol, 1.0 equiv); Pd$_2$(dba)$_3$ (5.3 mg, 5 mol%); *rac*-BINAP (6.2 mg, 10 mol%); NaO^{t}Bu (25.5 mg, 0.26 mmol, 3.0 equiv); and 8-bromoqunoline **4a** (41.8 mg, 0.2 mmol, 2.2 equiv) in 1 mL of toluene. The desired product was obtained (33.4 mg, 85% yield) as a white solid after purification by silica gel chromatography (PE/EA = 5/1). ^{1}H-NMR (400 MHz, Chloroform-*d*) δ 8.61 (dd, *J* = 4.1, 1.4 Hz, 2H), 8.06 (d, *J* = 4.0 Hz, 2H), 7.44 (d, *J* = 7.1 Hz, 2H), 7.40–7.13 (m, 10H), 7.06 (d, *J* = 8.1 Hz, 2H), 6.91 (d, *J* = 8.0 Hz, 2H), 6.18 (s, 2H), 4.63 (s, 2H), 3.97 (s, 2H).

(1R,2R)-1,2-Diphenyl-N^1,N^2-di(quinolin-8-yl)ethane-1,2-diamine (**L8**) [58]: Following the general procedure, the reaction was carried out with (1*R*,2*R*)-1,2-diphenylethane-1,2-diamine **3d** (1.06 g, 5.0 mmol, 1.0 equiv); Pd$_2$(dba)$_3$ (0.23 g, 0.25 mmol, 5 mol%); *rac*-BINAP (0.33 g, 0.5 mmol, 10 mol%); NaO^{t}Bu (1.47 g, 15 mmol, 3.0 equiv); and 8-bromoqunoline **4a** (2.51 g, 12 mmol, 2.4 equiv) in 90 mL of

toluene. The desired product was obtained (1.47 g, 63% yield) as a white solid after purification by silica gel chromatography (PE/DCM = 30/1 to PE/EA = 5/1). ^{1}H-NMR (400 MHz, Chloroform-*d*) δ 8.67 (dd, *J* = 4.3, 1.6 Hz, 2H), 8.05 (d, *J* = 8.4 Hz, 2H), 7.36 (m, 4H), 7.28–7.09 (m, 12H), 7.03 (d, *J* = 8.0 Hz, 2H), 6.50 (d, *J* = 7.6 Hz, 2H), 5.01 (s, 2H).

3.3. Preparation of Substituted Quinolines

General procedure for synthesis of substituted 8-bromoquinoline (Scheme 3): to a 50 mL round bottom flask was added 4-substituted 2-bromoaniline, glycerol (17.0 equiv), *m*-nitrobenzenesulfonate sodium (1.2 equiv), FeSO$_4$•7H$_2$O (0.05 equiv), and MsOH. The reaction mixture was heated at 125 °C for 24 h. After cooling to room temperature, aqueous NaOH solution (2.5 M) was added to the reaction mixture to adjust pH to 12. Then EtOH was added to form a black solution, which was extracted with EA or DCM (3 × 100 mL). The combined organic phase was washed with H$_2$O (100 mL), brine (100 mL), and dried with anhydrous Na$_2$SO$_4$. After removing the solvents, the residue was purified by silica gel chromatography.

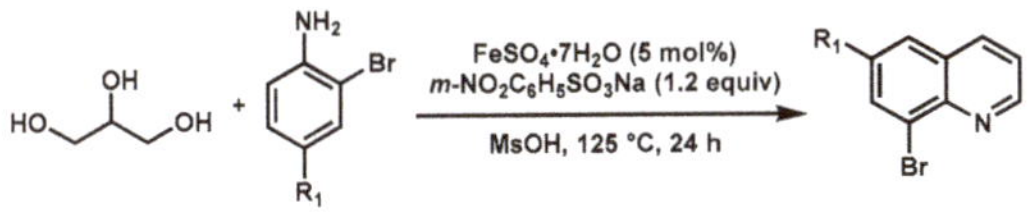

Scheme 3. Synthesis of Substituted Quinolines (Skraup Reaction).

8-Bromo-6-fluoroquinoline (**4b**) [59]: Following the general procedure, the reaction was carried out with 2-bromo-4-fluoroaniline (1.57 g, 8.3 mmol, 1.0 equiv); glycerol (11 mL, 149.0 mmol, 18.0 equiv); *m*-nitrobenzenesulfonate sodium (2.24 g, 10.0 mmol, 1.2 equiv); FeSO$_4$•7H$_2$O (0.12 g, 0.4 mmol, 0.05 equiv), and MsOH (11 mL). The desired product was obtained (0.86 g, 43% yield) as a pale yellow solid after purification by silica gel chromatography (PE/EA = 10/1 to 5/1). ^{1}H-NMR (400 MHz, Chloroform-*d*) δ 9.02 (dd, *J* = 4.2, 1.2 Hz, 1H), 8.14 (dd, *J* = 8.3, 1.6 Hz, 1H), 7.90 (dd, *J* = 8.1, 2.7 Hz, 1H), 7.50 (ddd, *J* = 8.3, 4.2, 0.6 Hz, 1H), 7.46 (dd, *J* = 8.3, 2.7 Hz, 1H).

8-Bromo-6-(trifluoromethyl)quinoline (**4c**): Following the general procedure, the reaction was carried out with 2-bromo-4-(trifluoromethyl)aniline (6.38 g, 26.6 mmol, 1.0 equiv); glycerol (20 mL, 271.5 mmol, 10.0 equiv); *m*-nitrobenzenesulfonate sodium (7.19 g, 32.0 mmol, 1.2 equiv); FeSO$_4$•7H$_2$O (0.37 g, 1.3 mmol, 0.05 equiv), and MsOH (35 mL). The desired product was obtained (1.57 g, 21% yield) as a pale orange solid after purification by silica gel chromatography (PE/EA = 10/1 to 5/1). ^{1}H-NMR (400 MHz, Chloroform-*d*) δ 9.16 (dd, *J* = 4.2, 1.7 Hz, 1H), 8.28 (dd, *J* = 8.3, 1.7 Hz, 1H), 8.24 (d, *J* = 1.9 Hz, 1H), 8.16–8.10 (m, 1H), 7.60 (dd, *J* = 8.3, 4.2 Hz, 1H) (Figure S12); ^{13}C-NMR (100 MHz, Chloroform-*d*) δ 153.4, 146.5, 137.7, 129.1(q, *J*$_{C-F}$ = 33.4 Hz), 129.0(q, *J*$_{C-F}$ = 3.1 Hz), 128.4, 126.3, 125.7 (q, *J*$_{C-F}$ = 4.3 Hz), 123.2, 123.2(q, *J*$_{C-F}$ = 272.8 Hz) (Figure S13); ^{19}F NMR (376 MHz, Chloroform-*d*) δ −62.5 (Figure S14); HRMS (ESI$^+$) calcd for C$_{10}$H$_6$BrF$_3$N [M + H]$^+$: 275.9630, found 275.9620; M. p. 58–62 °C.

8-Bromo-6-(tert-butyl)quinoline (**4d**) [60]: Following the general procedure, the reaction was carried out with 2-bromo-4-(*tert*-butyl)aniline (1.78 g, 7.8 mmol, 1.0 equiv); glycerol (10 mL, 135.7 mmol, 17.0 equiv); *m*-nitrobenzenesulfonate sodium (2.11 g, 9.4 mmol, 1.2 equiv); FeSO$_4$•7H$_2$O (0.11 g, 0.41 mmol, 0.05 equiv); and MsOH (10 mL). The desired product was obtained (1.73 g, 84% yield) as a yellow solid after purification by silica gel chromatography (PE/EA = 20/1). ^{1}H-NMR (400 MHz, Chloroform-*d*) δ 9.00 (dd, *J* = 4.2, 1.5 Hz, 1H), 8.18–8.12 (m, 2H), 7.70 (d, *J* = 2.0 Hz, 1H), 7.45 (dd, *J* = 8.2, 4.2 Hz, 1H), 1.42 (s, 9H).

4-Iodoacridine [**4e**] (Scheme 4)

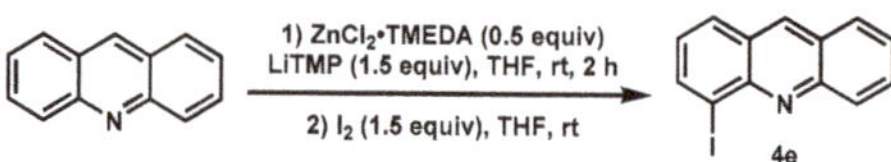

Scheme 4. Synthesis of 4-Iodoacridine.

To a 100 mL Schlenk flask were added TMP (1.14 g, 8.1 mmol, 1.5 equiv) and 20 mL of THF. The solution was cooled to 0 °C, and ^{n}BuLi (2.4 M, 4 mL, 9.6 mmol, 1.7 equiv) was added dropwise by syringe. Upon the completion of the addition, the mixture was stirred at 0 °C for another 30 min. Then ZnCl$_2$•TMEDA (0.68 g, 2.7 mmol, 0.5 equiv) was added at 0 °C, and the resultant mixture was stirred for 20 min before acridine (0.98 g, 5.5 mmol, 1.0 equiv) was added. After the reaction was warmed up to 25 °C, I$_2$ (2.17 g, 8.5 mmol, 1.5 equiv) in THF (20 mL) was added dropwise. The reaction mixture was stirred for 2 h and then quenched with saturated Na$_2$S$_2$O$_3$ solution and extracted with EA (3 × 30 mL). The combined organic phase was washed with brine and dried over anhydrous Na$_2$SO$_4$. The desired product was obtained (1.08 g, 64% yield) as a yellow solid after purification by silica gel chromatography (PE/DCM = 20/1 to 5/1). ^{1}H-NMR (400 MHz, Chloroform-*d*) δ 8.72 (s, 1H), 8.46 (dd, *J* = 7.1, 1.1 Hz, 1H), 8.39 (d, *J* = 8.8 Hz, 1H), 8.09–7.93 (m, 2H), 7.83 (ddd, *J* = 8.5, 6.6, 1.3 Hz, 1H), 7.63–7.52 (m, 1H), 7.31–7.19 (m, 1H) (Figure S15); ^{13}C-NMR (100 MHz, Chloroform-*d*) δ 149.8, 147.1, 141.0, 137.2, 130.9, 130.1, 129.4, 127.9, 127.2, 126.7, 126.63, 126.58, 104.0 (Figure S16); HRMS (ESI$^+$) calcd for C$_{13}$H$_9$IN [M + H]$^+$: 305.9774, found 305.9763; M.p. 100–104 °C.

2-Methoxynaphthalene [61]: Following the reported procedure, the reaction was carried out with 2-naphthol (**1.14** g, 10 mmol, 1.0 equiv); NaH (60% wt, 0.41 g, 17 mmol, 1.7 equiv); and MeI (1.76 g, 12 mmol, 1.2 equiv) in 10 mL of DMF. The desired product was obtained (1.28 g, 81% yield) as a white solid after purification by silica gel chromatography (DCM). ^{1}H-NMR (400 MHz, Chloroform-*d*) δ 7.78 (dd, *J* = 11.5, 8.4 Hz, 3H), 7.47 (ddd, *J* = 8.1, 6.8, 1.3 Hz, 1H), 7.36 (ddd, *J* = 8.1, 6.8, 1.2 Hz, 1H), 7.22–7.13 (m, 2H), 3.94 (s, 3H).

2-Bromo-3-methoxynaphthalene [62]: Following the reported procedure, the reaction was carried out with 2-methoxynaphthalene (0.79 g, 5.0 mmol, 1.0 equiv); ^{n}BuLi solution (1.67 M in hexane, 3 mL, 5.3 mmol, 1.1 equiv); and 1,2-dibromoethane (1.30 g, 6.9 mmol, 1.3 equiv) in 10 mL of THF. The desired product was obtained (0.87 g, 73% yield) as a white solid after recrystallization from hot hexane for 3 times. ^{1}H-NMR (400 MHz, Chloroform-*d*) δ 8.06 (s, 1H), 7.71 (dd, *J* = 13.0, 8.2 Hz, 2H), 7.51–7.42 (m, 1H), 7.40–7.32 (m, 1H), 7.16 (s, 1H), 4.01 (s, 3H).

3-Methoxynaphthalen-2-ol (**1b**) [63]: Following the reported procedure, the reaction was carried out with naphthalene-2, 3-diol (1.60 g, 10 mmol, 1.0 equiv); K$_2$CO$_3$ (1.81 g, 13 mmol, 1.3 equiv); and MeI (1.73 g, 12 mmol, 1.2 equiv) in 10 mL of acetone. The desired product was obtained (0.57 g, 33% yield) as a white solid after purification by silica gel chromatography (PE/EA = 50/1 to PE/EA = 20/1 to PE/EA = 10/1). ^{1}H-NMR (400 MHz, Chloroform-*d*) δ 8.06 (s, 1H), 7.71 (dd, *J* = 13.0, 8.2 Hz, 2H), 7.50–7.42 (m, 1H), 7.40–7.33 (m, 1H), 7.16 (s, 1H), 4.01 (s, 3H).

3-(Benzyloxy)naphthalen-2-ol (**1c**) [64]: Following the reported procedure, the reaction was carried out with naphthalene-2, 3-diol (1.61 g, 10 mmol, 1.0 equiv); K$_2$CO$_3$ (1.82 g, 13 mmol, 1.3 equiv); and BnBr (2.58 g, 15 mmol, 1.5 equiv) in 20 mL of DMF. The desired product was obtained (0.88 g, 35% yield) as a yellow solid after purification by silica gel chromatography (PE/EA = 50/1 to 20/1 to 10/1). ^{1}H-NMR (400 MHz, Chloroform-*d*) δ 7.71–7.64 (m, 2H), 7.52–7.31 (m, 7H), 7.29 (s, 1H), 7.22 (s, 1H), 5.97 (s, 1H), 5.24 (s, 2H).

General procedure (Scheme 5): To a Schlenk flask were added 2-bromo-3-methoxynaphthalene (1.0 equiv), aryl boronic acid (2.2 equiv), K_2CO_3 (3.0 equiv), $Pd(PPh_3)_4$ (2.5 mol%), and degassed EtOH/toluene/water (1/1/1) under Ar atmosphere. The mixture was heated at 90 °C until the completion of the reaction. Then the mixture was cooled to room temperature, and DCM was added. The mixture was washed with NaOH solution (20% wt), and the aqueous phase was extracted with DCM (2 × 20 mL). The combined organic phase was washed with brine (20 mL) and dried over anhydrous $MgSO_4$. After removing the solvent, the residue was dissolved in anhydrous DCM. The solution was cooled to −78 °C, and BBr_3 (1 M in DCM, 5.0 equiv) was added slowly by syringe. Then the mixture was warmed up to room temperature and stirred until the complete consumption of the starting material. The mixture was poured into the ice water (50 mL) and extracted with DCM (3 × 50 mL). The combined organic phase was washed with brine (100 mL) and dried over anhydrous Na_2SO_4. After removing the solvent, the residue was purified by silica gel chromatography to give the desired product.

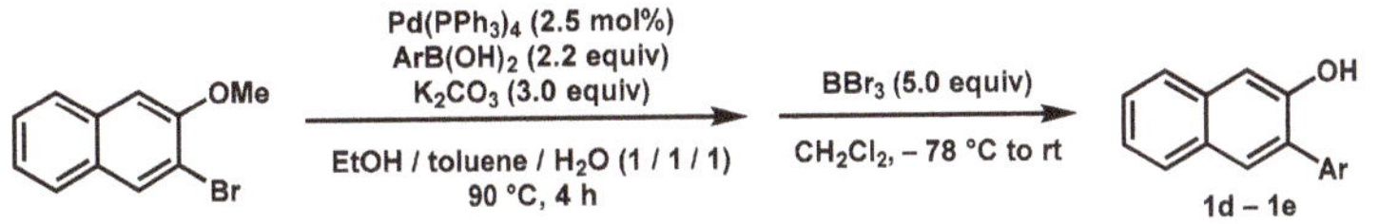

Scheme 5. Synthesis of Substrates **1d–1e**.

3-(o-Tolyl)naphthalen-2-ol (**1d**) [65]: Following the general procedure, the reaction was carried out with 2-bromo-3-methoxynaphthalene (236.2 mg, 1.0 mmol, 1.0 equiv); *o*-tolylboronic acid (304.8 mg, 2.2 mmol, 2.2 equiv); K_2CO_3 (417.7 mg, 3.0 mmol, 3.0 equiv); and $Pd(PPh_3)_4$ (29.9 mg, 2.5 mol%) in 6 mL of degassed solvents. Then BBr_3 (1 M in DCM, 5 mL, 5 mmol, 5.0 equiv) was used to remove the methyl group. The desired product was obtained (185.1 mg, 79% yield overall) as a brown sticky liquid after purification by silica gel chromatography (PE/DCM = 10/1). ^{1}H-NMR (400 MHz, Chloroform-*d*) δ 7.80–7.72 (m, 2H), 7.63 (s, 1H), 7.45 (ddd, *J* = 8.3, 6.9, 1.2 Hz, 1H), 7.44–7.28 (m, 6H), 4.92 (s, 1H), 2.20 (s, 3H).

3-Phenylnaphthalen-2-ol (**1e**) [66]: Following the general procedure, the reaction was carried out with 2-bromo-3-methoxynaphthalene (0.71 g, 3.0 mmol, 1.0 equiv); phenylboronic acid (0.55 g, 4.5 mmol, 1.5 equiv); K_2CO_3 (1.90 g, 13.8 mmol, 4.5 equiv); and $Pd(PPh_3)_4$ (0.09 g, 2.5 mol%) in 30 mL of degassed solvent. Then BBr_3 (1 M in DCM, 15 mL, 15 mmol, 5.0 equiv) was used to remove the methyl group. The desired product was obtained (0.63 g, 95% yield overall) as a pale brown solid after purification by silica gel chromatography (DCM). ^{1}H-NMR (400 MHz, Chloroform-*d*) δ 7.81–7.76 (m, 1H), 7.74 (d, *J* = 7.2 Hz, 2H), 7.60–7.51 (m, 4H), 7.49–7.41 (m, 2H), 7.39–7.32 (m, 2H), 5.30 (s, 1H).

3-Bromonaphthalen-2-ol (**1f**) [67]: Following the general procedure, the reaction was carried out with 2-bromo-3-methoxynaphthalene (240.6 mg, 1.0 mmol, 1.0 equiv) and BBr_3 (1 M in DCM, 5 mL, 5.0 mmol, 5.0 equiv). The desired product was obtained (226.0 mg, quantitative yield) as a white solid after purification by silica gel chromatography (DCM). ^{1}H-NMR (400 MHz, Chloroform-*d*) δ 8.03 (s, 1H), 7.69 (ddt, *J* = 7.4, 2.2, 1.2 Hz, 2H), 7.45 (ddd, *J* = 8.2, 6.8, 1.3 Hz, 1H), 7.39 (s, 1H), 7.35 (ddd, *J* = 8.2, 6.8, 1.2 Hz, 1H), 5.64 (s, 1H).

6-Phenylnaphthalen-2-ol (**1k**) [5]: Following the general procedure, the reaction was carried out with 6-bromonaphthalen-2-ol (1.12 g, 5.0 mmol, 1.0 equiv); phenylboronic acid (0.73 g, 6.0 mmol, 1.2 equiv); K_2CO_3 (3.00 g, 21.8 mmol, 4.4 equiv); and $Pd(PPh_3)_4$ (0.15 g, 2.5 mol%) in 30 mL of degassed solvent. The desired product was obtained (0.77 g, 70% yield) as a white solid after purification by silica gel chromatography (DCM). ^{1}H-NMR (400 MHz, Chloroform-*d*) δ 7.98 (d, *J* = 1.7 Hz, 1H), 7.82 (d, *J* =

(S)-7,7′-Bis(benzyloxy)-[1,1′-binaphthalene]-2,2′-diol (**2l**) [77]: The reaction was conducted with Fe(ClO$_4$)$_2$ (12.9 mg, 10 mol%) and **L1** (9.0 mg, 5 mol%); 7-(benzyloxy)naphthalen-2-ol (125.4 mg, 0.5 mmol, 1.0 equiv); and MS 4Å (160.6 mg). The desired product was obtained (88.5 mg, 71% yield) as a pale yellow solid after purification by silica gel chromatography (PE to PE/EA = 5/1). 58:42 er (HPLC: Chiralcel OD-H, hexane/propan-2-ol = 85/15, 1.0 mL/min, λ = 230 nm, t$_R$ (min): major = 14.8, minor = 26.7). ^{1}H-NMR (400 MHz, Chloroform-*d*) δ 7.89 (d, *J* = 8.8 Hz, 2H), 7.80 (d, *J* = 8.9 Hz, 2H), 7.22 (dt, *J* = 6.1, 2.2 Hz, 8H), 7.16 (dd, *J* = 6.6, 3.1 Hz, 4H), 7.11 (dd, *J* = 8.9, 2.5 Hz, 2H), 6.49 (d, *J* = 2.4 Hz, 2H), 4.99 (s, 2H), 4.80 (d, *J* = 11.7 Hz, 2H), 4.74 (d, *J* = 11.7 Hz, 2H).

(S)-7,7′-Dibutoxy-[1,1′-binaphthalene]-2,2′-diol (**2m**) [29]: The reaction was conducted with Fe(ClO$_4$)$_2$ (12.3 mg, 10 mol%) and **L1** (9.0 mg, 5 mol%); 7-butoxynaphthalen-2-ol (108.5 mg, 0.5 mmol, 1.0 equiv); and MS 4Å (150.0 mg). The desired product was obtained (106.5 mg, 99% yield) as a white solid after purification by silica gel chromatography (PE/EA = 10/1 to 5/1). 60:40 er (HPLC: Chiralpak AD-H, hexane/propan-2-ol = 90/10, 1.0 mL/min, λ = 254 nm, t$_R$ (min): major = 8.3, minor = 19.3). ^{1}H-NMR (400 MHz, Chloroform-*d*) δ 7.84 (d, *J* = 8.8 Hz, 2H), 7.76 (d, *J* = 8.9 Hz, 2H), 7.20 (d, *J* = 8.8 Hz, 2H), 7.03 (dd, *J* = 8.9, 2.4 Hz, 2H), 6.48 (d, *J* = 1.9 Hz, 2H), 5.08 (s, 2H), 3.71 (ddt, *J* = 27.6, 9.3, 6.5 Hz, 4H), 1.67–1.53 (m, 4H), 1.34 (tt, *J* = 16.4, 8.3 Hz, 4H), 0.86 (t, *J* = 7.4 Hz, 6H).

(S)-7,7′-Bis((tert-butyldimethylsilyl)oxy)-[1,1′-binaphthalene]-2,2′-diol (**2n**): The reaction was conducted with Fe(ClO$_4$)$_2$ (12.6 mg, 10 mol%) and **L1** (9.8 mg, 5 mol%); 7-((*tert*-butyldimethylsilyl)oxy) naphthalen-2-ol (138.2 mg, 0.5 mmol, 1.0 equiv); and MS 4Å (153.7 mg). The desired product was obtained (103.8 mg, 76% yield) as a brown solid after purification by silica gel chromatography (PE to PE/EA = 5/1). 68:32 er (HPLC: Chiralpak IC, hexane/propan-2-ol = 97/3, 1.0 mL/min, λ = 254 nm, t$_R$ (min): major = 6.0, minor = 8.5). ^{1}H-NMR (400 MHz, Chloroform-*d*) δ 7.86 (dd, *J* = 9.0, 0.7 Hz, 2H), 7.75 (d, *J* = 8.8 Hz, 2H), 7.21 (d, *J* = 8.9 Hz, 2H), 6.95 (dd, *J* = 8.8, 2.4 Hz, 2H), 6.46 (d, *J* = 2.4 Hz, 2H), 5.07 (s, 2H), 0.83 (s, 18H), −0.03 (s, 6H), −0.06 (s, 6H) (Figure S17); ^{13}C-NMR (100 MHz, Chloroform-*d*) δ 155.3, 153.2, 134.9, 131.1, 129.9, 125.1, 119.9, 115.3, 111.7, 109.8, 25.8, 18.3, −4.5 (Figure S18); HRMS (ESI$^-$), *m/z* calc'd for C$_{32}$H$_{41}$O$_4$Si$_2$ [M − H]$^-$: 545.2549, found 545.2578; $[\alpha]_D^{24}$ = 56.6 (*c* = 1.0, CHCl$_3$); M.p. 118–122 °C.

(S)-7,7′-Dimethoxy-[1,1′-binaphthalene]-2,2′-diol (**2o**) [78]: The reaction was conducted with Fe(ClO$_4$)$_2$ (12.4 mg, 10 mol%) and **L1** (9.4 mg, 5 mol%); 7-methoxynaphthalen-2-ol (87.3 mg, 0.5 mmol, 1.0 equiv); and MS 4Å (156.3 mg). The desired product was obtained (56.3 mg, 65% yield) as a white solid after purification by silica gel chromatography (PE to PE/EA = 5/1). 59:41 er (HPLC: Chiralcel OD-H, hexane/propan-2-ol = 85/15, 1.0 mL/min, λ = 230 nm, t$_R$ (min): major = 11.1, minor = 18.8). ^{1}H-NMR (400 MHz, Chloroform-*d*) δ 7.88 (dd, *J* = 9.0, 0.7 Hz, 2H), 7.78 (d, *J* = 8.9 Hz, 2H), 7.22 (d, *J* = 8.8 Hz, 2H), 7.03 (dd, *J* = 8.9, 2.5 Hz, 2H), 6.48 (d, *J* = 2.5 Hz, 2H), 3.57 (s, 6H).

4. Conclusions

In summary, we have developed an iron/bisquinolyldiamine-catalyzed asymmetric oxidative coupling of 2-naphthols. This method employs in situ-formed iron complexes from Fe(ClO$_4$)$_2$ and readily available ligand **L1** and uses 1 atm oxygen as the oxidant. The atom economy of this transformation, the easily available catalyst, and operationally simple procedure provide new applications of asymmetric iron catalysis. Further studies on synthesizing a library of nitrogen ligands and extending their applications are underway in our laboratory.

Supplementary Materials: The following are available online. NMR and HPLC spectra.

Author Contributions: L.-Y.W. performed the experiments; L.-Y.W., M.U., and W.-B.L. wrote the paper; and W.-B.L. conceived and designed the experiments. All authors have read and agreed to the published version of the manuscript.

Funding: We thank the National Natural Science Foundation of China (21602160 and 21772148) and Wuhan University for financial support.

Conflicts of Interest: The authors declare no conflicts of interest.

References

1. Wang, Y.-B.; Tan, B. Construction of Axially Chiral Compounds via Asymmetric Organocatalysis. *Acc. Chem. Res.* **2018**, *51*, 534–547. [CrossRef] [PubMed]

2. Wencel-Delord, J.; Panossian, A.; Leroux, F.R.; Colobert, F. Recent Advances and New Concepts for the Synthesis of Axially Stereoenriched Biaryls. *Chem. Soc. Rev.* **2015**, *44*, 3418–3430. [CrossRef] [PubMed]

3. Moliterno, M.; Cari, R.; Puglisi, A.; Antenucci, A.; Sperandio, C.; Moretti, E.; Sabato, A.D.; Salvio, R.; Bella, M. Quinine-Catalyzed Asymmetric Synthesis of 2,2'-Binaphthol-type Biaryls under Mild Reaction Conditions. *Angew. Chem. Int. Ed.* **2016**, *128*, 6635–6639. [CrossRef]

4. Wang, J.-Z.; Zhou, J.; Xu, C.; Sun, H.; Kürti, L.; Xu, Q.-L. Symmetry in Cascade Chirality-Transfer Processes: A Catalytic Atroposelective Direct Arylation Approach to BINOL Derivatives. *J. Am. Chem. Soc.* **2016**, *138*, 5202–5205. [CrossRef]

5. Chen, Y.-H.; Cheng, D.-J.; Zhang, J.; Wang, Y.; Liu, X.-Y.; Tan, B. Atroposelective Synthesis of Axially Chiral Biaryldiols via Organocatalytic Arylation of 2-Naphthols. *J. Am. Chem. Soc.* **2015**, *137*, 15062–15065. [CrossRef]

6. Wang, H. Recent Advances in Asymmetric Oxidative Coupling of 2-Naphthol and Its Derivatives. *Chirality* **2010**, *22*, 827–837. [CrossRef]

7. Brunel, J.M. BINOL: A Versatile Chiral Reagent. *Chem. Rev.* **2005**, *105*, 857–898. [CrossRef]

8. Chen, Y.; Yekta, S.; Yudin, A.K. Modified BINOL Ligands in Asymmetric Catalysis. *Chem. Rev.* **2003**, *103*, 3155–3212. [CrossRef]

9. Pu, L. 1,1'-Binaphthyl Dimers, Oligomers, and Polymers: Molecular Recognition, Asymmetric Catalysis, and New Materials. *Chem. Rev.* **1998**, *98*, 2405–2494. [CrossRef]

10. Tomino, I.; Tanimoto, Y.; Noyori, R. Virtually Complete Enantioface Differentiation in Carbonyl Group Reduction by a Complex Aluminum Hydride Reagent. *J. Am. Chem. Soc.* **1979**, *101*, 3129–3131.

11. Berthod, M.; Mignani, G.; Woodward, G.; Lemaire, M. Modified BINAP: The How and the Why. *Chem. Rev.* **2005**, *105*, 1801–1836. [CrossRef] [PubMed]

12. Kočovský, P.; Vyskočil, Š.; Smrčina, M. Non-Symmetrically Substituted 1,1'-Binaphthyls in Enantioselective Catalysis. *Chem. Rev.* **2003**, *103*, 3213–3245. [CrossRef] [PubMed]

13. Parmar, D.; Sugiono, E.; Raja, S.; Rueping, M. Complete Field Guide to Asymmetric BINOL-Phosphate Derived Brønsted Acid and Metal Catalysis: History and Classification by Mode of Activation; Brønsted Acidity, Hydrogen Bonding, Ion Pairing, and Metal Phosphates. *Chem. Rev.* **2014**, *114*, 9047–9153. [CrossRef] [PubMed]

14. Zhang, Z.-F.; Xie, F.; Yang, B.; Yu, H.; Zhang, W.-B. Chiral Phosphoramidite Ligand and Its Application in Asymmetric Catalysis. *Chin. J. Org. Chem.* **2011**, *31*, 429–442.

15. Hatano, M.; Ishihara, K. Chiral 1,1'-Binaphthyl-2,2'-Disulfonic Acid (BINSA) and Its Derivatives for Asymmetric Catalysis. *Asian J. Org. Chem.* **2014**, *3*, 352–365. [CrossRef]

16. Komanduri, V.; Krische, M.J. Enantioselective Reductive Coupling of 1,3-Enynes to Heterocyclic Aromatic Aldehydes and Ketones via Rhodium-Catalyzed Asymmetric Hydrogenation: Mechanistic Insight into the Role of Brønsted Acid Additives. *J. Am. Chem. Soc.* **2006**, *128*, 16448–16449. [CrossRef]

17. Costa, A.L.; Piazza, M.G.; Tagliavini, E.; Trombini, C.; Ronchi, A.U. Catalytic Asymmetric Synthesis of Homoallylic Alcohols. *J. Am. Chem. Soc.* **1993**, *115*, 7001–7002. [CrossRef]

18. Sakane, S.; Fujiwara, J.; Maruoka, K.; Yamamoto, H. Chiral Leaving Group. Biogenetic-Type Asymmetric Synthesis of Limonene and Bisabolenes. *J. Am. Chem. Soc.* **1983**, *105*, 6154–6155. [CrossRef]

19. Mukaiyama, T.; Inubushi, A.; Suda, S.; Hara, R.; Kobayashi, S. [1,1'-Bi-2-naphthalenediolato(2-)-O,O'] oxotitanium. An Efficient Chiral Catalyst for the Asymmetric Aldol Reaction of Silyl Enol Ethers with Aldehydes. *Chem. Lett.* **1990**, *19*, 1015–1018. [CrossRef]

20. Qi, L.-W.; Mao, J.-H.; Zhang, J.; Tan, B. Organocatalytic Asymmetric Arylation of Indoles Enabled by Azo Groups. *Nat. Chem.* **2018**, *10*, 58–64. [CrossRef]

21. Takizawa, S. Development of Dinuclear Vanadium Catalysts for Enantioselective Coupling of 2-Naphthols via a Dual Activation Mechanism. *Chem. Pharm. Bull.* **2009**, *57*, 1179–1188. [CrossRef]

22. Chu, C.-Y.; Hwang, D.R.; Wang, S.-K.; Uang, B.J. Chiral Oxovanadium Complex Catalyzed Enantioselective Oxidative Coupling of 2-Naphthols. *Chem. Commun.* **2001**, *37*, 980–981. [CrossRef]

23. Barhate, N.B.; Chen, C.-T. Catalytic Asymmetric Oxidative Couplings of 2-Naphthols by Tridentate *N*-Ketopinidene-Based Vanadyl Dicarboxylates. *Org. Lett.* **2002**, *4*, 2529–2532. [CrossRef]

24. Guo, Q.-X.; Wu, Z.-J.; Luo, Z.-B.; Liu, Q.-Z.; Ye, J.-L.; Luo, S.-W.; Cun, L.-F.; Gong, L.-Z. Highly Enantioselective Oxidative Couplings of 2-Naphthols Catalyzed by Chiral Bimetallic Oxovanadium Complexes with either Oxygen or Air as Oxidant. *J. Am. Chem. Soc.* **2007**, *129*, 13927–13938. [CrossRef]

25. Luo, Z.; Liu, Q.; Gong, L.-Z.; Cui, X.; Mi, A.; Jiang, Y. Novel Achiral Biphenol-Derived Diastereomeric Oxovanadium(IV) Complexes for Highly Enantioselective Oxidative Coupling of 2-Naphthols. *Angew. Chem. Int. Ed.* **2002**, *41*, 4532–4535. [CrossRef]

26. Luo, Z.-B.; Liu, Q.-Z.; Gong, L.-Z. The Rational Design of Novel Chiral Oxovanadium(IV) Complexes for Highly Enantioselective Oxidative Coupling of 2-Naphthols. *Chem. Commun.* **2002**, *38*, 914–915. [CrossRef] [PubMed]

27. Sako, M.; Takizawa, S.; Yoshida, Y.; Sasai, H. Enantioselective and Aerobic Oxidative Coupling of 2-Naphthol Derivatives Using Chiral Dinuclear Vanadium(V) Complex in Water. *Tetrahedron Asymmetry* **2015**, *26*, 613–616. [CrossRef]

28. Takizawa, S.; Katayama, T.; Sasai, H. Dinuclear Chiral Vanadium Catalysts for Oxidative Coupling of 2-Naphtholsvia a Dual Activation Mechanism. *Chem. Commun.* **2008**, *44*, 4113–4122. [CrossRef]

29. Takizawa, S.; Katayama, T.; Somei, H.; Asano, Y.; Yoshida, T.; Kameyama, C.; Rajesh, D.; Onitsuka, K.; Suzuki, T.; Mikami, M.; et al. Dual Activation in Oxidative Coupling of 2-Naphthols Catalyzed by Chiral Dinuclear Vanadium Complexes. *Tetrahedron* **2008**, *64*, 3361–3371. [CrossRef]

30. Nakajima, M.; Miyoshi, I.; Kanayama, K.; Hashimoto, S.; Noji, M.; Koga, K. Enantioselective Synthesis of Binaphthol Derivatives by Oxidative Coupling of Naphthol Derivatives Catalyzed by Chiral Diamine•Copper Complexes. *J. Org. Chem.* **1999**, *64*, 2264–2271. [CrossRef]

31. Gao, J.; Reibenspies, J.H.; Martell, A.E. Structurally Defined Catalysts for Enantioselective Oxidative Coupling Reactions. *Angew. Chem. Int. Ed.* **2003**, *42*, 6008–6012. [CrossRef] [PubMed]

32. Kozlowski, M.C.; Morgan, B.J.; Linton, E.C. Total Synthesis of Chiral Biaryl Natural Products by Asymmetric Biaryl Coupling. *Chem. Soc. Rev.* **2009**, *38*, 3193–3207. [CrossRef] [PubMed]

33. Hewgley, J.B.; Stahl, S.S.; Kozlowski, M.C. Mechanistic Study of Asymmetric Oxidative Biaryl Coupling: Evidence for Self-Processing of the Copper Catalyst to Achieve Control of Oxidase vs. Oxygenase Activity. *J. Am. Chem. Soc.* **2008**, *130*, 12232–12233. [CrossRef] [PubMed]

34. Podlesny, E.E.; Kozlowski, M.C. Enantioselective Total Synthesis of (*S*)-Bisoranjidiol, an Axially Chiral Bisanthraquinone. *Org. Lett.* **2012**, *14*, 1408–1411. [CrossRef]

35. O'Brien, E.M.; Morgan, B.J.; Mulrooney, C.A.; Carroll, P.J.; Kozlowski, M.C. Perylenequinone Natural Products: Total Synthesis of Hypocrellin A. *J. Org. Chem.* **2010**, *75*, 57–68. [CrossRef]

36. Li, X.; Hewgley, J.B.; Mulrooney, C.A.; Yang, J.; Kozlowski, M.C. Enantioselective Oxidative Biaryl Coupling Reactions Catalyzed by 1,5-Diazadecalin Metal Complexes: Efficient Formation of Chiral Functionalized BINOL Derivatives. *J. Org. Chem.* **2003**, *68*, 5500–5511. [CrossRef]

37. Mulrooney, C.A.; Li, X.; DiVirgilio, E.S.; Kozlowski, M.C. General Approach for the Synthesis of Chiral Perylenequinones via Catalytic Enantioselective Oxidative Biaryl Coupling. *J. Am. Chem. Soc.* **2003**, *125*, 6856–6857. [CrossRef]

38. Li, X.; Yang, J.; Kozlowski, M.C. Enantioselective Oxidative Biaryl Coupling Reactions Catalyzed by1,5-Diazadecalin Metal Complexes. *Org. Lett.* **2001**, *3*, 1137–1140. [CrossRef]

39. Temma, T.; Hatano, B.; Habaue, S. Cu(I)-Catalyzed Asymmetric Oxidative Cross-Coupling of 2-Naphthol Derivatives. *Tetrahedron* **2006**, *62*, 8559–8563. [CrossRef]

40. Temma, T.; Habaue, S. Highly Selective Oxidative Cross-Coupling of 2-Naphthol Derivatives with Chiral Copper(I)–Bisoxazoline Catalysts. *Tetrahedron Lett.* **2005**, *46*, 5655–5657. [CrossRef]

41. Prause, F.; Arensmeyer, B.; Fröhlich, B.; Breuning, M. In-Depth Structure–Selectivity Investigations on Asymmetric, Copper-Catalyzed Oxidative Biaryl Coupling in the Presence of 5-*cis*-Substituted Prolinamines. *Catal. Sci. Technol.* **2015**, *5*, 2215–2226. [CrossRef]

42. Kim, K.-H.; Lee, D.-W.; Lee, Y.-S.; Ko, D.-H.; Ha, D.-C. Enantioselective Oxidative Coupling of Methyl 3-Hydroxy-2-naphthoate Using Mono-*N*-alkylated Octahydrobinaphthyl-2,2′-Diamine Ligand. *Tetrahedron* **2004**, *60*, 9037–9042. [CrossRef]

43. Alamsetti, S.K.; Poonguzhali, E.; Ganapathy, D.; Sekar, G. Enantioselective Oxidative Coupling of 2-Naphthol Derivatives by Copper-(*R*)-1,1′-Binaphthyl-2,2′-Diamine-TEMPO Catalyst. *Adv. Synth. Catal.* **2013**, *355*, 2803–2808. [CrossRef]

44. Zhang, Q.; Cui, X.; Chen, L.; Liu, H.; Wu, Y. Syntheses of Chiral Ferrocenophanes and Their Application to Asymmetric Catalysis. *Eur. J. Org. Chem.* **2014**, *2014*, 7823–7829. [CrossRef]

45. Tian, J.-M.; Wang, A.-F.; Yang, J.-S.; Zhao, X.-J.; Tu, Y.-Q.; Zhang, S.-Y.; Chen, Z.-M. Copper-Complex-Catalyzed Asymmetric Aerobic Oxidative Cross-Coupling of 2-Naphthols: Enantioselective Synthesis of 3,3′-Substituted *C1*-Symmetric BINOLs. *Angew. Chem. Int. Ed.* **2019**, *58*, 11023–11027. [CrossRef] [PubMed]

46. Egami, H.; Katsuki, T. Iron-Catalyzed Asymmetric Aerobic Oxidation: Oxidative Coupling of 2-Naphthols. *J. Am. Chem. Soc.* **2009**, *131*, 6082–6083. [CrossRef] [PubMed]

47. Egami, H.; Matsumoto, K.; Oguma, T.; Kunisu, T.; Katsuki, T. Enantioenriched Synthesis of *C1*-Symmetric BINOLs: Iron-Catalyzed Cross-Coupling of 2-Naphthols and Some Mechanistic Insight. *J. Am. Chem. Soc.* **2010**, *132*, 13633–13635. [CrossRef]

48. Horibe, T.; Nakagawa, K.; Hazeyama, T.; Takeda, K.; Ishihara, K. An Enantioselective Oxidative Coupling Reaction of 2-Naphthol Derivatives Catalyzed by Chiral Diphosphine Oxide–iron(II) Complexes. *Chem. Commun.* **2019**, *55*, 13677–13680. [CrossRef]

49. Narute, S.; Parnes, R.; Toste, F.D.; Pappo, D. Enantioselective Oxidative Homocoupling and Cross-Coupling of 2-Naphthols Catalyzed by Chiral Iron Phosphate Complexes. *J. Am. Chem. Soc.* **2016**, *138*, 16553–16560. [CrossRef]

50. Shalit, H.; Dyadyuk, A.; Pappo, D. Selective Oxidative Phenol Coupling by Iron Catalysis. *J. Org. Chem.* **2019**, *84*, 1677–1686. [CrossRef]

51. Liu, W.; Zhong, D.-Y.; Yu, C.-L.; Zhang, Y.; Wu, D.; Feng, Y.-L.; Cong, H.-J.; Lu, X.-Q.; Liu, W.-B. Iron-Catalyzed Intramolecular Amination of Aliphatic C–H Bonds of Sulfamate Esters with High Reactivity and Chemoselectivity. *Org. Lett.* **2019**, *21*, 2673–2678. [CrossRef] [PubMed]

52. Zhong, D.-Y.; Wu, D.; Zhang, Y.; Lu, Z.-W.; Usman, M.; Liu, W.; Lu, X.-Q.; Liu, W.-B. Synthesis of Sultams and Cyclic *N*-Sulfonyl Ketimines via Iron-Catalyzed Intramolecular Aliphatic C-H Amidation. *Org. Lett.* **2019**, *21*, 5808–5812. [CrossRef] [PubMed]

53. Zang, C.; Liu, Y.-G.; Xu, Z.-J.; Tse, C.-T.; Guan, X.-G.; Wei, J.-H.; Huang, J.-S.; Che, C.-M. Highly Enantioselective Iron-Catalyzed *cis*-Dihydroxylation of Alkenes with Hydrogen Peroxide Oxidant via an Fe(III)-OOH Reactive Intermediate. *Angew. Chem. Int. Ed.* **2016**, *55*, 10253–10257. [CrossRef] [PubMed]

54. Wei, J.-H.; Cao, B.; Tse, C.-W.; Chang, X.-Y.; Zhou, C.-Y.; Che, C.-M. Chiral *cis*-Iron(II) Complexes with Metal- and Ligand-Centered Chirality for Highly Regio- and Enantioselective Alkylation of *N*-heteroaromatics. *Chem. Sci.* **2020**, *11*, 684–693. [CrossRef]

55. Ruiz-Castillo, P.; Buchwald, S.L. Applications of Palladium-Catalyzed C–N Cross-Coupling Reactions. *Chem. Rev.* **2016**, *116*, 12564–12649. [CrossRef]

56. Baumgartner, L.M.; Dennis, J.M.; White, N.A.; Buchwald, S.L.; Jensen, K.F. Use of a Droplet Platform to Optimize Pd-Catalyzed C–N Coupling Reactions Promoted by Organic Bases. *Org. Process Res. Dev.* **2019**, *23*, 1594–1601. [CrossRef]

57. Pangborn, A.B.; Giardello, M.A.; Grubbs, R.H.; Rosen, R.K.; Timmers, F.J. Safe and Convenient Procedure for Solvent Purification. *Organometallics* **1996**, *15*, 1518–1520. [CrossRef]

58. Michon, C.; Ellern, A.; Angelici, R.J. Chiral Tetradentate Amine and Tridentate Aminocarbene Ligands: Synthesis, Reactivity and X-Ray Structural Characterizations. *Inorganica Chimica Acta* **2006**, *39*, 4549–4556. [CrossRef]

59. Ramann, G.A.; Cowen, B.J. Quinoline Synthesis by Improved Skraup–Doebner–Von Miller Reactions Utilizing Acrolein Diethyl Acetal. *Tetrahedron Lett.* **2015**, *56*, 6436–6439. [CrossRef]

60. Chou, C.-C.; Hu, F.-C.; Yeh, H.-H.; Wu, H.-P.; Chi, Y.; Clifford, J.N.; Palomares, E.; Liu, S.-H.; Chou, P.-T.; Lee, G.-H. Highly Efficient Dye-Sensitized Solar Cells Based on Panchromatic Ruthenium Sensitizers with Quinolinyl Bipyridine Anchors. *Angew. Chem. Int. Ed.* **2014**, *53*, 178–183. [CrossRef]

61. Patra, T.; Agasti, S.; Akanksha; Maiti, D. Nickel-Catalyzed Decyanation of Inert Carbon–Cyano Bonds. *Chem. Commun.* **2013**, *49*, 69–71. [CrossRef] [PubMed]

62. Niimi, K.; Mori, H.; Miyazaki, E.; Osaka, I.; Kakizoe, H.; Takimiya, K.; Adachi, C. [2,2'] Bi[Naphtho [2,3-B] Furanyl]: A Versatile Organic Semiconductor with a Furan–Furan Junction. *Chem. Commun.* **2012**, *48*, 5892–5894. [CrossRef] [PubMed]

63. Bolchi, C.; Catalano, P.; Fumagalli, L.; Gobbi, M.; Pallavicini, M.; Pedretti, A.; Villa, L.; Vistoli, G.; Valoti, E. Structure–Affinity Studies for a Novel Series of Homochiral Naphthol and Tetrahydronaphthol Analogues of A1 Antagonist WB-4101. *Bioorganic Med. Chem.* **2004**, *12*, 4937–4951. [CrossRef]

64. Liu, Y.-C.; Trzoss, M.; Brimble, M.A. A Facile Synthesis of Aryl Spirodioxines Based on a 3H,3'H-2,2'-Spirobi (Benzo [*b*][1,4] Dioxine) Skeleton. *Synthesis* **2007**, *9*, 1392–1402.

65. Forkosh, H.; Vershinin, V.; Reiss, H.; Pappo, D. Stereoselective Synthesis of Optically Pure 2-Amino-2'-Hydroxy-1,1'-Binaphthyls. *Org. Lett.* **2018**, *20*, 2459–2463. [CrossRef] [PubMed]

66. Xiao, B.; Fu, Y.; Xu, J.; Gong, G.-T.; Dai, J.-J.; Yi, J.; Liu, L. Pd(II)-Catalyzed C–H Activation/Aryl-Aryl Coupling of Phenol Esters. *J. Am. Chem. Soc.* **2010**, *132*, 468–469. [CrossRef]

67. Ueta, Y.; Mikami, K.; Ito, S. Accessto Air-STable 1,3-Diphosphacyclobutane-2,4-Diyls by an Arylation Reaction with Arynes. *Angew. Chem. Int. Ed.* **2016**, *55*, 7525–7529. [CrossRef]

68. Buzard, D.J.; Olsson, C.; Noson, K.; Lipshutz, B.H. A Modular Route to Nonracemic Cyclo-Nobins. Preparation of the Parent Ligand for Homo- and Heterogeneous Catalysis. *Tetrahedron* **2004**, *60*, 4443–4449.

69. Schreiner, J.L.; Pirkle, W.H. Chiral High-Pressure Liquid Chromatographic Stationary Phases. Separation of the Enantiomers of Bi-β-naphthols and Analogues. *J. Org. Chem.* **1981**, *46*, 4988–4991.

70. Kingsbury, W.D. Synthesis of Structural Analogs of Leukotriene B$_4$ and Their Receptor Binding Activity. *J. Med. Chem.* **1993**, *36*, 3308–3320. [CrossRef]

71. Wolsey, W.C. Perchlorate Salts, Their Uses and Alternatives. *J. Chem. Educ.* **1973**, *50*, A335–A337. [CrossRef]

72. Theveau, L.; Bellini, R.; Dydio, P.; Szabo, Z.; Werf, A.; Sander, R.A.; Reek, J.N.K.; Moberg, C. Cofactor-Controlled Chirality of Tropoisomeric Ligand. *Organometallics* **2016**, *35*, 1956–1963. [CrossRef]

73. Ahmed, I.; Clark, D.-A. Rapid Synthesis of 3,3' Bis-Arylated BINOL Derivatives Using a C–H Borylation in Situ Suzuki–Miyaura Coupling Sequence. *Org. Lett.* **2014**, *16*, 4332–4335. [CrossRef] [PubMed]

74. Xu, X.-J.; Clarkson, G.C.; Docherty, G.; North, C.L.; Woodward, G.; Wills, M. Ruthenium (II) Complexes of Monodonor Ligands: Efficient Reagents for Asymmetric Ketone Hydrogenation. *J. Org. Chem.* **2005**, *70*, 8079–8087. [CrossRef] [PubMed]

75. Song, S.; Sun, X.; Li, X.-W.; Yuan, Y.-Z.; Jiao, N. Efficient and Practical Oxidative Bromination and Iodination of Arenes and Heteroarenes with DMSO and Hydrogen Halide: A Mild Protocol for Late-Stage Functionalization. *Org. Lett.* **2015**, *17*, 2886–2889. [CrossRef]

76. Meesala, Y.; Wu, H.-L.; Koteswararao, B.; Kuo, T.-S.; Lee, W.-Z. Aerobic Oxidative Coupling of 2-Naphthol Derivatives Catalyzed by a Hexanuclear Bis(μ-hydroxo) Copper(II) Catalyst. *Organometallics* **2014**, *33*, 4385–4393. [CrossRef]

77. Simonsen, K.B.; Gothelf, K.V.; Jørgensen, K.A. A Simple Synthetic Approach to 3,3'-Diaryl BINOLs. *J. Org. Chem.* **1998**, *63*, 7536–7538. [CrossRef]

78. Qu, B.; Haddad, N. Ligand-Accelerated Stereoretentive Suzuki–Miyaura Coupling of Unprotected 3,3'-Dibromo-BINOL. *J. Org. Chem.* **2016**, *81*, 745–750. [CrossRef]

Sample Availability: Not available.

Article

Iron-Catalyzed Synthesis, Structure, and Photophysical Properties of Tetraarylnaphthidines

Alexander Purtsas [1], Sergej Stipurin [1], Olga Kataeva [2] and Hans-Joachim Knölker [1,*]

1   Faculty of Chemistry, Technische Universität Dresden, Bergstraße 66, 01069 Dresden, Germany; alexander.purtsas@chemie.tu-dresden.de (A.P.); sergej.stipurin@chemie.tu-dresden.de (S.S.)
2   A. E. Arbuzov Institute of Organic and Physical Chemistry, FRC Kazan Scientific Center, Russian Academy of Sciences, Arbuzov Str. 8, 420088 Kazan, Russia; olga-kataeva@yandex.ru
*   Correspondence: hans-joachim.knoelker@tu-dresden.de; Fax: +49-351-463-37030

Received: 22 February 2020; Accepted: 29 March 2020; Published: 1 April 2020

Abstract: We describe the synthesis and photophysical properties of tetraarylnaphthidines. Our synthetic approach is based on an iron-catalyzed oxidative C–C coupling reaction as the key step using a hexadecafluorinated iron–phthalocyanine complex as a catalyst and air as the sole oxidant. The N,N,N',N'-tetraarylnaphthidines proved to be highly fluorescent with quantum yields of up to 68%.

Keywords: iron catalysis; oxidative coupling; naphthidines; fluorescence

---

1. Introduction

Triarylamines have been the subject of numerous studies due to their electron-rich character and the resulting low oxidation potentials [1–3]. These unique properties of triarylamine derivatives have induced extensive investigations towards their potential applications in organic films for optoelectronics [4] and as hole-transport materials in OLEDs (organic light emitting diodes) and solar cells [5–8]. Naphthidines (1,1'-binaphthalene-4,4'-diamines) have also been studied because of their ready oxidation to radical cations [9–12]. Starting from the corresponding 1-naphthylamines by an oxidative coupling, the synthesis of naphthidines can be induced electrochemically [9,10,13], by stoichiometric amounts of titanium tetrachloride [11], stoichiometric amounts of iron(III) chloride [14], chloranil [15], or (bis(trifluoroacetoxy)iodo)benzene (PIFA) [16]. It has been well-known for a long time that naphthidines can be separated into their stable atropisomers [17].

Herein, we report a simple synthesis of N,N,N',N'-tetraarylnaphthidines by an iron-catalyzed oxidative coupling of the corresponding 1-(diphenylamino)naphthalene precursor using iron(II)–hexadecafluorophthalocyanine (FePcF$_{16}$) [18] as the catalyst and air as the final oxidant (Figure 1). Moreover, we have investigated the photophysical properties of the obtained naphthidine derivatives.

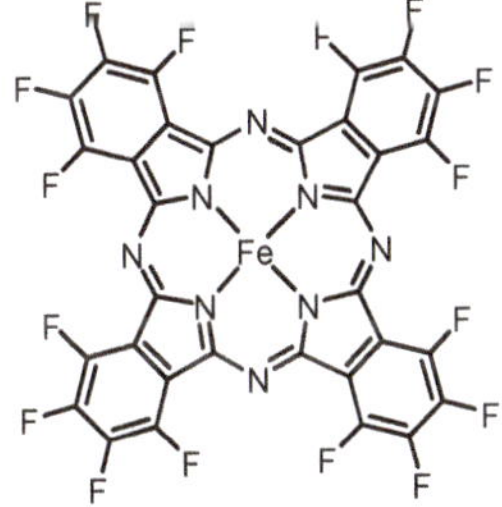

Figure 1. Structural formula of iron(II)–hexadecafluorophthalocyanine (FePcF$_{16}$).

2. Results and Discussion

2.1. Synthesis

Following a procedure reported by Nechaev et al. for the Buchwald–Hartwig reaction under solvent-free conditions [19], we have achieved the coupling of diphenylamine (**1**) and 1-bromonaphthalene (**2**) to afford 1-(diphenylamino)naphthalene (**3**) in high yield (Scheme 1).

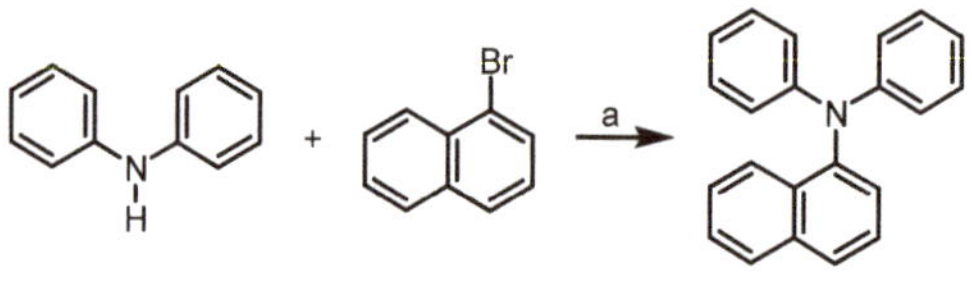

Scheme 1. Synthesis of 1-(diphenylamino)naphthalene (**3**). *Reagents and conditions:* (**a**) 1 mol% Pd(OAc)$_2$, 2 mol% RuPhos, 1.2 equivalents NaO*t*Bu, air, 110 °C, 12 h, 89% compound **3**.

Then, we turned our attention to the projected iron-catalyzed oxidative homocoupling of 1-(diphenylamino)naphthalene (**3**). Recently, we applied our iron-catalyzed oxidative coupling methodology to the C–C homocoupling of diarylamines, 1-, and 2-hydroxycarbazoles, as well as to the cross coupling of tertiary anilines with hydroxyarenes [20–22]. Iron as a first-row transition metal is environmentally safe and has become a powerful tool in synthetic organic chemistry with many applications for selective C–H bond activation [23–26]. According to a previous report by Yang, *N,N,N′,N′*-tetraphenylnaphthidine (**4**) could not be prepared by oxidative coupling of compound **3** with stoichiometric amounts of iron(III) chloride [14]. The oxidation of compound **3** with chloranil provides compound **4** only as the minor isomer in an inseparable mixture with the corresponding benzidine derivative [15]. Based on our previous studies [20–22], we envisaged to develop an iron-catalyzed homocoupling of the triarylamine **3** with air as the sole oxidant. Using catalytic amounts (2 mol%) of FePcF$_{16}$ and substoichiometric amounts (40 mol%) of methanesulfonic acid as an additive at room temperature provided *N,N,N′,N′*-tetraphenylnaphthidine (**4**) in 48% yield along with the *N,N,N′,N′*-tetraarylnaphthidine **5** as a by-product in 11% yield (Scheme 2). Obviously, compound **5** results from a further oxidative C–C coupling at the *p*-position of one of the phenyl rings of the initial coupling product compound **4**. The structural assignments for compounds **4** and **5** were based on their ^{1}H-NMR and ^{13}C-NMR spectroscopic data and an X-ray analysis of compound **4** (see Section 2.2).

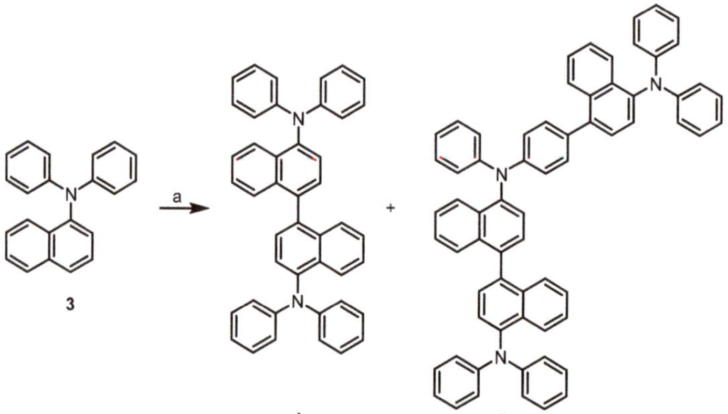

Scheme 2. Iron-catalyzed oxidative C–C coupling of 1-(diphenylamino)naphthalene (**3**). *Reagents and conditions:* (**a**) 2 mol% FePcF$_{16}$, 40 mol% MsOH, CH$_2$Cl$_2$, air, room temperature, 24 h, 48% compound **4**, 11% compound **5**.

2.2. Structure

The ¹H-NMR, ¹³C-NMR, and DEPT spectra of compound **4** displayed signals for a highly symmetrical compound. Signals for nine aromatic CH and five quaternary aromatic carbon atoms were identified. The COSY experiment revealed the presence of three spin systems (Supplementary Materials, COSY of compound **4**). The first is caused by coupling of H-1 with H-2 and H-3, which belong to the phenyl fragment. The second spin system consists of H-6 and H-7. The third system is formed by H-10, H-11, H-12, and H-13. The assignment of all ¹³C-NMR signals to the respective ¹H-NMR signals could be achieved by an HSQC measurement (Figure 2 and Table S1). The constitution of naphthidine **4** was confirmed by analysis of the HMBC spectrum (Supplementary Materials, HMBC of compound **4**, Table S1). Characteristic HMBC signals (C-4/H-2 and C-4/H-3) led to elucidation of the quaternary carbon atom C-4 by connecting the proton spin systems. The position of C-5 was established based on the HMBC cross-peaks with H-6, H-7, and H-13. Accordingly, the quaternary aromatic carbon atom C-8 was assigned based on the HMBC interactions with H-6, H-7, and H-10. HMBC cross-peaks between C-9/H-7, H-11, H-13 and C-14/H-6, H-10, H-12 unambiguously clarified the location of both naphthyl-based quaternary carbon atoms. The NOE interactions were exploited to confirm the positions of H-3 and H-13 (Supplementary Materials, NOESY of compound **4**). Due to the interactions of these fjord region protons, H-13 showed a substantial downfield shift to 8.08 ppm. Moreover, H-2, H-3, and H-6 exhibited an interaction with the ¹⁵N atom (Supplementary Materials, ¹H/¹⁵N HMBC of compound **4**).

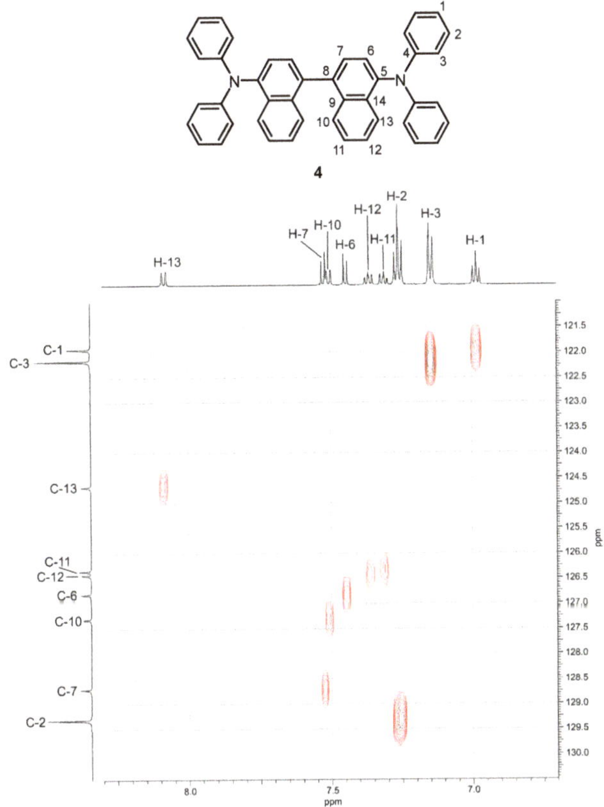

Figure 2. Assignment of the ¹³C-NMR signals of the *N,N,N′,N′*-tetraphenylnaphthidine (**4**) to the respective ¹H-NMR signals by the HSQC spectrum.

The structural assignment for *N,N,N′,N′*-tetraphenylnaphthidine (**4**) was confirmed by an X-ray crystal structure determination (Figure 3). The molecule of compound **4** adopts a *trans* conformation around the central biaryl bond (C1–C11) with a torsional angle of 99.4(2)° (Figure 4) according to the notation accepted for binaphthyl derivatives [27,28]. The geometry of the naphtho fragments exhibiting large variations of the bond lengths is in agreement with the crystal structure of naphthalene [29,30] and its structure obtained by quantum chemical calculations [31]. The orientation of the phenyl groups attached to the nitrogen atoms at the two sides of the molecule is different. The two phenyl rings at N21 are symmetrically oriented respective to the bisector plane drawn through the C4–N21 bond, while the two phenyl groups at N34 have different orientations. In the crystal, the molecules of compound **4** participate in multiple weak intermolecular C–H···π interactions between the naphtho groups (Figure 5). The distances between the hydrogen atoms and the sp^2-carbon atoms range from 2.7 to 2.8 Å and the angles are typical for these weak hydrogen bonds [32]. This sort of packing with face-to-edge interactions has been observed also in naphthalene itself [29,30] and in binaphthyl derivatives [27,28]. It is noteworthy that the naphtho groups interact mostly with each other and the phenyl groups interact predominantly with phenyl groups, whereas naphtho–phenyl contacts are rare.

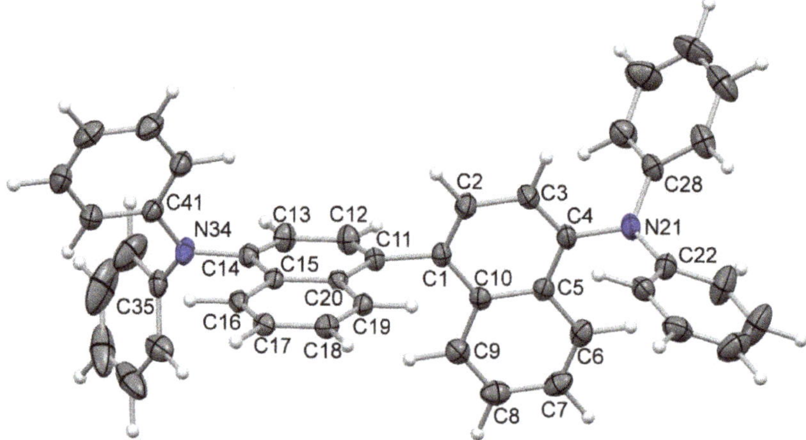

Figure 3. Molecular structure of *N,N,N′,N′*-tetraphenylnaphthidine (**4**) in the crystal (ORTEP plot showing thermal ellipsoids at the 50% probability level). Selected bond lengths (Å): C1–C2 1.370(3), C1–C10 1.424(2), C5–C10 1.431(2), C4–N21 1.436(2), C1–C11 1.500(2), C11–C12 1.370(2), C11–C20 1.428(2), C15–C20 1.432(2), C14–N34 1.430(2). Selected bond angles (°): C10–C1–C11 121.4(2), C2–C1–C11 118.8(2), C2–C1–C10 119.7(2), C1–C10–C9 122.2(2), C3–C4–N21 120.6(2), C5–C4–N21 120.0(2), C4–N21–C22 117.1(2), C4–N21–C28 118.3(2), and C22–N21–C28 120.6(2).

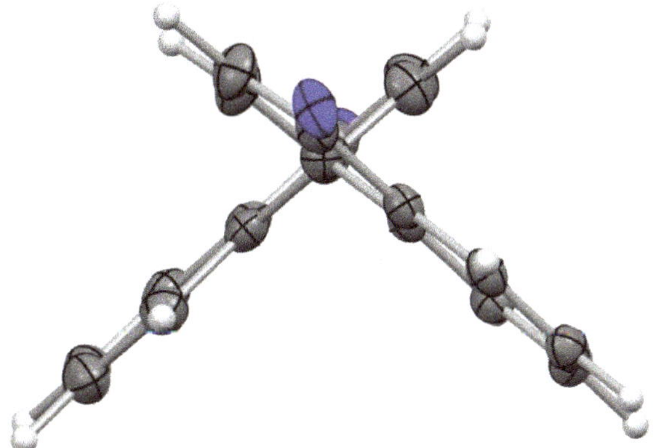

Figure 4. Crystal structure of compound **4** with a view along the N21–C4/C1–C11/C14–N34 axis (ORTEP plot showing thermal ellipsoids at the 50% probability level, the *N*–phenyl substituents have been omitted for clarity). Torsional angle (C2–C1–C11–C12) 99.4(2)°.

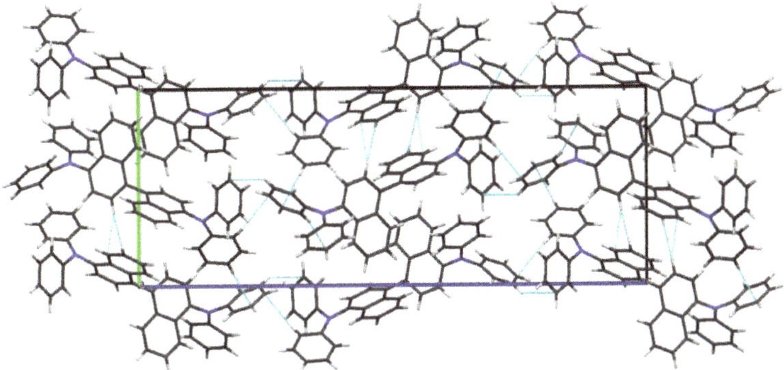

Figure 5. Fragment of the crystal packing of *N,N,N',N'*-tetraphenylnaphthidine (**4**) showing the unit cell and C–H···π interactions (view along the *a*-axis of the crystal).

The connectivity in compound **5** has been secured by a full set of 2D-NMR spectra (see below and the copies of the 2D-NMR spectra in the Supplementary Materials). The NMR spectra of **5** display signals for a non-symmetrical aromatic compound. The JRES experiment revealed the absence of a singlet. Thus, the additional oxidative coupling of the naphthidine **4** with compound **3** did not occur at positions 2, 6, 7, 11, or 12 (Supplementary Materials, JRES of compound **5**). The ^{1}H-NMR spectrum in combination with the HSQC (Figure 6) and the JRES spectra displayed signals for four substantially downfield-shifted doublets. On the basis of the fjord region effect observed for compound **4**, these four downfield-shifted signals in the ^{1}H-NMR spectrum of compound **5** are the naphthyl protons H 10, H 13, H 27, and H-38. The signals of the four corresponding carbon atoms are detected at δ_C = 124.66 (C-13), 124.70 (C-27), 124.71 (C-38), and 126.93 (C-10) ppm in the ^{13}C-NMR spectrum (HSQC in Figure 6, Table S2). The NOE interactions indicated the positions of the protons at the phenyl groups (Supplementary Materials, NOESY of compound **5**). The positions of the protons at C-3, C-30, and C-44 have been identified unambiguously by NOE interactions with H-13, H-27, and H-38. Support for this assignment of the protons H-3, H-30, and H-44 is obtained from an HMBC interaction with ^{15}N (Supplementary Materials, ^{1}H/^{15}N HMBC of compound **5**). A strong NOE correlation of H-10 with H-16 and the presence of a spin interaction between H-16 and H-17 in the COSY (Supplementary Materials, COSY of compound **5**) confirmed an oxidative coupling of compound **3** at C-1 of naphthidine **4**. Further support was obtained from HMBC correlations of H-16 with the quaternary carbon atoms

C-8 at δ_C = 138.64 ppm and C-18 at δ_C = 147.93 ppm and the HMBC correlation of H-17 with C-15 at δ_C = 133.51 ppm (Supplementary Materials, HMBC of compound **5**). The obtained NMR data are in excellent agreement with the structure assigned for compound **5**.

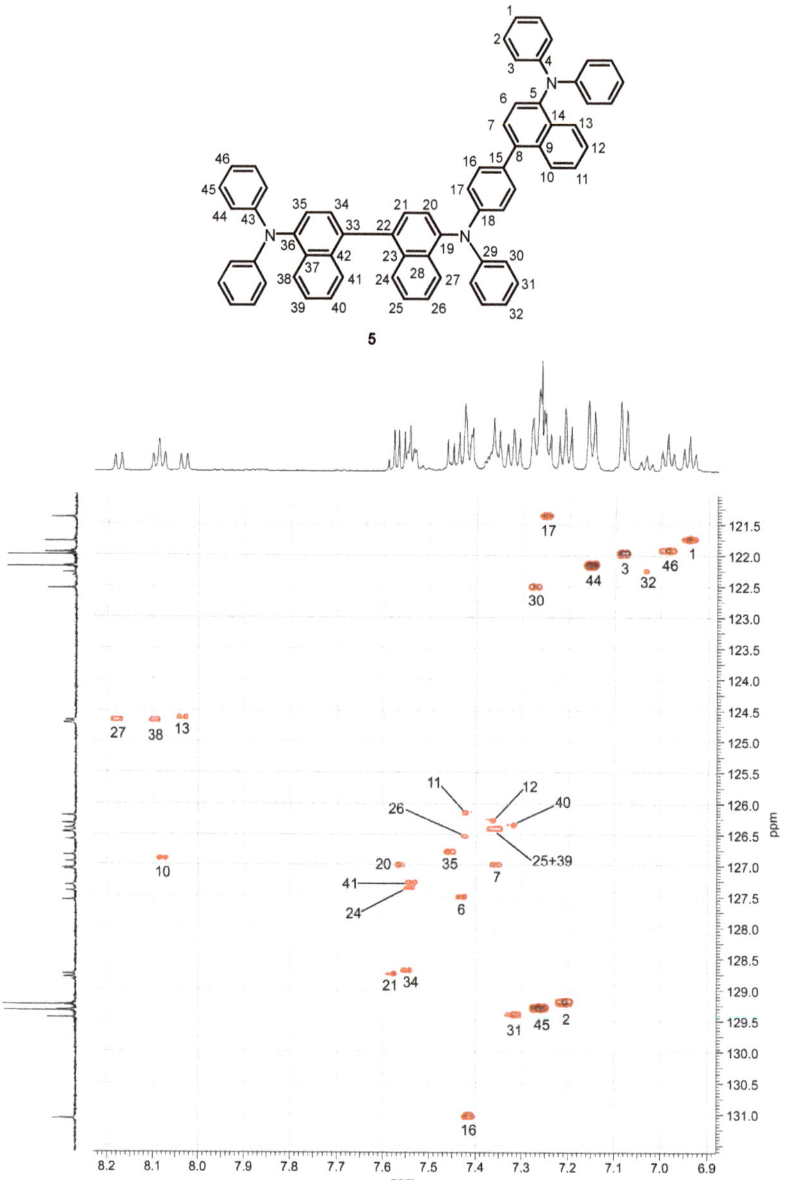

Figure 6. Assignment of the ^{13}C-NMR signals of the *N,N,N′,N′*-tetraarylnaphthidine (**5**) to the respective ^{1}H-NMR signals by the HSQC spectrum.

2.3. Photophysical Studies

On irradiation with UV-light (λ_{ex} = 254 nm) in methanolic solution, 1-(diphenylamino)naphthalene (3) and the resulting *N,N,N′,N′*-tetraarylnaphthidines **4** and **5** show a strong blue fluorescence (Figure 7). Therefore, we investigated the photophysical properties of compounds **3**, **4**, and **5** in more detail. The UV absorption and fluorescence emission data for all three compounds are listed in Table 1 and the normalized fluorescence emission spectra are displayed in Figure 8. The fluorophores **3**, **4**, and **5** exhibit large Stokes shifts of 149 to 162 nm. While the UV absorption maxima are identical, the fluorescence emission maxima show a bathochromic shift of only 7 nm for *N,N,N′,N′*-tetraphenylnaphthidine (4) and 14 nm for compound **5**, as compared to 1-(diphenylamino)naphthalene (3). Previously, even larger Stokes shifts have been observed for *N,N*-dimethylaminonaphthalenes [33] and strong bathochromic shifts have been reported for substituted triarylamines [34].

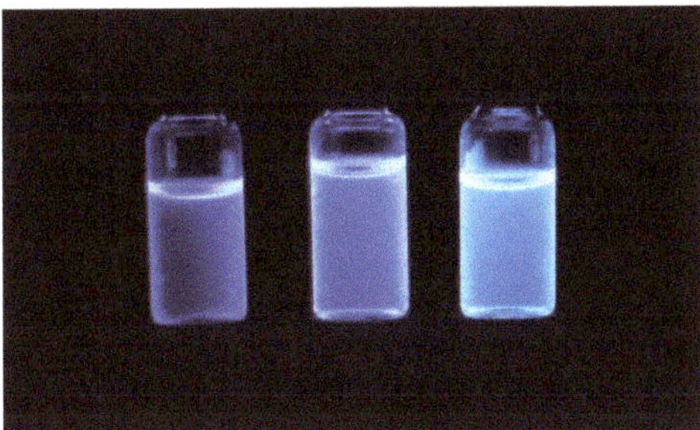

Figure 7. Fluorescence of the compounds **3** (left), **4** (center), and **5** (right) in methanol (concentration c = 2 mg mL^{-1}) by excitation with UV light (λ_{ex} = 254 nm).

Table 1. UV absorption and fluorescence emission data of the compounds **3**, **4**, and **5** in methanol.

Compound	λ_{abs} (nm) [a]	ε (M^{-1} cm^{-1}) [b]	λ_{em} (nm) [c]	$\Delta\lambda$ (nm) [d]
aminonaphthalene **3**	290	15,700	439	149
naphthidine **4**	291	11,400	446	155
naphthidine **5**	291	6000	453	162

[a] UV absorption maximum; [b] molar extinction coefficient; [c] fluorescence emission maximum (excitation at λ_{ex} = 291 nm); [d] Stokes shift, $\Delta\lambda = \lambda_{em} - \lambda_{abs}$.

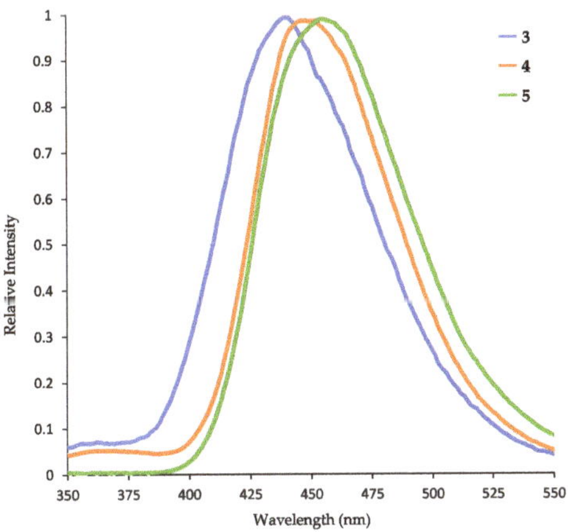

Figure 8. Normalized fluorescence emission spectra of 1-(diphenylamino)naphthalene (3) (blue), *N,N,N′,N′*-tetraphenylnaphthidine (4) (orange), and naphthidine **5** (green) in methanol (excitation at λ_{ex} = 291 nm).

For the naphthidines **4** and **5**, the UV absorption and fluorescence emission maxima along with fluorescence quantum yields, fluorescence lifetimes, and CIE-coordinates were determined (Table 2).

Table 2. Photophysical data of the naphthidines **4** and **5**.

Compound	λ_{abs} (nm) [a]	ε (M^{-1} cm^{-1}) [b]	λ_{em} (nm) [c]	Φ_{fl} [d,e]	τ(ns) [f,e]	CIE-coord. (x; y) [g,e]
naphthidine **4**	294	32,000	449	0.68	4.2	(0.153; 0.054)
naphthidine **5**	294	25,600	453	0.65	3.6	(0.152; 0.062)

[a] UV absorption maximum in CH_2Cl_2; [b] molar extinction coefficient in CH_2Cl_2; [c] fluorescence emission maximum in CH_2Cl_2 (λ_{ex} = 294 nm); [d] fluorescence quantum yield (λ_{ex} = 310 nm); [e] in a 2% PMMA film; [f] fluorescence lifetime (λ_{ex} = 360 nm); [g] Commission internationale de l'éclairage (CIE) (λ_{ex} = 310 nm).

The UV absorption maximum for the naphthidines **4** and **5** is at the same wavelength (294 nm), whereas the fluorescence emission maximum of compound **5** shows a bathochromic shift of 4 nm as compared with compound **4**. The fluorescence quantum yields of compounds **4** and **5** are relatively high and in the same range (68% and 65%, respectively, by excitation at 310 nm). Compound **5** exhibits a shorter fluorescence lifetime of 3.6 ns as compared with 4.2 ns for N,N,N',N'-tetraphenylnaphthidine (**4**). The CIE-coordinates for the blue light emissions of compounds **4** and **5** are indicative of a blue light which could be useful for applications in OLEDs [35–39]. The fluorescence quantum yields are very comparable with compounds already investigated as blue fluorescent OLEDs [37–39]. It has been demonstrated for other naphthalene fluorophores that a bathochromic shift of the emission can be achieved by increasing the size of the π-system and by the introduction of appropriate substituents [40,41]. Thus, application of these tools should easily allow an optimization of the present fluorophoric system.

In addition, we investigated the absorption and fluorescence properties of N,N,N',N'-tetraphenylnaphthidine (**4**) in various nonpolar and polar solvents. The fluorescence behavior of compound **4** on excitation at 254 nm was demonstrated in isohexane, dichloromethane, ethyl acetate, THF, and methanol (Figure 9). The corresponding UV absorption and fluorescence emission spectra are shown in the Supplementary Materials. The corresponding photophysical data of compound **4** are summarized in Table 3. For a series of N,N-dimethylaminonaphthalene fluorophores, Brummond et al. observed a significant red shift of the fluorescence emission maxima by increasing the solvent polarity [33], whereas in other systems the solvent dependency of the emission was significantly lower [37]. For N,N,N',N'-tetraphenylnaphthidine (**4**), this solvatochromic shift is much less pronounced (about 21 nm in dichloromethane as compared with the corresponding value in isohexane) (Table 3).

Figure 9. Fluorescence of N,N,N',N'-tetraphenylnaphthidine (**4**) by excitation with UV light (λ_{ex} = 254 nm) in different solvents (concentration c = 2 mg mL^{-1} and from left to right, isohexane, dichloromethane, ethyl acetate, THF, and methanol).

Table 3. UV absorption and fluorescence emission data of the naphthidine **4** in different solvents.

	Isohexane	CH_2Cl_2	Ethyl Acetate	THF	Methanol
λ_{abs} (nm) [a]	291	294	291	292	291
λ_{em} (nm) [b]	428	449	439	445	446

[a] UV absorption maximum and [b] fluorescence emission maximum (for each solvent $\lambda_{ex} = \lambda_{abs}$).

3. Materials and Methods

3.1. General

All reactions were performed in oven-dried glassware using anhydrous solvents under argon, unless stated otherwise. Pd(OAc)$_2$ was recrystallized from glacial AcOH. All other chemicals were used as received from commercial sources. The iron(III)-catalyzed reactions were carried out in non-dried solvents under air. Iron(II)–hexadecafluorophthalocyanine was prepared following a procedure reported previously [18]. Flash chromatography was performed using silica gel from Acros Organics (0.035 to 0.070 mm). Automated flash chromatography was performed on a Büchi Sepacore system equipped with an UV monitor on silica gel (Acros Organics, 0.035 to 0.070 mm). The TLC was performed with TLC plates from Merck (60 F254) using UV light for visualization. The melting points were measured on a Gallenkamp MPD 350 melting point apparatus. Ultraviolet spectra were recorded on a PerkinElmer 25 UV/Vis spectrometer. The fluorescence spectra were measured on a Varian Cary Eclipse spectrophotometer. The IR spectra were recorded on a Thermo Nicolet Avatar 360 FT-IR spectrometer using the ATR method (attenuated total reflectance). The NMR spectra were recorded on Bruker DRX 500 and Avance III 600 spectrometers. The chemical shifts δ were reported in ppm with the solvent signal as internal standard. Standard abbreviations were used to denote the multiplicities of the signals. EI mass spectra were recorded by GC/MS-coupling using an Agilent Technologies 6890 N GC System equipped with a 5973 Mass Selective Detector (70 eV). The ESI-MS spectra were recorded on an Esquire LC using an ion trap detector from Bruker. Positive and negative ions were detected. Elemental analyses were measured on a EuroVector EuroEA3000 elemental analyzer. The X-ray crystal structure analyses were performed with a Bruker–Nonius Kappa CCD and with a Bruker AXS Smart APEX diffractometer equipped with a 700 series Cryostream low temperature device from Oxford Cryosystems. SHELXS-97 [42], SADABS version 2.10 [43], SHELXL-97 [44], POV-Ray for Windows version 3.7.0.msvc10.win64, and ORTEP-3 for Windows [45] were used as software.

3.2. Photoluminescence Measurements

The 2 wt% emitter films were prepared by doctor blading a solution of an emitter in a 10 wt% poly(methyl methacrylate) solution in dichloromethane on a quartz substrate with a 60 µm doctor blade. The film was dried, and the emission measured under nitrogen atmosphere. Excitation was conducted in a wavelength range of 250–400 nm (Xe lamp with monochromator), and the emission was detected with a calibrated quantum yield detection system (Hamamatsu, model C11347-01). The fluorescence lifetime was measured with an Edinburgh Instruments mini-τ by excitation with pulses of an EPLED (Edinburgh picosecond Pulsed Light Emitting Diode) (360 nm, 20 kHz) and time-resolved photon counting (TCSPC).

3.3. Procedures

3.3.1. *N,N*-Diphenylnaphthalen-1-amine (1-(Diphenylamino)naphthalene) (3)

Compound **3** was prepared following a literature procedure [19]. To a screw-cap vial were added 1-bromonaphthalene (**2**) (307 µL, 455 mg, 2.20 mmol), diphenylamine (**1**) (338 mg, 2.00 mmol), Pd(OAc)$_2$ (5.4 mg, 24 µmol), RuPhos (19 mg, 41 µmol), and powdered NaO*t*Bu (236 mg, 2.46 mmol). The reaction mixture was stirred at 110 °C for 12 h. After cooling to room temperature, the mixture was dissolved in CH$_2$Cl$_2$/H$_2$O (1:1). The aqueous layer was extracted twice with dichloromethane

(5 mL). The combined organic layers were dried over MgSO$_4$, the solvent was evaporated in vacuo, and the crude product was purified by chromatography on silica gel with isohexane/dichloromethane (30:1) to afford compound **3** (523 mg, 1.77 mmol, 89%) as a colorless solid. M.p. 132.5 °C; UV (MeOH) λ = 219, 290, and 336 nm; fluorescence (MeOH) λ_{ex} = 291 and λ_{em} = 439 nm; IR (ATR) v = 3036, 1934, 1740, 1586, 1563, 1488, 1389, 1341, 1290, 1272, 1176, 1087, 1026, 1014, 891, 798, 773, 749, 693, and 624 cm^{-1}; ^{1}H-NMR (500 MHz, CDCl$_3$) δ = 6.93 (t, J = 7.5 Hz, 2H), 7.03 (dd, J = 8.7, 0.8 Hz, 4H), 7.15–7.22 (m, 4H), 7.30–7.38 (m, 2H), 7.42–7.50 (m, 2H), 7.77 (d, J = 8.2 Hz, 1H), 7.88 (d, J = 8.2 Hz, 1H), and 7.94 (d, J = 8.5 Hz, 1H); ^{13}C-NMR and DEPT at θ = 135° (125 MHz, CDCl$_3$) δ = 121.76 (2 CH), 121.97 (4 CH), 124.41 (CH), 126.25 (CH), 126.48 (CH), 126.51 (CH), 126.56 (CH), 127.39 (CH), 128.51 (CH), 129.21 (4 CH), 131.41 (C), 135.41 (C), 143.71 (C), 148.58 (2 C); MS (EI): *m/z* (%) = 295 (100, [M]$^+$), 294 (45), 293 (11), 217 (23), and 77 (10); elemental analysis calcd. for C$_{22}$H$_{17}$N, C 89.46, H 5.80, and N 4.74; found, C 89.61, H 6.05, and N 4.76.

3.3.2. Iron-Catalyzed Oxidative C–C Coupling

Dichloromethane (9 mL) was added to a mixture of iron(II)–hexadecafluorophthalocyanine (15.7 mg, 18.3 µmol, 2 mol%), methanesulfonic acid (34.8 mg, 362 µmol) and 1-(diphenylamino)naphthalene (**3**) (265 mg, 897 µmol). A 100 mL splash head was attached on the flask to prevent evaporation of the solvent, while ensuring sufficient gas exchange. The resulting solution was stirred at room temperature for 24 h under air. Then, 10 mL of a saturated aqueous solution of sodium hydrogen carbonate was added. The aqueous layer was extracted three times with dichloromethane. The combined organic layers were dried (magnesium sulfate). The solvent was evaporated and the residue was purified by automated flash chromatography (silica gel, isohexane/dichloromethane, 20% to 50% in 1.5 h) to provide *N,N,N′,N′*-tetraphenylnaphthidine (**4**) (128 mg, 217 µmol, 48%) as a colorless solid (less polar fraction) and naphthidine **5** (29.5 mg, 33.4 µmol, 11%) as a colorless solid (more polar fraction). Crystallization of compound **4** from isohexane afforded colorless crystals which were suitable for X-ray analysis.

N^4,N^4′,N^4′,N^4′-Tetraphenyl-[1,1′-binaphthalene]-4,4′-diamine (N,N,N′,N′-Tetraphenylnaphthidine) (**4**): M.p. 249–250 °C; UV (MeOH) λ = 217, 291, and 361 nm; fluorescence (MeOH) λ_{ex} = 291 and λ_{em} = 446 nm; IR (ATR) v = 3058, 3034, 1931, 1740, 1582, 1488, 1458, 1420, 1397, 1376, 1266, 1153, 1075, 1053, 1025, 920, 836, 746, 691, and 624 cm^{-1}; ^{1}H-NMR (600 MHz, CDCl$_3$) δ = 6.98 (t, J = 7.3 Hz, 4H), 7.14 (dd, J = 8.7, 1.1 Hz, 8H), 7.23–7.28 (m, 8H), 7.31 (ddd, J = 8.4, 6.9, 1.3 Hz, 2H), 7.36 (ddd, J = 8.4, 6.9, 1.3 Hz, 2H), 7.44 (d, J = 7.3 Hz, 2H), 7.50 (d, J = 7.9 Hz, 2H), 7.52 (d, J = 7.3 Hz, 2H), and 8.08 (d, J = 8.1 Hz, 2H); ^{13}C-NMR and DEPT at θ = 135° (150 MHz, CDCl$_3$) δ = 121.93 (4 CH), 122.17 (8 CH), 124.67 (2 CH), 126.35 (2 CH), 126.43 (2 CH), 126.81 (2 CH), 127.32 (2 CH), 128.71 (2 CH), 129.32 (8 CH), 131.40 (2 C), 134.72 (2 C), 136.76 (2 C), 143.55 (2 C), and 148.70 (4 C); MS (ESI, +10 V) *m/z* = 589.3 [M + H]$^+$; elemental analysis calcd. for C$_{44}$H$_{32}$N$_2$, C 89.76, H 5.48, and N 4.76; found, C 89.61, H 5.63, and N 4.73.

Crystallographic data for *N,N,N′,N′*-tetraphenylnaphthidine (**4**): C$_{44}$H$_{32}$N$_2$, crystal size 0.100 × 0.102 × 0.522 mm^3, M = 588.71 g mol^{-1}, orthorhombic, space group $P2_12_12_1$, a = 7.6835(4), b = 12.5020(6), c = 33.1785(18) Å, V = 3187(3) Å^3, Z = 4, $\rho_{calcd.}$ = 1.227 g cm^{-3}, μ = 0.071 mm^{-1}, T = 150(2) K, λ = 0.71073 Å, θ range 2.04–28.34°, 50987 reflections collected, 7939 independent (R_{int} = 0.0473), and 415 parameters. The structure was solved by direct methods and refined by full-matrix least-squares on F^2; R_1 = 0.0391 and wR_2 = 0.0890 [$I > 2\sigma(I)$]; maximal residual electron density 0.172 e Å^{-3}. The absolute structure was not determined. CCDC 1983355.

N^4-(4-(4-(Diphenylamino)naphthalen-1-yl)phenyl)-N^4,N^4′,N^4′-triphenyl-[1,1′-binaphthalene]-4,4′-diamine (**5**): M.p. 272–275 °C; UV (MeOH) λ = 215, 291, and 367 nm; fluorescence (MeOH) λ_{ex} = 291 and λ_{em} = 453 nm; IR (ATR) v = 3061, 3035, 2952, 2920, 2851, 1727, 1585, 1488, 1457, 1420, 1377, 1265, 1174, 1154, 1074, 1026, 952, 920, 827, 748, 692, and 625 cm^{-1}; ^{1}H-NMR (600 MHz, CDCl$_3$) δ = 6.94 (t, J = 7.3 Hz, 2H), 6.99 (t, J = 7.3 Hz, 2H), 7.03 (t, J = 7.3 Hz, 1H), 7.06–7.11 (m, 4H), 7.13–7.17 (m, 4H), 7.19–7.30 (m, 12H), 7.32 (t, J = 8.1 Hz, 3H), 7.34–7.39 (m, 4H), 7.40–7.48 (m, 6H), 7.49–7.61 (m, 5H), 8.03 (d, J = 8.3 Hz, 1H), 8.09

(t, J = 7.7 Hz, 2H), and 8.18 (d, J = 8.7 Hz, 1H); ^{13}C-NMR and DEPT at θ = 135° (150 MHz, CDCl$_3$) δ = 121.40 (2 CH), 121.78 (2 CH), 121.95 (2 CH), 122.00 (4 CH), 122.19 (4 CH), 122.28 (CH), 122.54 (2 CH), 124.66 (CH), 124.70 (CH), 124.71 (CH), 126.19 (CH), 126.31 (CH), 126.39 (CH), 126.45 (CH), 126.46 (CH), 126.57 (CH), 126.82 (CH), 126.93 (CH), 127.03 (CH), 127.05 (CH), 127.31 (CH), 127.41 (CH), 127.55 (CH), 128.74 (CH), 128.79 (CH), 129.23 (4 CH), 129.33 (4 CH), 129.44 (2 CH), 131.07 (2 CH), 131.41 (C), 131.51 (C), 131.67 (C), 133.51 (C), 133.72 (C), 134.72 (C), 134.80 (C), 136.72 (C), 137.03 (C), 138.64 (C), 142.80 (C), 143.39 (C), 143.61 (C), 147.93 (C), 148.46 (C), 148.61 (2 C), and 148.70 (2 C); MS (ESI, +10 V) *m/z* = 882.5 [M + H]$^+$; elemental analysis calcd. for C$_{66}$H$_{47}$N$_3$, C 89.87, H 5.37, and N 4.76; found, C 89.56, H 5.31, and N 4.79.

4. Conclusions

In conclusion, we have developed a two-step synthesis of *N*,*N*,*N*′,*N*′-tetraphenylnaphthidine (**4**). Starting from diphenylamine (**1**), Buchwald–Hartwig coupling with 1-bromonaphthalene (**2**) and subsequent iron-catalyzed oxidative homocoupling of the resulting 1-(diphenylamino)naphthalene (**3**) provides compound **4** as the major product and as a minor product compound **5**, resulting from an additional oxidative C–C coupling. Thus, we could demonstrate that our method of iron-catalyzed oxidative C–C coupling with air as the final oxidant can be applied to the regioselective homocoupling of triarylamines. Compounds **4** and **5** exhibit a strong blue-light fluorescence with quantum yields of up to 68% and fluorescence lifetimes of 4.2 and 3.6 ns, for compounds **4** and **5**, respectively. Further structural changes of the *N*,*N*,*N*′,*N*′-tetraarylnaphthidines by extension of the π-system or modification of the substitution pattern could lead to improved photophysical properties and fluorophores for various potential applications.

Supplementary Materials: The following data are available online. Copies of the ^{1}H-NMR, ^{13}C-NMR, DEPT (θ = 135°), UV, and fluorescence spectra of the compounds **3**, **4**, and **5**; copies of the 2D-NMR spectra (COSY, HSQC, HMBC, NOESY, ^{1}H/^{15}N-HMBC, JRES) of the compounds **4** and **5**.

Author Contributions: A.P. performed the chemical synthesis and characterized the compounds; A.P. and H.-J.K. designed the experiments and analyzed the data; S.S. determined the fluorescence quantum yields and the fluorescence lifetimes; O.K. performed the X-ray analysis; A.P. and H.-J.K. wrote the paper. All authors have read and agreed to the final version of the manuscript.

Funding: We thank the Deutsche Forschungsgemeinschaft (DFG) for the financial support of our project "Green and Sustainable Catalysts for Synthesis of Organic Building Blocks" (DFG grant KN 240/19-2).

Acknowledgments: We are indebted to Dr. Tilo Lübken (NMR facility, TU Dresden) for the measurement and assignment of the NMR spectra. We are grateful to Maximilian Georgi (Physical Chemistry, TU Dresden) for his support when taking the fluorescence pictures (Figures 7 and 9).

Conflicts of Interest: The authors declare no conflict of interest.

References

1. Lambert, C.; Nöll, G. One- and two-dimensional electron transfer processes in triarylamines with multiple redox centers. *Angew. Chem. Int. Ed.* **1998**, *37*, 2107–2110. [CrossRef]
2. Lambert, C.; Nöll, G. The class II/III transition in triarylamine redox systems. *J. Am. Chem. Soc.* **1999**, *121*, 8434–8442. [CrossRef]
3. Low, P.J.; Paterson, M.A.J.; Puschmann, H.; Goeta, A.E.; Howard, J.A.K.; Lambert, C.; Cherryman, J.C.; Tackley, D.R.; Leeming, S.; Brown, B. Crystal, molecular and electronic structure of *N*,*N*′-diphenyl-*N*,*N*′-bis(2,4-dimethylphenyl)-(1,1′-biphenyl)-4,4′-diamine and the corresponding radical cation. *Chem. Eur. J.* **2004**, *10*, 83–91. [CrossRef] [PubMed]
4. Kim, M.-J.; Seo, E.-M.; Vak, D.; Kim, D.-Y. Photodynamic properties of azobenzene molecular films with triphenylamines. *Chem. Mater.* **2003**, *15*, 4021–4027. [CrossRef]
5. Schumann, J.; Kanitz, A.; Hartmann, H. Synthesis and characterization of some heterocyclic analogues of *N*,*N*′-perarylated phenylene-1,4-diamines and benzidines as a new class of hole transport materials. *Synthesis* **2002**, *9*, 1268–1276. [CrossRef]

6.   Meng, H.; Herron, N. Organic small molecule materials for organic light-emitting diodes. In *Organic Light-Emitting Materials and Devices*, 1st ed.; Li, Z., Meng, H., Eds.; CRC Press, Taylor & Francis: Boca Raton, FL, USA, 2007; pp. 295–412.

7.   Calió, L.; Kazim, S.; Grätzel, M.; Ahmad, S. Hole-transport materials for perovskite solar cells. *Angew. Chem. Int. Ed.* **2016**, *55*, 14522–14545. [CrossRef] [PubMed]

8.   Wang, J.; Liu, K.; Ma, L.; Zhan, X. Triarylamine: Versatile platform for organic, dye-sensitized, and perovskite solar cells. *Chem. Rev.* **2016**, *116*, 14675–14725. [CrossRef] [PubMed]

9.   Miras, M.C.; Silber, J.J.; Sereno, L. Two-electron oxidation of *N,N,N′,N′*-tetramethylnaphthidine in acetonitrile–The reactivity of its dication to aromatic nucleophilic substitution by pyridine. *J. Electroanal. Chem.* **1986**, *201*, 367–386. [CrossRef]

10.   Zón, M.A.; Fernández, H. Study of the bielectronic electro-oxidation of *N,N,N′,N′*-tetramethylnaphthidine (TMN) in non-aqueous medium–Determination of the individual kinetic parameters by a Laplace space analysis. *J. Electroanal. Chem.* **1990**, *295*, 41–58. [CrossRef]

11.   Desmarets, C.; Champagne, B.; Walcarius, A.; Bellouard, C.; Omar-Amrani, R.; Ahajji, A.; Fort, Y.; Schneider, R. Facile synthesis and characterization of naphthidines as a new class of highly nonplanar electron donors giving robust radical cations. *J. Org. Chem.* **2006**, *71*, 1351–1361. [CrossRef]

12.   Desmarets, C.; Omar-Amrani, R.; Walcarius, A.; Lambert, J.; Champagne, B.; Fort, Y.; Schneider, R. Naphthidine di(radical cation)s-stabilized palladium nanoparticles for efficient catalytic Suzuki–Miyaura cross-coupling reactions. *Tetrahedron* **2008**, *64*, 372–381. [CrossRef]

13.   Hornback, J.M.; Gossage, H.E. Electrochemical oxidative dehydrodimerization of naphthylamines. An efficient synthesis of 8,8′-dianilino-5,5′-binaphthalene-1,1′-disulfonate. *J. Org. Chem.* **1985**, *50*, 541–543. [CrossRef]

14.   Li, X.-L.; Huang, J.-H.; Yang, L.-M. Iron(III)-promoted oxidative coupling of naphthylamines: Synthetic and mechanistic investigations. *Org. Lett.* **2011**, *13*, 4950–4953. [CrossRef] [PubMed]

15.   Maddala, S.; Mallick, S.; Venkatakrishnan, P. Metal-free oxidative C–C coupling of arylamines using a quinone-based organic oxidant. *J. Org. Chem.* **2017**, *82*, 8958–8972. [CrossRef]

16.   Morimoto, K.; Koseki, D.; Dohi, T.; Kita, Y. Oxidative biaryl coupling of *N*-aryl anilines by using a hypervalent iodine(III) reagent. *Synlett* **2017**, *28*, 2941–2945. [CrossRef]

17.   Theilacker, W.; Hopp, R. Spaltung des Naphthidins und des 2,3,2′,3′-Tetramethylbenzidins in optische Antipoden. *Chem. Ber.* **1959**, *92*, 2293–2301. [CrossRef]

18.   Jones, J.G.; Twigg, M.V. A fluorinated iron phthalocyanine. *Inorg. Chem.* **1969**, *8*, 2018–2019. [CrossRef]

19.   Topchiy, M.A.; Asachenko, A.F.; Nechaev, M.S. Solvent-free Buchwald–Hartwig reaction of aryl and heteroaryl halides with secondary amines. *Eur. J. Org. Chem.* **2014**, *2014*, 3319–3322. [CrossRef]

20.   Fritsche, R.F.; Theumer, G.; Kataeva, O.; Knölker, H.-J. Iron-catalyzed oxidative C–C and N–N coupling of diarylamines and synthesis of spiroacridines. *Angew. Chem. Int. Ed.* **2017**, *56*, 549–553. [CrossRef]

21.   Brütting, C.; Fritsche, R.F.; Kutz, S.K.; Börger, C.; Schmidt, A.W.; Kataeva, O.; Knölker, H.-J. Synthesis of 1,1′- and 2,2′-bicarbazole alkaloids by iron(III)-catalyzed oxidative coupling of 2- and 1-hydroxycarbazoles. *Chem. Eur. J.* **2018**, *24*, 458–470. [CrossRef]

22.   Purtsas, A.; Kataeva, O.; Knölker, H.-J. Iron-catalyzed oxidative C–C cross-coupling reaction of tertiary anilines with hydroxyarenes by using air as sole oxidant. *Chem. Eur. J.* **2020**, *26*, 2499–2508. [CrossRef] [PubMed]

23.   Bolm, C.; Legros, J.; Le Paih, J.; Zani, L. Iron-catalyzed reactions in organic synthesis. *Chem. Rev.* **2004**, *104*, 6217–6254. [CrossRef] [PubMed]

24.   Bauer, I.; Knölker, H.-J. Iron catalysis in organic synthesis. *Chem. Rev.* **2015**, *115*, 3170–3387. [CrossRef] [PubMed]

25.   Fürstner, A. Iron catalysis in organic synthesis: A critical assessment of what it takes to make this base metal a multitasking champion. *ACS Cent. Sci.* **2016**, *2*, 778–789. [CrossRef]

26.   Shang, R.; Ilies, L.; Nakamura, E. Iron-catalyzed C–H bond activation. *Chem. Rev.* **2017**, *117*, 9086–9139. [CrossRef]

27.   Kress, R.B.; Duesler, E.N.; Etter, M.C.; Paul, I.C.; Curtin, D.Y. Solid-state resolution of binaphthyl: Crystal and molecular structures of the chiral (A) form and racemic (B) form and the study of the rearrangement of single crystals. Requirements for development of hemihedral faces for enantiomer identification. *J. Am. Chem. Soc.* **1980**, *102*, 7709–7714. [CrossRef]

28. Roszak, K.; Katrusiak, A. High-pressure and temperature dependence of the spontaneous resolution of 1,1′-binaphthyl enantiomers. *Phys. Chem. Chem. Phys.* **2018**, *20*, 5305–5311. [CrossRef]

29. Brock, C.P.; Dunitz, J.D. Temperature dependence of thermal motion in crystalline naphthalene. *Acta Cryst.* **1982**, *B38*, 2218–2228. [CrossRef]

30. Capelli, S.C.; Albinati, A.; Mason, S.A.; Willis, B.T.M. Molecular motion in crystalline naphthalene: Analysis of multi-temperature X-ray and neutron diffraction data. *J. Phys. Chem. A* **2006**, *110*, 11695–11703. [CrossRef]

31. Baba, M.; Kowaka, Y.; Nagashima, U.; Ishimoto, T.; Goto, H.; Nakayama, N. Geometrical structure of benzene and naphthalene: Ultrahigh-resolution laser spectroscopy and *ab initio* calculation. *J. Chem. Phys.* **2011**, *135*, 054305. [CrossRef]

32. Nishio, M.; Umezawa, Y.; Honda, K.; Tsuboyama, S.; Suezawa, H. CH/π hydrogen bonds in organic and organometallic chemistry. *CrystEngComm* **2009**, *11*, 1757–1788. [CrossRef]

33. Benedetti, E.; Kocsis, L.S.; Brummond, K.M. Synthesis and photophysical properties of a series of cyclopenta[*b*]naphthalene solvatochromic fluorophores. *J. Am. Chem. Soc.* **2012**, *134*, 12418–12421. [CrossRef] [PubMed]

34. Li, R.; Gong, Z.-L.; Tang, J.-H.; Sun, M.-J.; Shao, J.-Y.; Zhong, Y.-W.; Yao, J. Triarylamines with branched multi-pyridine groups: Modulation of emission properties by structural variation, solvents, and tris(pentafluorophenyl)borane. *Sci. China Chem.* **2018**, *61*, 545–556. [CrossRef]

35. Lee, M.-T.; Chen, H.-H.; Liao, C.-H.; Tsai, C.-H.; Chen, C.H. Stable styrylamine-doped blue organic electroluminescent device based on 2-methyl-9,10-di(2-naphthyl)anthracene. *Appl. Phys. Lett.* **2004**, *85*, 3301–3303. [CrossRef]

36. Gao, Z.Q.; Mi, B.X.; Chen, C.H.; Cheah, K.W.; Cheng, Y.K.; Wen, W.-S. High-efficiency deep blue host for organic light-emitting devices. *Appl. Phys. Lett.* **2007**, *90*, 123506. [CrossRef]

37. Wu, T.-L.; Chou, H.-H.; Huang, P.-Y.; Cheng, C.-H.; Liu, R.-S. 3,6,9,12-Tetrasubstituted chrysenes: Synthesis, photophysical properties, and application as blue fluorescent OLED. *J. Org. Chem.* **2014**, *79*, 267–274. [CrossRef]

38. Cho, S.; Kim, C.; Lee, S.E.; Kim, Y.K.; Yoon, S.S. Naphthalene derivatives end-capped with 2-(diphenylamino)-9,9-diethylfluorenes for blue organic light-emitting diodes. *J. Nanosci. Nanotechnol.* **2018**, *18*, 1251–1255. [CrossRef]

39. Lee, K.H.; Kang, L.K.; Lee, J.Y.; Kang, S.; Jeon, S.O.; Yook, K.S.; Lee, J.Y.; Yoon, S.S. Molecular engineering of blue fluorescent molecules based on silicon end-capped diphenylaminofluorene derivatives for efficient organic light-emitting materials. *Adv. Funct. Mater.* **2010**, *20*, 1345–1358. [CrossRef]

40. Maeda, H.; Maeda, T.; Mizuno, K. Absorption and fluorescence spectroscopic properties of 1- and 1,4-silyl-substituted naphthalene derivatives. *Molecules* **2012**, *17*, 5108–5125. [CrossRef]

41. Koo, J.Y.; Heo, C.H.; Shin, Y.-H.; Kim, D.; Lim, C.S.; Cho, B.R.; Kim, H.M.; Park, S.B. Readily accessible and predictable naphthalene-based two-photon fluorophore with full visible-color coverage. *Chem. Eur. J.* **2016**, *22*, 14166–14170. [CrossRef]

42. Sheldrick, G.M. *SHELXS-97, Programs for Crystal Structure Solution*; University of Göttingen: Göttingen, Germany, 1997.

43. Sheldrick, G.M. *SADABS, v. 2.10, Bruker/Siemens Area Detector Absorption Correction Program*; Bruker AXS Inc.: Madison, WI, USA, 2002.

44. Sheldrick, G.M. *SHELXL-97, Programs for Crystal Structure Refinement*; University of Göttingen: Göttingen, Germany, 1997.

45. Farrugia, L.J. ORTEP-3 for Windows—A version of Ortep-III with a graphical user interface (GUI). *J. Appl. Cryst.* **1997**, *30*, 565. [CrossRef]

Sample Availability: Samples of the compounds **3–5** are available from the authors.

 *molecules* 

Article

Iron-Catalyzed Cross-Coupling of *Bis*-(aryl)manganese Nucleophiles with Alkenyl Halides: Optimization and Mechanistic Investigations

Lidie Rousseau [1,2], Alexandre Desaintjean [3], Paul Knochel [3] and Guillaume Lefèvre [1,*]

[1] Chimie ParisTech, PSL University, CNRS, Institute of Chemistry for Life and Health Sciences (i-CLeHS FRE2027), CSB2D, 75005 Paris, France; lidie.rousseau@cea.fr

[2] NIMBE, CEA, CNRS, Univ. Paris-Saclay, 91191 Gif-sur-Yvette, France

[3] Department of Chemistry, Ludwig-Maximilians-Universitat München, Butenandstr. 5-13, Haus F, 81377 Munich, Germany; Alexandre.Desaintjean@cup.uni-muenchen.de (A.D.); Paul.Knochel@cup.uni-muenchen.de (P.K.)

* Correspondence: guillaume.lefevre@chimieparistech.psl.eu; Tel.: +33-1-85-78-41-70

Received: 15 January 2020; Accepted: 6 February 2020; Published: 7 February 2020

Abstract: Various substituted *bis*-(aryl)manganese species were prepared from aryl bromides by one-pot insertion of magnesium turnings in the presence of LiCl and in situ trans-metalation with $MnCl_2$ in THF at $-5\,°C$ within 2 h. These *bis*-(aryl)manganese reagents undergo smooth iron-catalyzed cross-couplings using 10 mol% $Fe(acac)_3$ with various functionalized alkenyl iodides and bromides in 1 h at 25 °C. The aryl-alkenyl cross-coupling reaction mechanism was thoroughly investigated through paramagnetic ^1H-NMR, which identified the key role of *tris*-coordinated *ate*-iron(II) species in the catalytic process.

Keywords: iron catalysis; cross-coupling; *bis*-(aryl)manganese; alkenyl halides; *ate* iron(II) complex

---

1. Introduction

Transition-metal catalyzed cross-couplings are widely used in the development and production of pharmaceutical compounds [1]. The most versatile of them are palladium-catalyzed and nickel-catalyzed cross-couplings [2–5] as they tolerate a great variety of functionalities on both coupling partners. Yet, these metals have drawbacks such as toxicity [6,7] and high prices in the case of palladium [8]. That is one of the reasons why copper [9], iron [10–13], or cobalt [14] have been developed as alternative metal-catalysts.

Organomanganese species pioneered by Cahiez [15] often considerably reduce the amount of side reactions such as homo-coupling [16,17] and have proven to be excellent nucleophiles in various types of reactions [18–24] including cross-couplings [25,26]. Organomanganese species then constitute an interesting alternative to usual cross-coupling partners such as organomagnesium [27], organozinc [28], and organo-boronic esters, which may have genotoxic properties [29,30].

Recently, we have developed a two-step preparation of functionalized *bis*-(aryl)manganese reagents by oxidative insertion of magnesium into the C-Br bond of aryl bromides, which is followed by a trans-metalation with $MnCl_2 \cdot 2LiCl$ [31]. Herein, we wish to report an effective one-pot preparation of those functionalized *bis*-(aryl)manganese reagents ($Ar_2Mn \bullet 2MgX_2 \bullet 4LiCl$, denoted as Ar_2Mn (1), Scheme 1) starting from aryl bromides, which are followed by an iron-catalyzed cross-coupling of 1 with alkenyl iodides and bromides, and provide a range of polyfunctionalized alkenes (4, Scheme 1). These *bis*-(aryl)manganese reagents are generally stable at RT (25 °C) for several hours, which makes them suitable reagents for mild cross-coupling reactions [32].

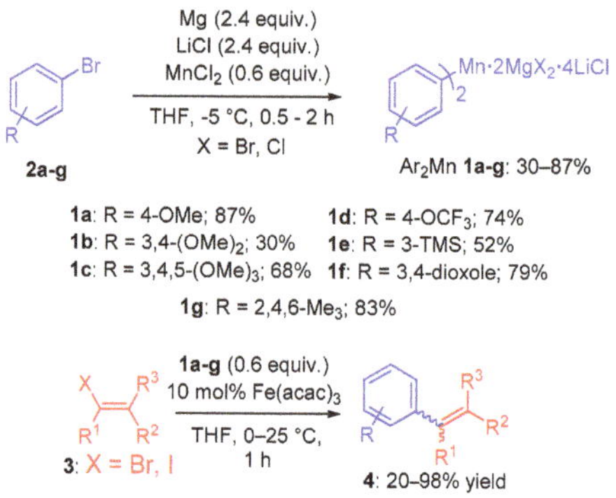

Scheme 1. One-pot preparation of *bis*-(aryl)manganese reagents by in situ trans-metalation followed by iron-catalyzed cross-couplings with alkenyl iodides and bromides.

2. Results

In preliminary experiments, the *bis*-(aryl)manganese reagent **1a** was conveniently prepared by treating 4-bromoanisole (**2a**, 1.0 equiv.) in THF at −5 °C with magnesium turnings and LiCl (2.4 equiv.) in the presence of MnCl$_2$ (0.6 equiv.) within 1 h. Titration [31] with iodine led to a yield for **1a** of 87%. Cross-coupling methodologies involving those *bis*-(aryl)manganese species were then investigated (see Supporting Information file for details).

In the absence of any iron catalyst, the cross-coupling of **1a** with (*Z*)-ethyl 3-iodoacrylate (**3a**; 25 °C, 1 h) produced a *Z/E* = 50:50 mixture of the desired cross-coupling product **4a** in 60% yield (Table 1, entry 1). Although the cross-coupling performed with FeBr$_2$ gave a moderate yield of 42%, the use of FeCl$_3$, FeBr$_3$, or FeCl$_2$ afforded the *E* isomer of **4a** in 54–64% yield (entries 2–5). Using Fe(acac)$_2$ proved to be more effective since the yield increased to 67% (entry 6). Our best result was obtained with Fe(acac)$_3$ (>99% purity) as a catalyst, producing the *E* isomer of **4a** in a 79% yield (entry 7).

Table 1. Catalyst screening of the reaction between the *bis*-(aryl)manganese reagent **1a** and (*Z*)-ethyl 3-iodoacrylate (**3a**).

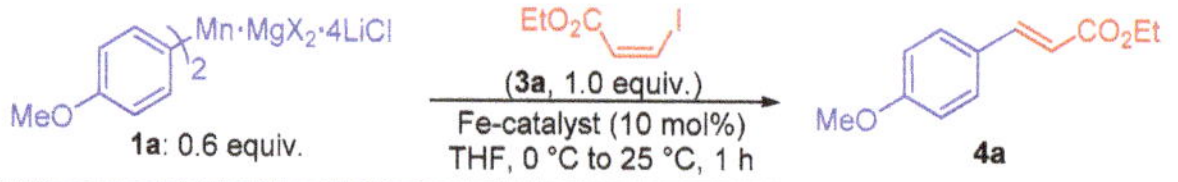

Entry	Catalyst (10 mol%)	Yield (%) [a]
1	none	60 [b]
2	FeBr$_2$	42
3	FeCl$_3$	54
4	FeBr$_3$	57
5	FeCl$_2$	64
6	Fe(acac)$_2$	67
7	Fe(acac)$_3$ (>99% purity)	79

[a] Yield of analytically pure product. [b] A *Z/E* = 50:50 mixture of **4a** was obtained.

Furthermore, the cross-coupling of 1a with (2-bromovinyl)trimethylsilane (**3b**; *Z/E* = 10:90) gave the olefin **4b** in 98% yield with complete *E*-selectivity (*Z/E* = 1:99) whereas the yield without iron salt was 24% (*Z/E* = 20:80, Table 2, entry 1). When the electron-rich *bis*-(3,4-dimethoxyphenyl)manganese (**1b**) was mixed with **3a**, the *E*-acrylate **4c** was generated in 69% yield and a *Z/E* = 69:31 mixture of products was obtained in 58% yield without an iron catalyst (entry 2). The tri-substituted *bis*-(3,4,5-trimethoxyphenyl)manganese (**1c**) underwent smooth cross-coupling with **3b** to afford the *E*-alkene **4d** in 80% (8% were obtained without a catalyst, entry 3). Additionally, **1c** reacted with

3a and 2-bromostyrene (**3c**; *Z/E* = 18:82) to give the acrylate **4e** and **4f** (*Z/E* = 1:99) in 57% and 82% yield whereas 66% (*Z/E* = 72:28) and 80% (*Z/E* = 53:47) were, respectively, obtained without a catalyst (entries 4–5). In the last experiment, **1c** reacted with (*E*)-1-iodooctene (**3d**) to provide the alkene **4g** in 87% yield (*Z/E* = 9:91) when 77% yield (*Z/E* = 4:96) was obtained without a catalyst (entry 6). Furthermore, *bis*-(4-(trifluoromethoxy)phenyl)manganese (**1d**) reacted with **3a** to provide the acrylate **4h** (*Z/E* = 1:99) in 77% yield (entry 7). The reaction without Fe(acac)$_3$ gave a similar yield but a mix of the two isomers (74%, *Z/E* = 81:19, entry 7). The silicon-containing *bis*-(aryl)manganese reagent **1e** could also react with **3a**, which produces **4i** (*Z/E* = 1:99) in 64% yield (51%, *Z/E* = 57:43 were obtained without Fe(acac)$_3$, entry 8). Some good yields could be achieved in the absence of the iron catalyst (entries 2, 4–8), which could be attributed to the catalytic activity of the manganese(II) itself. For example, manganese salts proved to efficiently catalyze several couplings of organomagnesium reagents with alkenyl electrophiles in the past [32]. The *bis*-benzo[d][1,3]dioxol-5-ylmanganese (**1f**) also reacted with **3a** and **3b** to yield the *E*-alkenes **4j** and **4k** in 78–84% yield (entries 9–10). The bulkier *bis*-mesitylmanganese **1g** reacted with **3b** to afford **4l** with a small 20% yield (91% after 18 h, entry 11). This method also proved to tolerate nitriles, since 4-(2-bromovinyl)benzonitrile **3e** (*Z/E* = 98:2) could be used as a coupling partner with **1d** and **1e** in good yields (entries 12–13). Moreover, the coupling generally proceeds with an excellent *E*-selectivity when iodoalkenes are used. Total isomerization is observed for *Z*-starting iodoalkenes (entries 2, 4, 7, 8, 10). A similar tendency is observed for bromoalkenes, with the exception of 4-(2-bromovinyl)benzonitrile **3e** (*Z/E* = 98:2), which, intriguingly, did not lead to isomerization of the starting *Z* bond when coupled with nucleophile **1d**, entry 12, or to a slight isomerization when coupled with **1e**, entry 13. In all cases, efficient transference of both aryl groups from the starting *bis*-(aryl)manganese species has been observed.

Table 2. Iron-catalyzed couplings of *bis*-(aryl)manganese (**1a–g**) [a] with alkenyl electrophiles (**3a–e**).

Reaction scheme: Ar$_2$Mn (**1a–g**, 0.6 equiv.) + alkenyl electrophile (**3a–e**, 1.0 equiv.), Fe(acac)$_3$ (10 mol%), THF, 0 °C to 25 °C, 1 h → **4a-n**: 20-98%. X = Br, I. R = 3-TMS; 4-OMe; 4-OCF$_3$; 3,4-(OMe)$_2$; 3,4,5-(OMe)$_3$; 3,4-dioxole; 2,4,6-Me$_3$.

Entry	Ar$_2$Mn	Electrophile	Yield (%) [b]	Entry	Ar$_2$Mn	Electrophile	Yield (%) [b]
1	**1a**	**3b**: *Z/E* = 10:90	**4b**: 98, *Z/E* = 1:99 (24, *Z/E* = 20:80) [c]	8	**1e**	**3a**	**4i**: 64, *Z/E* = 1:99 (51, *Z/E* = 57:43) [c]
2	**1b**	**3a**	**4c**: 69, *Z/E* = 1:99 (58, *Z/E* = 69:31) [c]	9	**1f**	**3b**: *Z/E* = 10:90	**4j**: 78, *Z/E* = 1:99 (39, *Z/E* = 20:80) [c]
3	**1e**	**3b**: *Z/E* = 10:90	**4d**: 80, *Z/E* = 1:99 (8, *Z/E* = 1:99) [c]	10	**1f**	**3a**	**4k**: 84, *Z/E* = 1:99 (48, *Z/E* = 99:1) [c]
4	**1c**	**3a**	**4e**: 57, *Z/E* = 1:99 (66, *Z/E* = 72:28) [c]	11	**1g**	**3b**: *Z/E* = 10:90	**4l**: [d] 20, *Z/E* = 21:79 91, *Z/E* = 5:95 [e] (traces, *Z/E* = 50:50) [c]

Table 2. *Cont.*

Entry	Ar$_2$Mn	Electrophile	Yield (%) [b]	Entry	Ar$_2$Mn	Electrophile	Yield (%) [b]
5	1c	3c: Z/E = 18:82	4f: 82, Z/E = 1:99 (80, Z/E = 53:47) [c]	12	1d	3e: Z/E = 98:2	4m: 74, Z/E = 98:2 (19, Z/E = 73:27) [c]
6	1c	3d	4g: 87, Z/E = 9:91 (77, Z/E = 4:96) [c]	13	1e	3e: Z/E = 98:2	4n: 66, Z/E = 70:30 (41, Z/E = 72:28) [c]
7	1d	3a	4h: 77, Z/E = 1:99 (74, Z/E = 81:19) [c]				

[a] For clarity reasons, the magnesium salt has been omitted. [b] Yield of analytically pure product. [c] In parentheses, yield and Z/E ratio obtained without catalysis. [d] Yields determined by GC and ^{1}H-NMR. [e] After 18 h.

3. Discussion

In order to rationalize some of the mechanistic features of the transformations reported in the first section, we focused our efforts on the coupling system involving various *bis*-(aryl)manganese nucleophiles (*bis*-mesitylmanganese and *bis*-phenylmanganese) with (2-bromovinyl)-trimethylsilane (**3b**). This choice has been motivated by the low cross-coupling yields obtained with this electrophile in the absence of the iron catalyst, which ascertains the requirement of an Fe-based catalysis for this coupling (see Table 2, entries 1, 3, 9 and 11).

The *bis*-(mesityl)manganese reagent was prepared by adding MesMgBr (2.0 equiv.) into a solution of MnCl$_2$•LiCl (1.0 equiv.) in THF at −5 °C within 1 h. ^{1}H-NMR showed no free MesMgBr left. The spectra presented high signal-to-noise ratios and broad signals, due to the high paramagnetism of manganese(II) species, which could not be attributed to a specific molecule (Figure 1a) [33]. High-spin organomanganese compounds are often reported to be NMR silent [34] or without NMR characterization at all [35–37]. Yet, after addition of a catalytic load of FeCl$_2$ (0.10 equiv.) to the Mes$_2$Mn solution, the *ate* complex [Mes$_3$Fe^{II}]$^-$ was detected by ^{1}H-NMR from the three signals at 127 ppm (s, 6H, *meta*-H of the Mes group), 110 ppm (s, 9H, *para*-CH$_3$), and 26 ppm (bs, 18 H, *ortho*-CH$_3$), as shown in Figure 1b. These signals attest to a strong paramagnetism, due to the high-spin (S = 2) configuration of this complex [38]. This proves a fast trans-metalation of the aryl groups from the manganese toward the iron(II) center. A similar result was obtained while adding an excess of the Mes$_2$Mn solution onto Fe(acac)$_3$ (0.10 equiv.) as an iron(III) precursor. [Mes$_3$Fe^{II}]$^-$ was detected by ^{1}H-NMR, showing that, when an iron(III) precursor is used, a first 1-electron reduction of the latter by the nucleophile can take place, affording an iron(II) species. This is in agreement with recent reports by Neidig [39] and by some of us [40] regarding the reduction of iron(III) salts by Grignard reagents as MeMgBr and PhMgBr. Accordingly, all the mechanistic studies discussed thereafter were performed using an iron(II) precursor.

Upon addition of the electrophile **3b** ((2-bromovinyl)trimethylsilane) to a mixture of FeCl$_2$ and Mes$_2$Mn at 25 °C, the signals corresponding to [Mes$_3$Fe^{II}]$^-$ were observed to slowly decrease, affording [Mes$_2$BrFe^{II}]$^-$, characterized by new signals at 128 ppm (s, 4H, *meta*-H of the Mes group), 104 ppm (s, 6H, *para*-CH$_3$), and 29 ppm (bs, 12 H, *ortho*-CH$_3$) (see Figure 1c) (this tricoordinate *ate* species also presents a high-spin S = 2 configuration) [38]. The same reaction was run at 25 °C for 1 h, then quenched, and analyzed by GC-MS, which proved formation of the desired cross-coupling product with a low conversion (ca. 20%). This is in fair agreement with the result given in Table 2, entry 11 (due to the high paramagnetism of the NMR-analyzed solution and due to the presence of non-deuterated solvents (THF solutions of organometallics), NMR monitoring of the coupling product formation could not be efficiently performed).

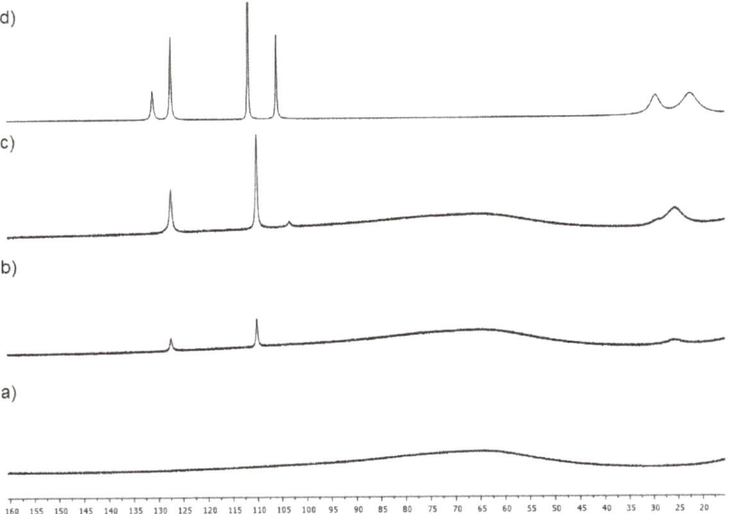

Figure 1. ^{1}H-NMR spectra (recorded at 25 °C in d_8-THF) of a 0.08 M solution of (**a**) Mes$_2$Mn, (**b**) Mes$_2$Mn + 0.10 equiv. FeCl$_2$, (**c**) Mes$_2$Mn + 0.10 equiv. FeCl$_2$ + 1.0 equiv. **3b**, (**d**) FeCl$_2$ + 3.0 equiv. MesMgBr + 10 equiv. **3b**.

The same observations were made while performing the coupling of **3b** with MesMgBr as a sole nucleophilic partner in a Kumada-type reaction using FeCl$_2$, as [Mes$_3$Fe^{II}]$^-$ and [Mes$_2$BrFe^{II}]$^-$ were also detected under these cross-coupling conditions (Figure 1d, in the absence of manganese, the signals of [Mes$_2$BrFe^{II}]$^-$ shifted to 131, 106, and 30 ppm). These series of experiments, therefore, confirm that both [Ar$_3$Fe^{II}]$^-$ and [Ar$_2$BrFe^{II}]$^-$ *ate* complexes are part of a coupling catalytic cycle and [Ar$_3$Fe^{II}]$^-$ can be involved in the activation step of the electrophile. The following catalytic cycle (Scheme 2) can be suggested, which echoes recent reports by Neidig on the Fe-catalyzed alkyl-alkenyl coupling reactions [39], and by ourselves on the benzyl-alkenyl coupling [26].

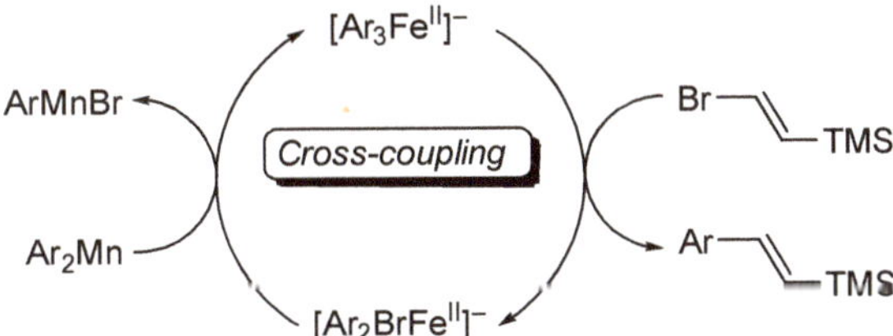

Scheme 2. Proposed catalytic cycle for the aryl-alkenyl cross-coupling between Ar$_2$Mn and an alkenyl bromide under Fe-catalytic conditions.

Thanks to the steric hindrance in the *ortho* positions, the *ate* [Mes$_3$Fe^{II}]$^-$ species remains stable for hours at 25 °C [38,41]. Thus, the use of a mesityl nucleophile in the mechanistic experiments discussed earlier prevents any degradation of the Fe^{II} catalyst toward lower oxidation states. In order to delineate the influence of the formation of lower oxidation states on the system, additional investigations were, therefore, carried out using PhMgBr as a less hindered nucleophile [42].

First, $[Ph_3Fe^{II}]^-$ was generated at $-20\ ^\circ C$ by stoichiometric trans-metalation between $FeCl_2$ and 3.0 equiv. of PhMgBr, and characterized by its ^1H-NMR signals at 116 and -41 ppm. As we recently reported, $[Ph_3Fe^{II}]^-$ is stable for more than 1 h at this temperature [40,41]. Its fate upon addition of an excess of the electrophile **3b** (10 equiv.) was then monitored by paramagnetic ^1H-NMR. $[Ph_3Fe^{II}]^-$ reacted rapidly, as attested by the decrease of its resonances (ca. 75% of the starting $[Ph_3Fe^{II}]^-$ reacted after 10 min). The reaction was quenched after 1 h, and the GC-MS confirmed formation of the cross-coupling product, which confirms that $[Ph_3Fe^{II}]^-$ was able to react with **3b** in a cross-coupling process, akin to $[Mes_3Fe^{II}]^-$. Moreover, several transient resonances in the $-15/-5$ ppm area could also be detected in the course of the reaction (see Figure 2). These elusive resonances quickly disappeared, and were not detected after 30 min at $-20\ ^\circ C$. These signals echo the formation of $(\eta^2\text{-alkene})_n\text{-Fe}^0$ intermediates, as recently reported by Deng, which exhibit similar resonances [43]. This suggests that Fe^0 species are formed in situ by 2-electron reductive elimination from $[Ph_3Fe^{II}]^-$, which is in agreement with a recent report by some of us demonstrating that the evolution of $[Ph_3Fe^{II}]^-$ led to the formation of a distribution of Fe^0 and Fe^I oxidation states (identified as $(\eta^4\text{-arene})_2Fe^0$ and $[(\eta^6\text{-arene})Fe^I(Ph)_2]^-$, "arene" being an aromatic ligand present in the bulk medium (e.g., C_6H_6 or $C_6H_5\text{-}C_6H_5$ coming from the oxidation of PhMgBr). Fe^0 being preponderantly formed [40,41]). Those Fe^0 intermediates would then be trapped by alkene ligands present in the bulk medium, which leads to the observed resonances.

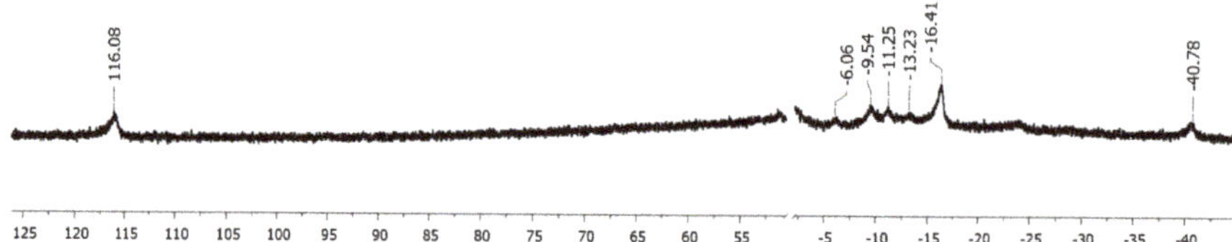

Figure 2. ^1H-NMR spectrum (recorded at $-20\ ^\circ C$ in d_8-THF) of a 0.08 M solution of $[Ph_3Fe^{II}]^-$ followed by addition of **3b** (10 equiv.). Spectrum recorded 20 min after addition of **3b**.

Then, the reactivity of the low valent Fe^0 and Fe^I oxidation states in the reaction medium was investigated. Following one of our recent procedures, reduction of Fe^{II} into a distribution of Fe^0 and Fe^I species was performed, by fast trans-metalation between $FeCl_2$ and PhMgBr (2.0 equiv.) at room temperature [40,41]. After 10 min, 1.0 equiv. of MesMgBr was added. The ^1H-NMR spectrum showed no signal in the 100–150 ppm area, which attests to the absence of any Mes-Fe^{II} species, which shows that all the starting Fe^{II} was reduced by PhMgBr (Figure 3a).

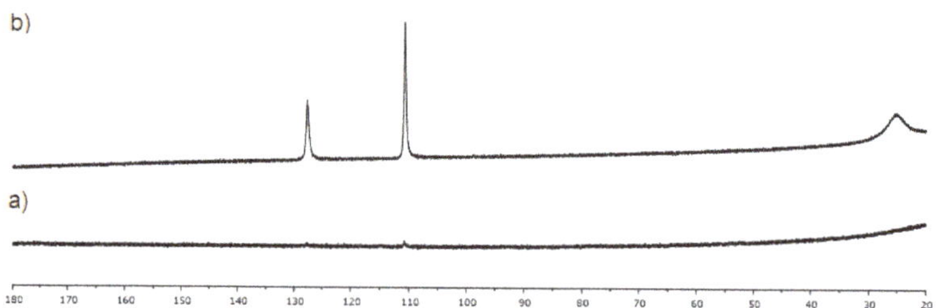

Figure 3. ^1H-NMR spectra (recorded at $25\ ^\circ C$ in d_8-THF) of a 0.08 M solution of (**a**) $FeCl_2$ + 2.0 equiv. PhMgBr, + 1.0 equiv. MesMgBr after 10 min agitation at RT. (**b**) Same tube, + 3.0 equiv. **3b**, after 30 min.

The addition of 3.0 equiv. of the electrophile **3b** to the in situ generated solution of Fe^0 and Fe^I species led to a color change of the sample, which turned from dark brown to yellow. The ^1H-NMR spectrum showed that, after 30 min, ca. 20% of the iron contained in the solution was converted into $[Mes_3Fe^{II}]^-$ (Figure 3b). The presence of **3b**, therefore, allows a re-oxidation of the reduced Fe^0 and/or Fe^I species to the Fe^{II} oxidation state, the latter being trapped by trans-metalation with

MesMgBr to afford [Mes$_3$Fe^{II}]$^-$. The re-oxidation of low iron oxidation states by **3b** to the Fe^{II} stage was also confirmed by the observation of *bis*(trimethylsilyl)butadienes TMS-CH=CH-CH=CH-TMS (TMS-(CH)$_4$-TMS, *E/E*; *Z/E*; *Z/Z*) in GC-MS, after catalytic reactions involving 10 mol% of FeCl$_2$, PhMgBr, and **3b** as coupling partners. Formation of TMS-(CH)$_4$-TMS undoubtedly comes from the sacrificial monoelectronic reduction of the electrophile that permits re-oxidation of the low Fe^0 and/or Fe^I oxidation states. TMS-(CH)$_4$-TMS, moreover, also appears as a suitable ligand for Fe^0 oxidation state, and a (η^4-TMS-(CH)$_4$-TMS)Fe^0 complex might, thus, contribute to the group of high field resonances in the in the $-15/-5$ ppm area (Figure 2). Quantity of detected TMS-(CH)$_4$-TMS represents ca. 10% of the quantity of a detected coupling product, which shows that this off-cycle sacrificial reduction pathway is not preponderant. By comparison, no traces of TMS-(CH)$_4$-TMS were detected when using mesityl nucleophiles (conditions of Figure 1), attesting that no sacrificial 1-electron reduction of **3b** by oxidation states lower than Fe^{II} formed in situ occurred.

Scheme 3 presents a summary of the competitive reactions that were observed in this work during the Fe-catalyzed coupling of Ph$_2$Mn with (2-bromovinyl)-trimethylsilane (**3b**), taking into account the possibility of an off-cycle process involving in situ formed low iron oxidation states.

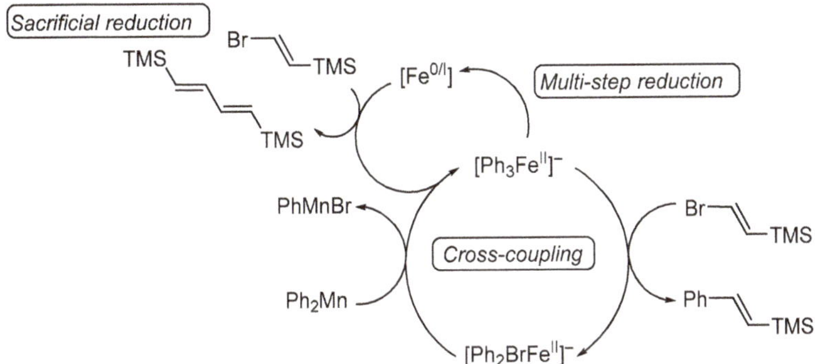

Scheme 3. Proposed catalytic cycle and side reactions for the aryl-alkenyl cross-coupling between Ph$_2$Mn and an alkenyl bromide under Fe-catalytic conditions.

Kinetic studies will further be pursued in order to determine the global kinetics of the aryl-alkenyl cross-coupling reaction, and to examine the possibility for the low Fe^0 and Fe^I oxidation states involved in a cross-coupling catalytic cycle, in addition to the off-cycle sacrificial reduction of the alkenyl electrophile evidenced herein. Such studies will also help analyze the mechanism of the activation of the C-X bond of the alkenyl halide. Isomerization toward the sterically more stable *E* coupling products may suggest the implication of an iron-based radical activation of the alkenyl halide, as observed in the case of alkyl electrophiles [44], albeit formation of the C$_{sp2}$-centered radical is generally more energetically-demanding. Additionally, it cannot be excluded in an alternative scenario that isomerization of the C=C bond occurs after the coupling step, akin to the observations made by Jacobi von Wangelin for the iron-catalyzed isomerization of *Z*-olefins to their *E* analogues [45].

4. Materials and Methods

4.1. Materials and Instruments

All reactions, except otherwise noted, were carried out in flame-dried glassware equipped with magnetic stirring under an argon atmosphere using standard Schlenk techniques. To transfer solvents or reagents, syringes were used, which were purged three times with argon prior to use. After purification by flash column chromatography, products were concentrated using a rotary evaporator and, subsequently, dried under high vacuum. Indicated yields are isolated yields of compounds estimated to be >95% pure, as determined by ^{1}H-NMR (25 °C) and capillary GC.

To examine the reaction progress of the performed reactions, GC-analysis of quenched hydrolyzed and iodolyzed reaction aliquots relative to an internal standard was used. For this purpose, small amounts of the reaction mixture were hydrolyzed using a saturated aqueous solution of NH_4Cl, subsequently extracted with EtOAc, dried over $MgSO_4$ and gaschromatographically quantified. To monitor the process of directed metalations and oxidative insertion reactions, small amounts of the reaction mixture were iodolyzed. A small quantity of iodine was dissolved in freshly distilled THF (0.50 mL), charged with the reaction mixture, and added to a solution of $Na_2S_2O_3$. The mixture was extracted with EtOAc, dried over $MgSO_4$, and was then gas chromatographically measured.

To determine the concentration of the different synthesized metallic reagents, iodometric titration was used. For this purpose, a known amount of iodine was charged with freshly distilled THF (1.00 mL) to give a deep red solution. The metallic reagent was added dropwise at 2 °C to the iodine solution until the red coloration went colorless. The concentration of the organometallic reagent could be calculated via the consumed volume of the reaction mixture and the amount of used iodine.

Thin layer chromatography (TLC) was implemented on alumina plates coated with SiO_2 (Merck 60, F-254, Merck, Darmstadt, Germany). To visualize the spots of the different products, UV light was used.

Flash column chromatography was performed using SiO_2 (0.04–0.06 mm, 230–400 mesh) from Merck.

^{1}H-NMR, ^{13}C-NMR, ^{19}F-NMR, and 2D-NMR spectra were recorded on VARIAN Mercury 200, BRUKER ARX 300, VARIAN VXR 400 S, and BRUKER AMX 600 instruments (Bruker, Billerica, MA, USA). Chemical shifts are reported as δ-values in ppm relative to tetramethylsilane. The following abbreviations were used to characterize signal multiplicities: s (singlet), d (doublet), t (triplet), q (quartet), m (multiplet), and bs (broad singlet).

Mass spectroscopy: High resolution (HRMS) and low resolution (MS) spectra were recorded on a FINNIGAN MAT 95Q instrument (now Thermo Fisher company, Waltham, MA, USA). Electron impact ionization (EI) was conducted with an ionization energy of 70 eV. For coupled gas chromatography/mass spectrometry, a HEWLETT-PACKARD HP 6890 /MSD 5973 GC/MS system was used. Molecular fragments are reported starting at a relative intensity of 10%.

Infrared spectra (IR) were recorded from 4500 cm^{-1} to 650 cm^{-1} on a PERKIN ELMER Spectrum BX59343 instrument (Perkin Elmer, Wellesley, MA, USA). For detection, a SMITHS DETECTION DuraSamplIR IIDiamond ATR sensor (Smiths Detection, Hemel Hempstead, UK) was used. Wavenumbers are reported in cm^{-1} starting at an absorption of 10%.

Melting points (m.p.) were determined on a BÜCHI B-540 melting point apparatus (BÜCHI LabortechnikAG, Flawil, Switzerland) and are uncorrected. Compounds decomposing upon melting are indicated by (decomp.).

Gas chromatography was executed with machines of type Agilent Technologies 7890A GC-Systems with 6890 GC inlets, detectors (Agilent, Santa Clara, CA, USA), a GC oven, and a column of type HP 5 (Hewlett-Packard, 5% phenylmethylpolysiloxane; length: 10 m, diameter: 0.25 mm, film thickness: 0.2 μm).

Gas chromatography-Mass spectra were recorded on a networking system called Hewlett-Packard 6890/MSD 5973 GC/MS (Hewlett Packard, Palo Alto, CA, USA) with a column of type HP 5 (Hewlett-Packard, 5% phenylmethylpolysiloxane; length: 10m, diameter: 0.25 mm, film thickness: 0.2 μm).

4.2. Chemicals, Solvents, and Typical Procedures

All chemicals were purchased from commercial sources and were used without any further purification unless otherwise noted.

THF was continuously refluxed and freshly distilled from benzophenone ketyl under nitrogen. The freshly distilled THF was stored over a molecular sieve (4 Å) under argon. Solvents for column chromatography were distilled prior to use.

4.2.1. Typical Procedure for the One-Pot Preparation of *Bis*-(aryl)manganese Reagents **1a–g**

A dry and argon-flushed Schlenk-tube, equipped with a magnetic stirring bar and a rubber septum, was charged with LiCl (0.610 g, 14.4 mmol, 2.4 equiv.), heated to 450 °C under high vacuum, and then cooled to room temperature. After being switched to argon, the same procedure was applied after $MnCl_2$ was added (453 mg, 3.60 mmol, 0.6 equiv.). After cooling to room temperature, magnesium turnings were added (0.350 g, 14.4 mmol, 2.4 equiv.), which was followed by freshly distilled THF (12 mL). After the reaction mixture was cooled to −5 °C, the aryl bromides **2a–g** were then added dropwise using 1 mL syringes (6.0 mmol, 1.0 equiv., addition time: 1 min) and the reaction mixture was stirred until a complete conversion of the starting material was observed. The reaction progress was monitored by GC-analysis of hydrolyzed and iodolyzed aliquots.

When the metalation was completed, the concentration of the *bis*-(aryl)manganese species was determined by titration against iodine in freshly distilled THF. The black solutions of the aryl reagents **1a–g** were then separated from the magnesium turnings using a syringe and, subsequently, transferred into another pre-dried and argon-flushed Schlenk-tube, which was cooled to −5 °C. After a titration against iodine in freshly distilled THF was performed, the reagent was ready to use for Cross-Couplings.

4.2.2. Typical Procedure for the Cross-Coupling Reactions of *Bis*-(aryl)manganese Reagents **1a–g** with Different Electrophiles **3a–e**

A pre-dried and argon-flushed Schlenk-tube equipped with a magnetic stirring bar and a rubber septum was charged with $Fe(acac)_3$ (35 mg, 0.10 mmol, 10 mol%), the corresponding electrophile (**3a–e**, 1.0 mmol, 1.0 equiv.), tetradecane as internal standard (50 µL) and freshly distilled THF (1.0 mL) as solvent. The reaction mixture was cooled to 0 °C and the *bis*-(aryl)manganese solution (**1a–g**, 0.6 equiv.) was added dropwise whereupon a color change to dark brown could be recognized. After the addition was complete, the reaction mixture was stirred for a given time at room temperature and the completion of the cross-coupling reaction was monitored by GC-analysis of hydrolyzed aliquots. Thereupon, a saturated aqueous solution of NH_4Cl was added and the aqueous layer was extracted with EtOAc (3×100 mL). The combined organic layers were dried over $MgSO_4$, filtered, and concentrated under reduced pressure. Purification of the crude products by flash column chromatography afforded the desired cross-coupling reaction products (**4a–k**, **4m**, **4n**).

4.3. Studies on the Catalytically Active Species and Catalytic Cycle

All the samples were prepared in a recirculating JACOMEX inert atmosphere (Ar) drybox and vacuum Schlenk lines. Glassware was dried overnight at 120 °C before use. NMR spectra were obtained using a Bruker DPX 400 MHz spectrometer (Bruker, Billerica, MA, USA). Chemical shifts for ^{1}H-NMR spectra were referenced to solvent impurities (herein, THF). NMR tubes were equipped with a J. Young valves were used for all ^{1}H-NMR experiments and catalytic tests. The GC-MS analysis was performed using *n*-decane as an internal standard. The reaction media aliquots were quenched by the addition of distilled water under air. The organic products were extracted using DCM, and injected into the GC-MS. Mass spectra were recorded on a Hewlett-Packart HP 5973 mass spectrometer (Hewlett Packard, Palo Alto, CA, USA) via a GC-MS coupling with a Hewlett-Packart HP 6890 chromatograph (Hewlett Packard, Palo Alto, CA, USA) equipped with a capillary column HP-5MS (50 m × 0.25 mm × 0.25 µm, Hewlett Packard, Palo Alto, CA, USA). Ionisation was due to an electronic impact (EI, 70 eV).

5. Conclusions

In summary, various functionalized *bis*-(aryl)manganese species have been readily prepared in one-pot conditions from the corresponding aryl bromides by inserting magnesium in the presence of

LiCl and in situ trans-metalation with $MnCl_2$ in THF at $-5\ °C$ within 2 h. These *bis*-(aryl)manganese reagents have been allowed to undergo smooth iron-catalyzed cross-couplings using 10 mol% $Fe(acac)_3$ and various functionalized alkenyl iodides and bromides at 25 °C for 1 h. Mechanistic investigations carried out by ^{1}H-NMR showed that *ate*-iron(II) species $[Ar_3Fe^{II}]^-$ are formed by trans-metalation of the *bis*-(aryl)manganese reagent with the iron catalyst, and that they can react with alkenyl bromides to afford the expected cross-coupling product. Low-valent Fe^0 and Fe^I oxidation states can also be formed by the reduction of the *ate*-iron(II) catalyst under these conditions. This leads to the sacrificial reduction of the alkenyl electrophile via an off-cycle pathway, which partly regenerates the Fe^{II} oxidation state, where the latter is able to enter a new catalytic cycle.

Supplementary Materials: The following are available online. Additional synthesis and characterization (NMR, IR, HRMS, m.p.) data are available online.

Author Contributions: Conceptualization, G.L. and P.K. Methodology, A.D. and L.R. Writing—original draft preparation, G.L., L.R., and A.D. Writing—review and editing, G.L. and P.K. Supervision, G.L. and P.K. Project administration, G.L. and P.K. Funding acquisition, G.L. and P.K. All authors have read and agreed to the published version of the manuscript.

Funding: This work was supported by the CNRS, Chimie ParisTech (Paris), and LMU (Munich) in the framework of the International Associated Laboratory IrMaCar.

Acknowledgments: We thank the DFG (SFB749) for financial support, and Albemarle (Germany) and BASF (Ludwigshafen, Germany) for the generous gift of chemicals. G.L. also thanks the ANR research program (project JCJC SIROCCO). Aurélien Adenot (CEA Saclay) is warmly thanked for his kind technical assistance.

Conflicts of Interest: The authors declare no conflict of interest.

References and Notes

1. Crawley, M.L.; Trost, B.M. *Applications of Transition Metal Catalysis in Drug Discovery and Development: An Industrial Perspective*; John Wiley & Sons: Hoboken, NJ, USA, 2012.
2. Miyaura, N. Cross-Coupling Reactions. A Practical Guide. *Org. Process Res. Dev.* **2003**, *7*, 1084. [CrossRef]
3. Phapale, V.B.; Cárdenas, D.J. Nickel-Catalysed Negishi cross-coupling reactions: Scope and mechanisms. *Chem. Soc. Rev.* **2009**, *38*, 1598–1607. [CrossRef] [PubMed]
4. Hartwig, J.F. Organotransition Metal Chemistry. From Bonding to Catalysis. *Angew. Chem. Int. Ed.* **2010**, *49*, 7622. [CrossRef]
5. Jana, R.; Pathak, T.P.; Sigman, M.S. Advances in Transition Metal (Pd,Ni,Fe)-Catalyzed Cross-Coupling Reactions Using Alkyl-organometallics as Reaction Partners. *Chem. Rev.* **2011**, *111*, 1417–1492. [CrossRef] [PubMed]
6. $LD_{50}(FeCl_2$, rat oral) = 900 mg/kg; $LD_{50}(NiCl_2$, rat oral) = 186 mg/kg.
7. Egorova, K.S.; Ananikov, V.P. Which Metals are Green for Catalysis? Comparison of the Toxicities of Ni, Cu, Fe, Pd, Pt, Rh, and Au Salts. *Angew. Chem. Int. Ed.* **2016**, *55*, 12150–12162. [CrossRef] [PubMed]
8. $FeCl_2$ ca. 332 €/mol, $PdCl_2$ ca. 6164 €/mol; prices retrieved from Alfa Aesar in August 2019.
9. Thapa, S.; Shrestha, B.; Gurung, S.K.; Giri, R. Copper-catalysed cross-coupling: An untapped potential. *Org. Biomol. Chem.* **2015**, *13*, 4816–4827. [CrossRef]
10. Fürstner, A.; Leitner, A.; Méndez, M.; Krause, H. Iron-Catalyzed Cross-Coupling Reactions. *J. Am. Chem. Soc.* **2002**, *124*, 13856–13863. [CrossRef]
11. Bedford, R.B.; Brenner, P.B. *Iron Catalysis II*; Bauer, E., Ed.; Springer: Berlin, Germany, 2015.
12. Cahiez, G.; Moyeux, A.; Cossy, J. Grignard Reagents and Non-Precious Metals: Cheap and Eco-Friendly Reagents for Developing Industrial Cross-Couplings. A Personal Account. *Adv. Synth. Catal.* **2015**, *357*, 1983–1989. [CrossRef]
13. Bauer, I.; Knölker, H.-J. Iron Catalysis in Organic Synthesis. *Chem. Rev.* **2015**, *115*, 3170–3387. [CrossRef]
14. Cahiez, G.; Moyeux, A. Cobalt-Catalyzed Cross-Coupling Reactions. *Chem. Rev.* **2010**, *110*, 1435–1462. [CrossRef]
15. Cahiez, G.; Duplais, C.; Buendia, J. Chemistry of Organomanganese(II) Compounds. *Chem. Rev.* **2009**, *109*, 1434–1476. [CrossRef] [PubMed]

16. Peng, Z.; Li, N.; Sun, X.; Wang, F.; Xu, L.; Jiang, C.; Song, L.; Yan, Z.-F. The transition-metal-catalyst-free oxidative homocoupling of organomanganese reagents prepared by the insertion of magnesium into organic halides in the presence of $MnCl_2 \cdot 2LiCl$. *Org. Biomol. Chem.* **2014**, *12*, 7800–7809. [CrossRef] [PubMed]

17. Haas, D.; Hammann, J.M.; Moyeux, A.; Cahiez, G.; Knochel, P. Oxidative Homocoupling of Diheteroaryl- or Diarylmanganese Reagents Generated via Directed Manganation Using TMP_2Mn. *Synlett* **2015**, *26*, 1515. [CrossRef]

18. Cahiez, G.; Masuda, A.; Bernard, D.; Normant, J.F. Reactivite des derives organo-manganeux.I-action sur les chlorures d'acides. synthese de cetones. *Tetrahedron Lett.* **1976**, *36*, 3155–3156. [CrossRef]

19. Friour, G.; Alexakis, A.; Cahiez, G.; Normant, J.F. Reactivite des derives organomanganeux—VIII: Préparation de cétones par acylation d'organomanganeux. Influence de la nature de l'agent acylant, des solvants et des ligands. *Tetrahedron* **1984**, *40*, 683–693. [CrossRef]

20. Kauffmann, T.; Bisling, M. Zwei neue reaktionsweisen von alkyl-Mn(II)-verbindungen: 1,4-addition an cyclohex-2-enon(1) und desoxygenierung von oxiranen (1). *Tetrahedron Lett.* **1984**, *25*, 293–296. [CrossRef]

21. Cahiez, G.; Alami, M. Organomanganese (II) reagents XI.: A study of their reactions with cyclic conjugated enones: Conjugate addition and reductive dimerization. *Tetrahedron Lett.* **1986**, *27*, 569–572. [CrossRef]

22. Cahiez, G.; Alami, M. Organomanganese (II) reagents XVI1: Copper-catalyzed 1,4-addition of organomanganese chlorides to conjugated enones. *Tetrahedron Lett.* **1989**, *30*, 3541–3544. [CrossRef]

23. Cahiez, G.; Alami, M. Composés organomanganeux XXI. Réaction d'un composé organomanganeux avec une énone conjuguée: Influence d'un acide de Lewis, d'un sel de fer ou d'un sel de nickel. *J. Organomet. Chem.* **1990**, *397*, 291–298. [CrossRef]

24. Quinio, P.; Benischke, A.D.; Moyeux, A.; Cahiez, G.; Knochel, P. New Preparation of Benzylic Manganese Chlorides by the Direct Insertion of Magnesium into Benzylic Chlorides in the Presence of $MnCl_2 \cdot 2LiCl$. *Synlett* **2015**, *26*, 514. [CrossRef]

25. Cahiez, G.; Marquais, S. Highly Chemo- and Stereoselective Fe-Catalyzed Alkenylation of Organomanganese Reagents. *Tetrahedron Lett.* **1996**, *37*, 1773–1776. [CrossRef]

26. Desaintjean, A.; Belrhomari, S.; Rousseau, L.; Lefèvre, G.; Knochel, P. Iron-Catalyzed Cross-Coupling of Functionalized Benzylmanganese Halides with Alkenyl Iodides, Bromides, and Triflates. *Org. Lett.* **2019**, *21*, 8684–8688. [CrossRef] [PubMed]

27. Tamao, K.; Sumitani, K.; Kumada, M. Selective carbon-carbon bond formation by cross-coupling of Grignard reagents with organic halides. Catalysis by nickel-phosphine complexes. *J. Am. Chem. Soc.* **1972**, *94*, 4374–4376. [CrossRef]

28. Haas, D.; Hammann, J.F.; Greiner, R.; Knochel, P. Recent Developments in Negishi Cross-Coupling Reactions. *ACS Catal.* **2016**, *6*, 1540–1552. [CrossRef]

29. O'Donovan, M.R.; Mee, C.D.; Fenner, S.; Teasdale, A.; Phillips, D.H. Boronic acids-a novel class of bacterial mutagen. *Mutat. Res.* **2011**, *724*, 1–6. [CrossRef]

30. Hansen, M.M.; Jolly, R.A.; Linder, R.J. Boronic Acids and Derivatives-Probing the Structure-Activity Relationships for Mutagenicity. *Org. Process Res. Dev.* **2015**, *19*, 1507–1516. [CrossRef]

31. Benischke, A.D.; Desaintjean, A.; Juli, T.; Cahiez, G.; Knochel, P. Nickel-Catalyzed Cross-Coupling of Functionalized Organo-manganese Reagents with Aryl and Heteroaryl Halides Promoted by 4-Fluorostyrene. *Synthesis* **2017**, *49*, 5396. [CrossRef]

32. Cahiez, G.; Gager, O.; Lecomte, F. Manganese-Catalyzed Cross-Coupling Reaction between Aryl Grignard Reagents and Alkenyl Halides. *Org. Lett.* **2008**, *10*, 5255–5256. [CrossRef]

33. Additionally, ^{55}Mn nucleus is quadrupolar (100 % natural abundance, I = 5/2).

34. Morris, R.J.; Girolami, G.S. High-valent organomanganese chemistry. 2. Synthesis and characterization of manganese(III) aryls. *Organometallics* **1991**, *10*, 799–804. [CrossRef]

35. Krasovskiy, A.; Knochel, P. Convenient Titration Method for Organometallic Zinc, Magnesium, and Lanthanide Reagents. *Synthesis* **2006**, *5*, 890. [CrossRef]

36. Fischer, R.; Görls, H.; Friedrich, M.; Westerhausen, M. Reinvestigation of arylmanganese chemistry—Synthesis and molecular structures of $[(thf)_4Mg(\mu-Cl)_2Mn(Br)Mes]$, $[Mes(thf)Mn(\mu-Mes)]_2$, and $(MnPh_2)^{\infty}$ ($Ph=C_6H_5$; Mes= mesityl, $2,4,6-Me_3C_6H_2$). *J. Organomet. Chem.* **2009**, *694*, 1107–1111. [CrossRef]

37. Meyer, R.M.; Hanusa, T.P. Structural organomanganese chemistry. In *The Chemistry of Organomanganese Compounds*; Rappoport, Z., Marek, I., Eds.; Wiley: Hoboken, NJ, USA, 2011; p. 45.

38. Bedford, R.B.; Brenner, P.B.; Carter, E.; Cogswell, P.M.; Haddow, M.F.; Harvey, J.N.; Murphy, D.M.; Nunn, J.; Woodall, C.H. TMEDA in Iron-Catalyzed Kumada Coupling: Amine Adduct versus Homoleptic "ate" Complex Formation. *Angew. Chem. Int. Ed.* **2014**, *53*, 1804–1808. [CrossRef] [PubMed]

39. Muñoz, S.B.; Daifuku, S.L.; Sears, J.D.; Baker, T.M.; Carpenter, S.H.; Brennessel, W.W.; Neidig, M.L. The N-Methylpyrrolidone (NMP) Effect in Iron-Catalyzed Cross-Coupling with Simple Ferric Salts and MeMgBr. *Angew. Chem. Int. Ed.* **2018**, *57*, 6496–6500. [CrossRef] [PubMed]

40. Clémancey, M.; Cantat, T.; Blondin, G.; Latour, J.-M.; Dorlet, P.; Lefèvre, G. Structural insights into the nature of Fe^0 and Fe^I low-valent species obtained upon reduction of Iron salts by Aryl Grignard reagents. *Inorg. Chem.* **2017**, *56*, 3834–3848. [CrossRef]

41. Rousseau, L.; Herrero, C.; Clémancey, M.; Imberdis, A.; Blondin, G.; Lefèvre, G. Evolution of ate organoiron(II) species towards lower oxidation states: Role of the steric and electronic factors. *Chem. Eur. J.* **2019**. [CrossRef]

42. The mechanistic studies were pursued using only aryl Grignard nucleophiles, as the transmetalation of the latter with iron was proceeding in the same way than that of manganese, and as the NMR spectra allowed a much more precise and accurate interpretation thanks to the absence of highly paramagnetic Mn-containing species.

43. Cheng, J.; Chen, Q.; Leng, X.; Ye, S.; Deng, L. Three-Coordinate Iron(0) complexes with N-Heterocyclic Carbene and Vinyltrimethylsilane Ligation: Synthesis, Characterization, and Ligand Substitution Reactions. *Inorg. Chem.* **2019**, *58*, 13129–13141. [CrossRef]

44. Mo, Z.; Deng, L. Open-shell iron hydrocarbyls. *Coord. Chem. Rev.* **2017**, *350*, 285–299. [CrossRef]

45. Mayer, M.; Welther, A.; Jacobi von Wangelin, A. Iron-Catalyzed Isomerizations of Olefins. *ChemCatChem* **2011**, *3*, 1567–1571. [CrossRef]

Sample Availability: Samples of all the compounds described in this article are available from the authors.

Article

Iron-Catalyzed C(sp^2)–C(sp^3) Cross-Coupling of Aryl Chlorobenzoates with Alkyl Grignard Reagents

Elwira Bisz [1,*] and Michal Szostak [1,2,*]

[1] Department of Chemistry, Opole University, 48 Oleska Street, 45-052 Opole, Poland
[2] Department of Chemistry, Rutgers University, 73 Warren Street, Newark, NJ 07102, USA
* Correspondence: ebisz@uni.opole.pl (E.B.); michal.szostak@rutgers.edu (M.S.); Tel.: +1-973-353-5329 (E.B.); +48-77-452-7160 (M.S.)

Academic Editor: Hans-Joachim Knölker
Received: 13 December 2019; Accepted: 31 December 2019; Published: 6 January 2020

Abstract: Aryl benzoates are compounds of high importance in organic synthesis. Herein, we report the iron-catalyzed C(sp^2)–C(sp^3) Kumada cross-coupling of aryl chlorobenzoates with alkyl Grignard reagents. The method is characterized by the use of environmentally benign and sustainable iron salts for cross-coupling in the catalytic system, employing benign urea ligands in the place of reprotoxic NMP (NMP = N-methyl-2-pyrrolidone). It is notable that high selectivity for the cross-coupling is achieved in the presence of hydrolytically-labile and prone to nucleophilic addition phenolic ester C(acyl)–O bonds. The reaction provides access to alkyl-functionalized aryl benzoates. The examination of various O-coordinating ligands demonstrates the high activity of urea ligands in promoting the cross-coupling versus nucleophilic addition to the ester C(acyl)–O bond. The method showcases the functional group tolerance of iron-catalyzed Kumada cross-couplings.

Keywords: iron; cross-coupling; aryl esters; C–O activation; Fe-catalysis; Kumada cross-coupling

---

1. Introduction

Iron catalyzed cross-couplings have recently emerged as an extremely valuable platform for organic synthesis [1–19]. Of particular interest is the high natural abundance of iron [11–13], which in combination with the low toxicity of iron salts and their facile removal from post-reaction mixtures makes it attractive for large-scale industrial processes [19]. The beneficial effect of iron for cross-coupling reactions extends far beyond its economical and sustainable ecological profile, and it is demonstrated by the establishment of the traditionally challenging C(sp^2)–C(sp^3) cross-couplings employing alkyl Grignard reagents possessing β-hydrogens that are not easily accomplished using other transition metals [14–18]. In this regard, the iron-NMP (NMP = N-methyl-2-pyrrolidone) system elegantly pioneered by Fürstner and co-workers represents by far the most viable option for iron cross-coupling chemistry [20–33]. The success of the iron-NMP reagent relies in large part on the outstanding functional group tolerance of this catalyst, the high toxicity of NMP notwithstanding [34,35]. In this vein, our laboratory has reported iron-catalyzed cross-couplings with alkyl Grignard reagents using benign urea ligands that represent an effective alternative to the reprotoxic NMP [36–43].

In this Special Issue on *Recent Advances in Iron Catalysis*, we detail our findings on the development of the iron-catalyzed cross-coupling of aryl chlorobenzoates with alkyl Grignard reagents (Scheme 1). The reaction is notable for several reasons: (1) the method allows for the synthesis of alkyl-functionalized aryl benzoates, which represent compounds of high importance in organic synthesis (Scheme 2); (2) the method demonstrates the exceptional functional group tolerance of the iron system, wherein the selective Kumada cross-coupling is achieved in the presence of the hydrolytically labile and prone to nucleophilic addition C(acyl)–O ester moiety. This model system is well suited for the examination of various O-coordinating ligands in promoting the cross-coupling versus nucleophilic addition to the ester bond. More broadly, the reaction showcases the functional group tolerance in the industrially important iron-catalyzed Kumada cross-couplings.

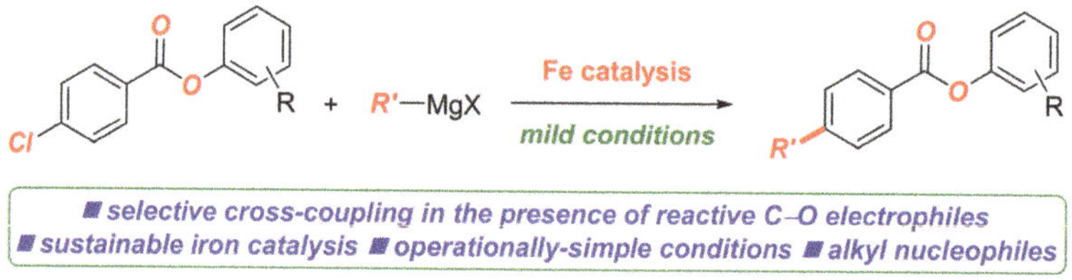

Scheme 1. Iron-catalyzed C(sp^2)–C(sp^3) cross-coupling of aryl chlorobenzoates with alkyl Grignard reagents (this study).

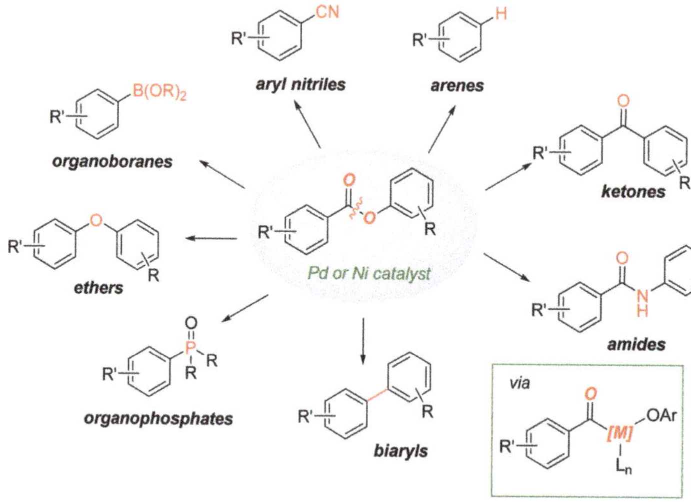

Scheme 2. The important transformations via substituted aryl esters, the products of this study.

2. Results

We became interested in developing the iron-catalyzed cross-coupling of aryl chlorobenzoates as part of our program in iron catalysis [36–43] and the cross-coupling of C(acyl)–X (X = N, O) electrophiles [44,45]. Recently, several groups have reported methods for the nickel and palladium-catalyzed C(acyl)–O bond activation of aryl benzoates, leading to the selective formation of acyl-metal intermediates (Scheme 2, box) [46–64]. While aryl benzoates have long been established as electrophiles in nucleophilic addition to the ester bond via tetrahedral intermediates owing to the increased electrophilicity of the ester bond due to O$_{lp}$ to Ar conjugation [65], the recent advances in accessing acyl metals from aryl benzoates significantly expand the utility of this class of carboxylic acid derivatives in organic synthesis. Thus, the direct iron-catalyzed Kumada cross-coupling would provide an attractive method for the functionalization of the aromatic ring; however, perhaps not surprisingly given the high reactivity of the C(acyl)–O bond, generally applicable methods for the C(sp^2)–C(sp^3) Kumada cross-coupling of aryl benzoates have been elusive.

At the outset, we probed the model reaction between phenyl 4-chlorobenzoate (**1a**) and ethylmagnesium chloride in the presence of benign DMI (DMI = 1,3-dimethyl-2-imidazolidinone) (Table 1). Under standard conditions, the cross-coupling proceeded in 27% yield with the remaining mass balance corresponding to the alcohol product (Table 1, entry 1). Lowering the equivalents of the Grignard reagent had no impact on the reaction efficiency (Table 1, entry 2). After experimentation, we found that the slow addition of the close to equimolar quantity of the Grignard reagent afforded the desired cross-coupling product in 65% yield (Table 1, entry 3). Interestingly, using an excess of DMI led to lower cross-coupling efficiency, which was likely due to facilitating the nucleophilic addition to the carbonyl group (Table 1, entry 4). DMI improves the coupling efficiency; however, this additive is not required, as demonstrated by the cross-coupling in its absence (Table 1, entries 3–6). Furthermore, using Grignard as the limiting reagent as well as extending the addition time had a deleterious effect on the cross-coupling (Table 1, entries 7–8). Likewise, increasing the iron loading to generate the active organoferrate in excess gave no observable increase in the reaction efficiency (Table 1, entries 9–10). Further, we note that an efficient reaction ensues at −40 °C (Table 1, entry 11), while negligible conversion was observed at −78 °C (Table 1, entry 12). Importantly, control reactions in the absence of iron, with and without DMI (Table 1, entries 13–14), resulted in no cross-coupling with the alcohol formed as the sole reaction product, thereby highlighting the key role of iron to promote the cross-coupling. Finally, for comparison purposes, we tested NMP as the additive (Table 1, entry 15). Interestingly, NMP resulted in lower cross-coupling efficiency than DMI (vide infra), highlighting the beneficial effect of this ligand beyond its favorable toxicological profile (cf. NMP).

Table 1. Optimization of iron-catalyzed cross-coupling.[1] DMI: 1,3-dimethyl-2-imidazolidinone.

Entry	Fe(acac)$_3$ (mol%)	Ligand	mol %	Addition Time (min)	Time (min)	Yield (%) [2]
1 [3]	5	DMI	200	0	10	27
2	5	DMI	200	0	10	27
3	5	DMI	200	60	180	65
4	5	DMI	600	60	180	52
5	5	DMI	20	60	180	48
6	5	-	-	60	180	44
7 [4]	5	DMI	200	60	180	57
8	5	DMI	200	180	60	52
9	10	DMI	200	60	180	60
10	50	DMI	200	60	180	28
11 [5]	5	DMI	200	0	180	52
12 [6]	5	DMI	200	0	180	<10
13	-	-	-	60	180	0
14	-	DMI	200	60	180	0
15	5	NMP	200	60	180	57

[1] Conditions: **1** (0.50 mmol), Fe(acac)$_3$ (5 mol%), THF (0.15 M), C$_2$H$_5$MgCl (1.05 equiv, 2.0 M, THF), 0 °C, 180 min. RMgCl added dropwise over 60 min. [2] Yields determined by ^{1}H-NMR and/or GC-MS. [3] C$_2$H$_5$MgCl (1.20 equiv). [4] C$_2$H$_5$MgCl (0.83 equiv). [5] −40 °C. [6] −78 °C.

Then, we examined the scope of the optimized iron catalytic system as outlined in Table 2. We were pleased to find that neutral as well as electron-rich aryl 4-chlorobenzoates, such as 4-*tert*-butyl and 4-methoxy, enabled the chemoselective cross-coupling in good yields (Table 2, entries 1–3). Furthermore, electron-deficient aryl 4-chlorobenzoates, such as 4-fluoro, are also tolerated, albeit the cross-coupling product is obtained in lower yield (Table 2, entry 4). As expected, the reactivity trend mirrors the electronic properties of the aryl ester in that electron-deficient aryl substituents increase O$_{lp}$ to Ar conjugation, leading to the lower yield in the cross-coupling. Pleasingly, we found that both sterically-hindered 2-methyl and 2,6-dimethyl aryl 4-chlorobenzoates are well-tolerated (Table 2, entries 5–6) and result in significantly improved yields for the cross-coupling as a result of steric shielding of the C(acyl)–O bond. Thus, we recommend that electron-rich or sterically hindered aryl benzoates are used for the cross-coupling to minimize the formation of the alcohol side products. 4-Chlorophenyl 4-chlorobenzoate is not a suitable substrate due to nucleophilic addition. The scope of Grignard reagents was also briefly examined (Table 2, entries 7–10). As such, longer primary alkyl Grignard reagents such as hexyl or tetradecyl gave the cross-coupling products in high yields (Table 2, entry 7–8). The cross-coupling of more sterically hindered secondary Grignard reagents is feasible; however, it leads to modest yield (Table 2, entry 9). Finally, we were pleased to find that the challenging phenethyl Grignard reagent that is prone to β-hydride elimination is also a competent nucleophile for this cross-coupling protocol (Table 2, entry 10), attesting to the efficiency of the cross-coupling. At present, cross-coupling of 3-substituted aryl chlorobenzoates is not feasible due to facile hydrolysis.

Table 2. Iron-catalyzed C(sp^2)–C(sp^3) cross-coupling of aryl chlorobenzoates with alkyl Grignards.[1.]

Entry	Substrate	2	Product	Yield (%)
1		2a		63
2		2b		68
3		2c		81
4		2d		51
5		2e		80
6		2f		90
7		2g		83

Table 2. *Cont.*

8	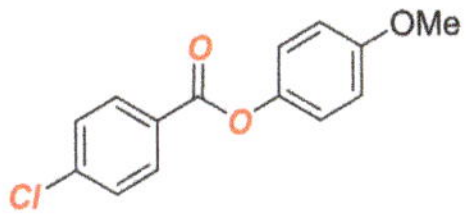	**2h**	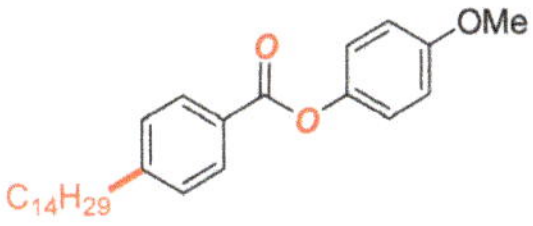	76
9 [2]		**2i**	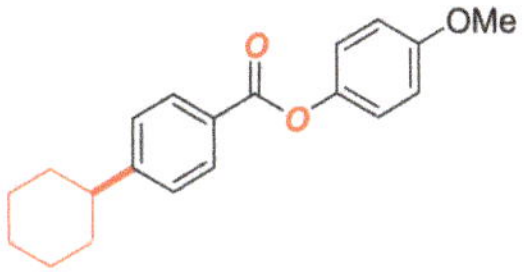	37
10 [2]	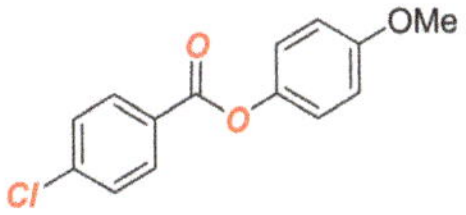	**2j**	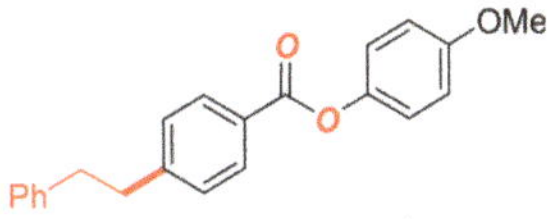	82

[1] Conditions: **1** (0.50 mmol), Fe(acac)$_3$ (5 mol%), THF (0.15 M), DMI (200 mol%), RMgX (1.05 equiv, THF), 0 °C, 180 min. RMgX added dropwise over 60 min. Isolated yields. [2] RMgX (1.20 equiv), 15 h. See the Supplementary Materials for details.

Next, intermolecular competition studies were performed to gain insight into the selectivity of this cross-coupling (Scheme 3). (A) Competition experiments between phenyl and methyl ester (OPh:OMe = 2.5:1.0) revealed that aryl esters are more reactive than their alkyl counterparts, which is consistent with the facility of oxidative addition. (B) Similarly, competition between electron-rich and electron-deficient aryl esters (4-MeO:4-F = 1.0:1.25) revealed that electron-deficient arenes are more reactive. This observation is in agreement with the O-aryl ester activating the aromatic ring for the cross-coupling; however, its increased electrophilicity leads to a competing nucleophilic addition to give the alcohol products. The formation of the alcohol could be minimized by using sterically hindered or electron-rich aromatic esters.

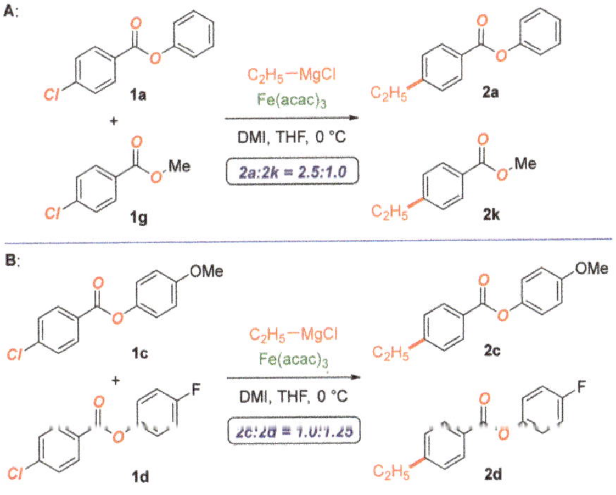

Scheme 3. Competition experiments. (**A**) Competition experiments between phenyl and methyl ester (OPh:OMe = 2.5:1.0) revealed that aryl esters are more reactive than their alkyl counterparts, which is consistent with the facility of oxidative addition. (**B**) Similarly, competition between electron-rich and electron-deficient aryl esters (4-MeO:4-F = 1.0:1.25) revealed that electron-deficient arenes are more reactive

Finally, we have probed the effect of various additives on the cross-coupling (Table 3 and Figure 1). At present, one of the major challenges in iron-catalyzed C(sp^2)–C(sp^3) cross-coupling is replacing the reprotoxic NMP by benign yet effective additives. The present system compares the cross-coupling efficiency versus nucleophilic addition, thereby indirectly measuring the ligand effect. Our study demonstrates that urea ligands such as DMI, DMPU (DMPU = 1,3-dimethyl-3,4,5,6-tetrahydro-2(1*H*)-pyrimidinone) and TMU (TMU = 1,1,3,3-tetramethylurea) are more reactive than NMP in the cross-coupling (Table 3, entries 1–4), while *N*-methylcaprolactam shows comparable reactivity to NMP (Table 3, entry 5). In contrast, the recently reported by our group *N,N*-bis(2-methoxyethyl)benzamide (Table 3, entry 6) and phenyl(piperidin-1-yl)methanone (Table 3, entry 7) appear to be less reactive than NMP [10]; however, ester hydrolysis is not observed in these cases, which may lead to unusual selectivity in the cross-coupling.

Table 3. Ligand effect on iron-catalyzed cross-coupling of aryl chlorobenzoates: cross-coupling vs. nucleophilic addition.[1] DMPU: 1,3-dimethyl-3,4,5,6-tetrahydro-2(1*H*)-pyrimidinone); NMP: *N*-methyl-2-pyrrolidone, TMU: 1,1,3,3-tetramethylurea.

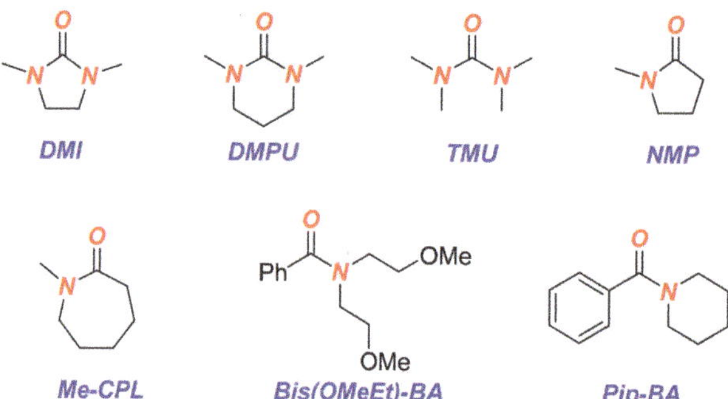

Entry	Fe(acac)$_3$ (mol%)	Ligand	mol%	Time (min)	Yield (%) [2]
1	5	DMI	200	180	98 (2)
2	5	DMPU	200	180	>98 (1)
3	5	TMU	200	180	>98 (<1)
4	5	NMP	200	180	95 (4)
5	5	*N*-Methylcaprolactam	200	180	92 (7)
6 [3]	5	Bis(OMeEt)-BA	200	180	57 (<1)
7 [4]	5	Pip-BA	200	180	75 (<1)

[1] Conditions: **1** (0.50 mmol), Fe(acac)$_3$ (5 mol%), THF (0.15 M), C$_2$H$_5$MgCl (1.0 equiv, 2.0 M, THF), 0 °C, 180 min. RMgCl added dropwise over 60 min. [2] Determined by ^{1}H-NMR and/or GC-MS. The number in brackets corresponds to the alcohol addition product. Entries 6–7: **1**: 43% and 25%. [3] Ligand: *N,N*-bis(2-methoxyethyl)benzamide. [4] Ligand: Phenyl(piperidin-1-yl)methanone. See Figure 1 for structures.

Figure 1. Structures of ligands used.

3. Discussion

In summary, we have reported the iron-catalyzed C(sp^2)–C(sp^3) Kumada cross-coupling of aryl chlorobenzoates with alkyl Grignard reagents. The iron-catalyzed cross-coupling reactions have gained significant momentum due to the beneficial environmental and sustainability profile compared to precious metals. However, what is equally important is the fact that iron catalysis enables cross-coupling reactions that are difficult or impossible to achieve with other metals, the prime example being the industrially-relevant C(sp^2)–C(sp^3) Kumada cross-coupling. The present study expands the scope of benign iron-catalyzed cross-couplings with urea ligands as replacements for toxic NMP to embrace the functional group tolerance of highly reactive aryl benzoates without cleavage of the sensitive C(acyl)–O bond. Future studies will be focused on expanding the scope of iron-catalyzed cross-couplings and the design of new amide-based ligands for iron catalysis.

4. Materials and Methods

4.1. General Information

All compounds reported in the manuscript are commercially available or have been previously described in the literature unless indicated otherwise. All experiments involving iron were performed using standard Schlenk techniques under argon or nitrogen atmosphere unless stated otherwise. All esters have been prepared by standard methods [66]. All yields refer to yields determined by ^{1}H-NMR and/or GC/MS using an internal standard (optimization) and isolated yields (preparative runs) unless stated otherwise. ^{1}H-NMR and ^{13}C-NMR data are given for all compounds in the Experimental section for characterization purposes. ^{1}H-NMR, ^{13}C-NMR, and HRMS data are reported for all new compounds. All products have been previously reported, unless stated otherwise. Spectroscopic data matched literature values. General methods have been published [36–43]. All new compounds have been characterized by established guidelines by ^{1}H-NMR, ^{13}C-NMR, HRMS, and Mp as appropriate.

4.2. General Procedure for Iron-Catalyzed C(sp^2)–C(sp^3) Cross-Coupling

An oven-dried vial equipped with a stir bar was charged with an ester substrate (neat, typically, 0.50 mmol, 1.0 equiv) and Fe(acac)$_3$ (typically, 5 mol%), which was placed under a positive pressure of argon and subjected to three evacuation/backfilling cycles under vacuum. Tetrahydrofuran (0.15 M) and ligand were sequentially added with vigorous stirring at room temperature, the reaction mixture was cooled to 0 °C, a solution of Grignard reagent (typically, 1.05 equiv) was added dropwise over 60 min with vigorous stirring, and the reaction mixture was stirred for the indicated time at 0 °C. After the indicated time, the reaction mixture was diluted with HCl (1.0 N, 1.0 mL) and Et$_2$O (1 × 30 mL), and the organic layer was extracted with HCl (1.0 N, 2 × 10 mL), dried, and concentrated. The sample was analyzed by ^{1}H-NMR (CDCl$_3$, 400 MHz) and GC-MS to obtain the conversion, yield and, selectivity using an internal standard and comparison with authentic samples. Purification by chromatography on silica gel afforded the title product.

4.3. General Procedure for Determination of Relative Reactivity

According to the general procedure, an oven-dried vial equipped with a stir bar was charged with two chloride substrates (each 0.50 mmol, 1.0 equiv) and Fe(acac)$_3$ (5 mol%), which was placed under a positive pressure of argon and subjected to three evacuation/backfilling cycles under vacuum. Tetrahydrofuran (0.15 M) and DMI (neat, 200 mol%) were sequentially added with vigorous stirring at room temperature, the reaction mixture was cooled to 0 °C, a solution of C$_2$H$_5$MgCl (2.0 M in THF, 0.25 mmol, 0.50 equiv) was added dropwise over 60 min with vigorous stirring, and the reaction mixture was stirred for 180 min at 0 °C. Following the standard work up, the sample was analyzed by ^{1}H-NMR (CDCl$_3$, 400 MHz) and GC-MS to obtain the conversion, yield, and selectivity using an internal standard and comparison with authentic samples.

4.4. Characterization Data for Starting Materials

Phenyl 4-chlorobenzoate (**1a**) [67]. Yield 95% (2.20 g). White solid. ^{1}H-NMR (400 MHz, CDCl$_3$) δ 8.13 (d, *J* = 8.8 Hz, 2H), 7.47 (d, *J* = 8.8 Hz, 2H), 7.45–7.39 (m, 2H), 7.30–7.25 (m, 1H), and 7.22–7.18

(m, 2H). ^{13}C{1H} NMR (100 MHz, CDCl$_3$) δ 164.47, 150.94, 140.26, 131.69, 129.69, 129.10, 128.19, 126.19, and 121.77.

4-(Tert-Butyl)phenyl 4-chlorobenzoate (**1b**). *New compound.* Yield 98% (2.84 g). White solid. Mp = 114–116 °C. ^{1}H-NMR (400 MHz, CDCl$_3$) δ 8.13 (d, *J* = 8.7 Hz, 2H), 7.46 (d, *J* = 8.6 Hz, 2H), 7.43 (d, *J* = 8.8 Hz, 2H), 7.12 (d, *J* = 8.8 Hz, 2H), and 1.34 (s, 9H). ^{13}C{1H} NMR (100 MHz, CDCl$_3$) δ 164.63, 149.01, 148.55, 140.16, 131.68, 129.06, 128.30, 126.59, 121.05, 34.67, and 31.58. HRMS (ESI/Q-TOF) *m/z*: [M + Na]$^+$ calcd for C$_{17}$H$_{17}$ClO$_2$Na 311.0815 found 311.0822.

4-Methoxyphenyl 4-chlorobenzoate (**1c**) [68]. Yield 95% (2.50 g). White solid. ^{1}H-NMR (400 MHz, CDCl$_3$) δ 8.12 (d, *J* = 8.6 Hz, 2H), 7.47 (d, *J* = 8.6 Hz, 2H), 7.12 (d, *J* = 9.0 Hz, 2H), 6.93 (d, *J* = 9.1 Hz, 2H), and 3.82 (s, 3H). ^{13}C{1H} NMR (100 MHz, CDCl$_3$) δ 164.85, 157.60, 144.42, 140.20, 131.67, 129.08, 128.30, 122.52, 114.74, and 55.78.

4-Fluorophenyl 4-chlorobenzoate (**1d**) [69]. Yield 98% (2.45 g). White solid. ^{1}H-NMR (400 MHz, CDCl$_3$) δ 8.12 (d, *J* = 8.8 Hz, 2H), 7.48 (d, *J* = 8.8 Hz, 2H), 7.19–7.14 (m, 2 H), and 7.14–7.08 (m, 2 H). ^{13}C{1H}-NMR (100 MHz, CDCl$_3$) δ 164.52, 161.74, 159.31, 146.71, 140.45, 131.70, 129.17, 127.89, 123.20 (d, J^F = 8.4 Hz), and 116.38 (d, J^F = 23.5 Hz).

o-Tolyl 4-chlorobenzoate (**1e**). *New compound.* Yield 97% (2.40 g). Colorless oil. ^{1}H-NMR (400 MHz, CDCl$_3$) δ 8.13 (d, *J* = 8.7 Hz, 2H), 7.45 (d, *J* = 8.7 Hz, 2H), 7.27–7.09 (m, 4H), and 2.21 (s, 3H). ^{13}C{1H}-NMR (100 MHz, CDCl$_3$) δ 164.07, 149.45, 140.18, 131.61, 131.32, 130.27, 129.06, 127.97, 127.13, 126.32, 122.00, and 16.31. HRMS (ESI/Q-TOF) *m/z*: [M + Na]$^+$ calcd for C$_{14}$H$_{11}$ClO$_2$Na 269.0345 found 269.0342.

2,6-Dimethylphenyl 4-chlorobenzoate (**1f**). *New compound.* Yield 98% (2.56 g). Colorless oil. ^{1}H-NMR (400 MHz, CDCl$_3$) δ 8.16 (d, *J* = 8.5 Hz, 2H), 7.46 (d, *J* = 8.4 Hz, 2H), 7.11–7.04 (m, 3H), and 2.17 (s, 6H). ^{13}C{1H}-NMR (100 MHz, CDCl$_3$) δ 163.56, 148.30, 140.20, 131.61, 130.32, 129.10, 128.76, 127.80, 126.13, and 16.43. HRMS (ESI/Q-TOF) *m/z*: [M + Na]$^+$ calcd for C$_{15}$H$_{13}$ClO$_2$Na 283.0502 found 283.0509.

4.5. Characterization Data for Cross-Coupling Products

Phenyl 4-ethylbenzoate (Table 2, **2a**) [70]. Prepared according to the general procedure using phenyl 4-chlorobenzoate (0.50 mmol), Fe(acac)$_3$ (5 mol%), DMI (200 mol%), THF (0.15 M), and C$_2$H$_5$MgCl (2.0 M in THF, 1.05 equiv). The reaction mixture was stirred for 180 min at 0 °C. Yield 63% (71.3 mg). White solid. ^{1}H-NMR (400 MHz, CDCl$_3$) δ 8.12 (d, *J* = 8.4 Hz, 2H), 7.45–7.40 (m, 2H), 7.33 (d, *J* = 8.5 Hz, 2H), 7.29–7.24 (m, 1H), 7.23–7.18 (m, 2H), 2.74 (q, *J* = 7.6 Hz, 2H), and 1.28 (t, *J* = 7.6 Hz, 3H). ^{13}C{1H}-NMR (100 MHz, CDCl$_3$) δ 165.44, 151.18, 150.78, 130.51, 129.64, 128.29, 127.17, 125.96, 121.95, 29.22, and 15.45.

4-(Tert-Butyl)phenyl 4-ethylbenzoate (Table 2, **2b**). *New compound.* Prepared according to the general procedure using 4-(tert-butyl)phenyl 4-chlorobenzoate (0.50 mmol), Fe(acac)$_3$ (5 mol%), DMI (200 mol%), THF (0.15 M), and C$_2$H$_5$MgCl (2.0 M in THF, 1.05 equiv). The reaction mixture was stirred for 180 min at 0 °C. Yield 68% (96.1 mg). Colorless oil. ^{1}H-NMR (400 MHz, CDCl$_3$) δ 8.12 (d, *J* = 8.4 Hz, 2H), 7.43 (d, *J* = 8.8 Hz, 2H), 7.32 (d, *J* = 8.5 Hz, 2H), 7.13 (d, *J* = 8.8 Hz, 2H), 2.74 (q, *J* = 7.6 Hz, 2H), 1.34 (s, 9H), 1.28 (t, *J* = 7.6 Hz, 3H). ^{13}C{1H}-NMR (100 MHz, CDCl$_3$) δ 165.57, 150.66, 148.81, 148.73, 130.49, 128.25, 127.31, 126.53, 121.21, 34.66, 31.61, 29.21, and 15.46. HRMS (ESI/Q-TOF) *m/z*: [M + Na]$^+$ calcd for C$_{19}$H$_{22}$O$_2$Na 305.1518 found 305.1519.

4-Methoxyphenyl 4-ethylbenzoate (Table 2, **2c**). *New compound.* Prepared according to the general procedure using 4-methoxyphenyl 4-chlorobenzoate (0.50 mmol), Fe(acac)$_3$ (5 mol%), DMI (200 mol%), THF (0.15 M), and C$_2$H$_5$MgCl (2.0 M in THF, 1.05 equiv). The reaction mixture was stirred for 180 min at 0 °C. Yield 81% (103.8 mg). White solid. Mp = 101–103 °C. ^{1}H-NMR (400 MHz, CDCl$_3$) δ 8.11 (d, *J* = 8.3 Hz, 2H), 7.32 (d, *J* = 8.5 Hz, 2H), 7.12 (d, *J* = 9.1 Hz, 2H), 6.93 (d, *J* = 9.1 Hz, 2H), 3.82 (s, 3H), 2.74 (q, *J* = 7.6 Hz, 2H), 1.28 (t, *J* = 7.6 Hz, 3H). ^{13}C{1H}-NMR (100 MHz, CDCl$_3$) δ 165.79, 157.39, 150.67,

144.63, 130.46, 128.25, 127.22, 122.67, 114.65, 55.77, 29.21, and 15.44. HRMS (ESI/Q-TOF) *m/z*: [M + Na]$^+$ calcd for $C_{16}H_{16}O_3Na$ 279.0997 found 279.0997.

4-Fluorophenyl 4-ethylbenzoate (Table 2, **2d**). *New compound*. Prepared according to the general procedure using 4-fluorophenyl 4-chlorobenzoate (0.50 mmol), Fe(acac)$_3$ (5 mol%), DMI (200 mol%), THF (0.15 M), and C_2H_5MgCl (2.0 M in THF, 1.05 equiv). The reaction mixture was stirred for 180 min at 0 °C. Yield 51% (62.4 mg). White solid. Mp = 38–40 °C. ^{1}H-NMR (400 MHz, CDCl$_3$) δ 8.11 (d, *J* = 8.4 Hz, 2H), 7.33 (d, *J* = 8.1 Hz, 2H), 7.19–7.14 (m, 2H), 7.13–7.07 (m, 2H), 2.75 (q, *J* = 7.6 Hz, 2H), 1.28 (t, *J* = 7.6 Hz, 3H). ^{13}C{1H}-NMR (100 MHz, CDCl$_3$) δ 165.45, 161.63, 159.20, 150.96, 146.97 (d, J^F = 2.9 Hz), 130.51, 128.33, 126.86, 123.33 (d, J^F = 8.5 Hz), 116.28 (d, J^F = 23.5 Hz), 29.23, and 15.43. HRMS (ESI/Q-TOF) *m/z*: [M + Na]$^+$ calcd for $C_{15}H_{13}FO_2Na$ 267.0797 found 267.0794.

o-Tolyl 4-ethylbenzoate (Table 2, **2e**). *New compound*. Prepared according to the general procedure using o-tolyl 4-chlorobenzoate (0.50 mmol), Fe(acac)$_3$ (5 mol%), DMI (200 mol%), THF (0.15 M), and C_2H_5MgCl (2.0 M in THF, 1.05 equiv). The reaction mixture was stirred for 180 min at 0 °C. Yield 80% (96.1 mg). Colorless oil. ^{1}H-NMR (400 MHz, CDCl$_3$) δ 8.14 (d, *J* = 8.3 Hz, 2H), 7.34 (d, *J* = 8.4 Hz, 2H), 7.30–7.22 (m, 2H), 7.20–7.11 (m, 2H), 2.75 (q, *J* = 7.6 Hz, 2H), 2.23 (s, 3H), and 1.29 (t, *J* = 7.6 Hz, 3H). ^{13}C{1H}-NMR (100 MHz, CDCl$_3$) δ 165.10, 150.77, 149.76, 131.31, 130.51, 128.32, 127.12, 127.09, 126.17, 122.23, 29.23, 16.43, and 15.47. HRMS (ESI/Q-TOF) *m/z*: [M + Na]$^+$ calcd for $C_{16}H_{16}O_2Na$ 263.1048 found 263.1044.

2,6-Dimethylphenyl 4-ethylbenzoate (Table 2, **2f**). *New compound*. Prepared according to the general procedure using 2,6-dimethylphenyl 4-chlorobenzoate (0.50 mmol), Fe(acac)$_3$ (5 mol%), DMI (200 mol%), THF (0.15 M), and C_2H_5MgCl (2.0 M in THF, 1.05 equiv). The reaction mixture was stirred for 180 min at 0 °C. Yield 90% (114.2 mg). Colorless oil. ^{1}H-NMR (400 MHz, CDCl$_3$) δ 8.17 (d, *J* = 8.3 Hz, 2H), 7.34 (d, *J* = 8.4 Hz, 2H), 7.12–7.06 (m, 3H), 2.74 (q, *J* = 7.6 Hz, 2H), 2.19 (s, 6H), and 1.29 (t, *J* = 7.6 Hz, 3H). ^{13}C{1H}-NMR (100 MHz, CDCl$_3$) δ 164.56, 150.73, 148.53, 130.54, 130.50, 128.72, 128.32, 126.88, 125.97, 29.20, 16.54, and 15.44. HRMS (ESI/Q-TOF) *m/z*: [M + Na]$^+$ calcd for $C_{17}H_{18}O_2Na$ 277.1205 found 277.1209.

4-Methoxyphenyl 4-hexylbenzoate (Table 2, **2g**). *New compound*. Prepared according to the general procedure using 4-methoxyphenyl 4-chlorobenzoate (0.50 mmol), Fe(acac)$_3$ (5 mol%), DMI (200 mol%), THF (0.15 M), and $C_6H_{13}MgCl$ (2.0 M in THF, 1.05 equiv). The reaction mixture was stirred for 180 min at 0 °C. Yield 83% (129.8 mg). White solid. Mp = 64–66 °C. ^{1}H-NMR (400 MHz, CDCl$_3$) δ 8.10 (d, *J* = 8.3 Hz, 2H), 7.30 (d, *J* = 8.4 Hz, 2H), 7.12 (d, *J* = 9.1 Hz, 2H), 6.93 (d, *J* = 9.1 Hz, 2H), 3.82 (s, 3H), 2.69 (t, *J* = 7.6 Hz, 2H), 1.70–1.60 (m, 2H), 1.37–1.28 (m, 6H), and 0.89 (t, *J* = 6.9 Hz, 3H). ^{13}C{1H}-NMR (100 MHz, CDCl$_3$) δ 165.81, 157.40, 149.45, 144.65, 130.37, 128.80, 127.20, 122.67, 114.65, 55.78, 36.26, 31.84, 31.30, 29.09, 22.76, and 14.27. HRMS (ESI/Q-TOF) *m/z*: [M + Na]$^+$ calcd for $C_{20}H_{24}O_3Na$ 335.1623 found 335.1614.

4-Methoxyphenyl 4-tetradecylbenzoate (Table 2, **2h**). *New compound*. Prepared according to the general procedure using 4-methoxyphenyl 4-chlorobenzoate (0.50 mmol), Fe(acac)$_3$ (5 mol%), DMI (200 mol%), THF (0.15 M), and $C_{14}H_{29}MgCl$ (1.0 M in THF, 1.05 equiv). The reaction mixture was stirred for 180 min at 0 °C. Yield 76% (161.7 mg). White solid. Mp = 63–65 °C. ^{1}H-NMR (400 MHz, CDCl$_3$) δ 8.10 (d, *J* = 8.3 Hz, 2H), 7.30 (d, *J* = 8.3 Hz, 2H), 7.12 (d, *J* = 9.1 Hz, 2H), 6.93 (d, *J* = 9.1 Hz, 2H), 3.82 (s, 3H), 2.69 (t, *J* = 7.7 Hz, 2H), 1.69–1.60 (m, 2H), 1.35–1.24 (m, 22H), and 0.88 (t, *J* = 6.8 Hz, 3H). ^{13}C{1H}-NMR (100 MHz, CDCl$_3$) δ 165.80, 157.39, 149.46, 144.64, 130.37, 128.79, 127.19, 122.67, 114.64, 55.77, 36.26, 32.11, 31.35, 29.84, 29.75, 29.65, 29.55, 29.44, 22.88, and 14.32. HRMS (ESI/Q-TOF) *m/z*: [M + H]$^+$ calcd for $C_{28}H_{41}O_3$ 425.3056 found 425.3056.

4-Methoxyphenyl 4-cyclohexylbenzoate (Table 2, **2i**). *New compound*. Prepared according to the general procedure using 4-methoxyphenyl 4-chlorobenzoate (0.50 mmol), Fe(acac)$_3$ (5 mol%), DMI (200 mol%), THF (0.15 M), and c-$C_6H_{11}MgCl$ (1.0 M in THF, 1.20 equiv). The reaction mixture was stirred for 15 h at 0 °C. Yield 37% (57.8 mg). White solid. Mp = 131–133 °C. ^{1}H-NMR (400 MHz, CDCl$_3$) δ 8.11

(d, *J* = 8.4 Hz, 2H), 7.33 (d, *J* = 8.2 Hz, 2H), 7.11 (d, *J* = 9.1 Hz, 2H), 6.93 (d, *J* = 9.1 Hz, 2H), 3.82 (s, 3H), 2.65–2.54 (m, 1H), 1.94–1.82 (m, 4H), 1.81–1.73 (m, 1H), 1.51–1.34 (m, 4H), and 1.33–1.23 (m, 1H). ^{13}C{1H}-NMR (100 MHz, CDCl$_3$) δ 165.77, 157.38, 154.38, 144.64, 130.45, 127.31, 127.23, 122.67, 114.64, 55.77, 44.93, 34.30, 26.89, and 26.19. HRMS (ESI/Q-TOF) *m/z*: [M + Na]$^+$ calcd for C$_{20}$H$_{22}$O$_3$Na 333.1467 found 333.1474.

4-Methoxyphenyl 4-phenethylbenzoate (Table 2, **2j**). *New compound*. Prepared according to the general procedure using 4-methoxyphenyl 4-chlorobenzoate (0.50 mmol), Fe(acac)$_3$ (5 mol%), DMI (200 mol%), THF (0.15 M), and PhCH$_2$CH$_2$MgCl (1.0 M in THF, 1.2 equiv). The reaction mixture was stirred for 15 h at 0 °C. Yield 82% (136.1 mg). White solid. Mp = 116–118 °C. ^{1}H-NMR (400 MHz, CDCl$_3$) δ 8.09 (d, *J* = 8.3 Hz, 2H), 7.31–7.26 (m, 4H), 7.22–7.14 (m, 3H), 7.11 (d, *J* = 9.1 Hz, 2H), 6.93 (d, *J* = 9.1 Hz, 2H), 3.80 (s, 3H), 3.04–2.98 (m, 2H), and 2.97–2.92 (m, 2H). ^{13}C{1H}-NMR (100 MHz, CDCl$_3$) δ 165.71, 157.38, 148.03, 144.58, 141.19, 130.40, 128.88, 128.61, 128.57, 127.49, 126.27, 122.63, 114.63, 55.73, 38.07, and 37.59. HRMS (ESI/Q-TOF) *m/z*: [M + Na]$^+$ calcd for C$_{22}$H$_{20}$O$_3$Na 355.1310 found 355.1308.

Supplementary Materials: ^{1}H and ^{13}C-NMR spectra are available online at http://www.mdpi.com/1420-3049/25/1/230/s1.

Author Contributions: E.B. conducted experimental work and analyzed the data; E.B. and M.S. initiated the project, designed experiments to develop this reaction, and wrote the paper. All authors have read and agreed to the published version of the manuscript.

Funding: We gratefully acknowledge Narodowe Centrum Nauki (grant no. 2014/15/D/ST5/02731), Rutgers University and the NSF (CAREER CHE-1650766) for generous financial support.

Conflicts of Interest: The authors declare no conflict of interest.

References

1. Fürstner, A.; Martin, R. Advances in Iron Catalyzed Cross Coupling Reactions. *Chem. Lett.* **2005**, *34*, 624–629. [CrossRef]
2. Sherry, B.D.; Fürstner, A. The Promise and Challenge of Iron-Catalyzed Cross Coupling. *Acc. Chem. Res.* **2008**, *41*, 1500–1511. [CrossRef]
3. Czaplik, W.M.; Mayer, M.; Cvengros, J.; Jacobi von Wangelin, A. Coming of Age: Sustainable Iron-Catalyzed Cross-Coupling Reactions. *ChemSusChem* **2009**, *2*, 396–417. [CrossRef] [PubMed]
4. Plietker, B. Topic in Organometallic Chemistry Also Available Electronically. In *Iron Catalysis–Fundamentals and Applications*; Springer: Berlin Heidelberg, Germany, 2011; Volume 33.
5. Bauer, E.B. *Iron Catalysis II. Top. Organomet. Chem.*; Springer: Berlin Heidelberg, Germany, 2015; Volume 50.
6. Marek, I.; Rappoport, Z. *The Chemistry of Organoiron Compounds*; Wiley: Weinheim, Germany, 2014.
7. Bauer, I.; Knölker, H.J. Iron Catalysis in Organic Synthesis. *Chem. Rev.* **2015**, *115*, 3170–3387. [CrossRef] [PubMed]
8. Legros, J.; Fidegarde, B. Iron-promoted C-C bond formation in the total synthesis of natural products and drugs. *Nat. Prod. Rep.* **2015**, *32*, 1541–1555. [CrossRef] [PubMed]
9. Bisz, E.; Szostak, M. Iron-Catalyzed C-O Bond Activation: Opportunity for Sustainable Catalysis. *ChemSusChem* **2017**, *10*, 3964–3981. [CrossRef] [PubMed]
10. Fürstner, A. Discussion Addendum for: 4-Nonylbenzoic Acid. *Org. Synth.* **2019**, *96*, 1–15. [CrossRef]
11. Fürstner, A. Iron Catalysis in Organic Synthesis: A Critical Assessment of What It Takes To Make This Base Metal a Multitasking Champion. *ACS Cent. Sci.* **2016**, *2*, 778–789. [CrossRef]
12. Fürstner, A. Base-Metal Catalysis Marries Utilitarian Aspects with Academic Fascination. *Adv. Synth. Catal.* **2016**, *358*, 2362–2363. [CrossRef]
13. Ludwig, J.R.; Schindler, C.S. Catalyst: Sustainable Catalysis. *Chem* **2017**, *2*, 313–316. [CrossRef]
14. Molander, G.A.; Wolfe, J.P.; Larhed, M. (Eds.) *Science of Synthesis: Cross-Coupling and Heck-Type Reactions*; Thieme: Stuttgart, Germany, 2013.
15. de Meijere, A.; Bräse, S.; Oestreich, M. (Eds.) *Metal-Catalyzed Cross-Coupling Reactions and More*; Wiley: New York, NY, USA, 2014.
16. Colacot, T.J. (Ed.) *New Trends in Cross-Coupling*; The Royal Society of Chemistry: Cambridge, UK, 2015.

17. Jana, R.; Pathak, T.P.; Sigman, M.S. Advances in Transition Metal (Pd,Ni,Fe)-Catalyzed Cross-Coupling Reactions Using Alkyl-organometallics as Reaction Partners. *Chem. Rev.* **2011**, *111*, 1417–1492. [CrossRef] [PubMed]

18. Giri, R.; Thapa, S.; Kafle, A. Palladium- Catalysed, Directed C-H Coupling with Organometallics. *Adv. Synth. Catal.* **2014**, *356*, 1395–1411. [CrossRef]

19. Piontek, A.; Bisz, E.; Szostak, M. Iron-Catalyzed Cross-Coupling in the Synthesis of Pharmaceuticals: In Pursuit of Sustainability. *Angew. Chem. Int. Ed.* **2018**, *57*, 11116–11128. [CrossRef] [PubMed]

20. Fürstner, A.; Leitner, A. Iron-Catalyzed Cross-Coupling Reactions of Alkyl-Grignard Reagents with Aryl Chlorides, Tosylates, and Triflates. *Angew. Chem. Int. Ed.* **2002**, *41*, 609–612. [CrossRef]

21. Fürstner, A.; Leitner, A.; Mendez, M.; Krause, H. Iron-Catalyzed Cross-Coupling Reactions. *J. Am. Chem. Soc.* **2002**, *124*, 13856–13863. [CrossRef]

22. Fürstner, A.; Leitner, A. A Catalytic Approach to (R)-(+)-Muscopyridine with Integrated "Self-Clearance". *Angew. Chem. Int. Ed.* **2003**, *42*, 308–311. [CrossRef]

23. Fürstner, A.; De Souza, D.; Parra-Rapado, L.; Jensen, J.T. Catalysis-Based Total Synthesis of Latrunculin, B. *Angew. Chem. Int. Ed.* **2003**, *42*, 5358–5360. [CrossRef]

24. Czaplik, W.M.; Mayer, M.; Jacobi von Wangelin, A. Domino Iron Catalysis: Direct Aryl-Alkyl Cross-Coupling. *Angew. Chem. Int. Ed.* **2009**, *48*, 607–610. [CrossRef]

25. Gülak, S.; Jacobi von Wangelin, A. Chlorostyrenes in Iron-Catalyzed Biaryl Coupling Reactions. *Angew. Chem. Int. Ed.* **2012**, *51*, 1357–1361. [CrossRef]

26. Gärtner, D.; Stein, A.L.; Grupe, S.; Arp, J.; Jacobi von Wangelin, A. Iron-Catalyzed Cross-Coupling of Alkenyl Acetates. *Angew. Chem. Int. Ed.* **2015**, *54*, 10545–10549. [CrossRef]

27. Kuzmina, O.M.; Steib, A.K.; Markiewicz, J.T.; Flubacher, D.; Knochel, P. Ligand-Accelerated Iron- and Cobalt-Catalyzed Cross-Coupling Reactions between N-Heteroaryl Halides and Aryl Magnesium Reagents. *Angew. Chem. Int. Ed.* **2013**, *52*, 4945–4949. [CrossRef] [PubMed]

28. Fürstner, A.; Martin, R.; Krause, H.; Seidel, G.; Goddard, R.; Lehmann, C.W. Preparation, Structure, and Reactivity of Nonstabilized Organoiron Compounds. Implications for Iron-Catalyzed Cross Coupling Reactions. *J. Am. Chem. Soc.* **2008**, *130*, 8773–8787. [CrossRef] [PubMed]

29. Cassani, C.; Bergonzini, G.; Wallentin, C.J. Active Species and Mechanistic Pathways in Iron-Catalyzed C–C Bond-Forming Cross-Coupling Reactions. *ACS Catal.* **2016**, *6*, 1640–1648. [CrossRef]

30. Casitas, A.; Krause, H.; Goddard, R.; Fürstner, A. Elementary Steps in Iron Catalysis: Exploring the Links between Iron Alkyl and Iron Olefin Complexes for their Relevance in C–H Activation and C–C Bond Formation. *Angew. Chem. Int. Ed.* **2015**, *54*, 1521–1526. [CrossRef] [PubMed]

31. Casitas, A.; Rees, J.A.; Goddard, R.; Bill, E.; DeBeer, D.; Fürstner, A. Two Exceptional Homoleptic Iron(IV) Tetraalkyl Complexes. *Angew. Chem. Int. Ed.* **2017**, *56*, 10108–10113. [CrossRef]

32. Muñoz, S.B., III; Daifuku, S.L.; Sears, J.D.; Baker, T.M.; Carpenter, S.H.; Brennessel, W.W.; Neidig, M.L. The N-Methylpyrrolidone (NMP) Effect in Iron-Catalyzed Cross-Coupling with Simple Ferric Salts and MeMgBr. *Angew. Chem. Int. Ed.* **2018**, *57*, 6496–6500. [CrossRef]

33. Sears, J.D.; Muñoz, S.B.; Daifuku, S.L.; Shaps, A.A.; Carpenter, S.H.; Brennessel, W.W.; Neidig, M.L. The Effect of β-Hydrogen Atoms on Iron Speciation in Cross-Couplings with Simple Iron Salts and Alkyl Grignard Reagents. *Angew. Chem. Int. Ed.* **2019**, *58*, 2769–2773. [CrossRef]

34. Åkesson, B. *N-Methyl-2-Pyrrolidone*; WHO: Geneva, Switzerland, 2001.

35. NMP is Classified as A Chemical of "Very High Concern" and A Proposal has been put forward to restrict the Use of NMP. Available online: https://echa.europa.eu/candidate-list-table (accessed on 10 December 2019).

36. Bisz, E.; Szostak, M. Cyclic Ureas (DMI, DMPU) as Efficient, Sustainable Ligands in Iron-Catalyzed C(sp^2)–C(sp^3) Coupling of Aryl Chlorides and Tosylates. *Green Chem.* **2017**, *19*, 5361–5366. [CrossRef]

37. Bisz, E.; Szostak, M. 2-Methyltetrahydrofuran: A Green Solvent for Iron-Catalyzed Cross-Coupling Reactions. *ChemSusChem* **2018**, *11*, 1290–1294. [CrossRef]

38. Piontek, A.; Szostak, M. Iron-Catalyzed C(sp^2)-C(sp^3) Cross-Coupling of Alkyl Grignard Reagents with Polyaromatic Tosylates. *Eur. J. Org. Chem.* **2017**, *48*, 7271–7276. [CrossRef]

39. Bisz, E.; Szostak, M. Iron-Catalyzed C(sp^2)–C(sp^3) Cross-Coupling of Chlorobenzamides with Alkyl Grignard Reagents: Development of Catalyst System, Synthetic Scope and Application. *Adv. Synth. Catal.* **2019**, *361*, 85–95. [CrossRef]

40. Bisz, E.; Szostak, M. Iron-Catalyzed C(sp^2)–C(sp^3) Cross-Coupling of Chlorobenzenesulfonamides with Alkyl Grignard Reagents: Entry to Alkylated Aromatics. *J. Org. Chem.* **2019**, *84*, 1640–1646. [CrossRef] [PubMed]

41. Bisz, E.; Podchorodecka, P.; Szostak, M. N-Methylcaprolactam as a Dipolar Aprotic Solvent for Iron-Catalyzed Cross-Coupling Reactions: Matching Efficiency with Safer Reaction Media. *ChemCatChem* **2019**, *11*, 1196–1199. [CrossRef]

42. Bisz, E.; Kardela, M.; Piontek, A.; Szostak, M. Iron-Catalyzed C(sp^2)–C(sp^3) Cross-Coupling at Low Catalyst Loading. *Catl. Sci. Technol.* **2019**, *9*, 1092–1097. [CrossRef]

43. Bisz, E.; Kardela, M.; Szostak, M. Ligand Effect on Iron-Catalyzed Cross-Coupling Reactions: Evaluation of Amides as O-Coordinating Ligands. *ChemCatChem* **2019**, *11*, 5733–5737. [CrossRef]

44. Shi, S.; Nolan, S.P.; Szostak, M. Well-Defined Palladium(II)-NHC (NHC = N-Heterocyclic Carbene) Precatalysts for Cross- Coupling Reactions of Amides and Esters by Selective Acyl CO–X. (X. = N., O) Cleavage. *Acc. Chem. Res.* **2018**, *51*, 2589–2599. [CrossRef]

45. Meng, G.; Szostak, M. N-Acyl-Glutarimides: Privileged Scaffolds in Amide N-C Bond Cross-Coupling. *Eur. J. Org. Chem.* **2018**, *20–21*, 2352–2365. [CrossRef]

46. Takise, R.; Muto, K.; Yamaguchi, J. Cross-Coupling of Aromatic Esters and Amides. *Chem. Soc. Rev.* **2017**, *46*, 5864–5888. [CrossRef]

47. Guo, L.; Rueping, M. Decarbonylative Cross-Couplings: Nickel Catalyzed Functional Group Interconversion Strategies for the Construction of Complex Organic Molecules. *Acc. Chem. Res.* **2018**, *51*, 1185–1195. [CrossRef]

48. Liu, C.; Szostak, M. Decarbonylative Cross-Coupling of Amides. *Org. Biomol. Chem.* **2018**, *16*, 7998–8010. [CrossRef]

49. Amaike, K.; Muto, K.; Yamaguchi, J.; Itami, K. Decarbonylative C-H Coupling of Azoles and Aryl Esters: Unprecedented Nickel Catalysis and Application to the Synthesis of Muscoride, A. *J. Am. Chem. Soc.* **2012**, *134*, 13573–13576. [CrossRef] [PubMed]

50. Muto, K.; Yamaguchi, J.; Musaev, D.G.; Itami, K. Decarbonylative Organoboron Cross-Coupling of Esters by Nickel Catalysis. *Nat. Commun.* **2015**, *6*, no. 7508. 1–8. [CrossRef] [PubMed]

51. Takise, R.; Isshiki, R.; Muto, K.; Itami, K.; Yamaguchi, J. Decarbonylative Diaryl Ether Synthesis by Pd and Ni Catalysis. *J. Am. Chem. Soc.* **2017**, *139*, 3340–3343. [CrossRef] [PubMed]

52. Isshiki, R.; Muto, K.; Yamaguchi, J. Decarbonylative C–P Bond Formation Using Aromatic Esters and Organophosphorus Compounds. *Org. Lett.* **2018**, *20*, 1150–1153. [CrossRef]

53. Halima, T.B.; Zhang, W.; Yalaoui, I.; Hong, X.; Yang, Y.-F.; Houk, K.N.; Newman, S.G. Palladium-Catalyzed Suzuki-Miyaura Coupling of Aryl Esters. *J. Am. Chem. Soc.* **2017**, *139*, 1311–1318. [CrossRef]

54. Halima, T.B.; Kishore, J.; Shkoor, V.M.; Newman, S.G. A Cross-Coupling Approach to Amide Bond Formation from Esters. *ACS Catal.* **2017**, *7*, 2176–2180. [CrossRef]

55. Masson-Makdissi, J.; Vandavasi, J.; Newman, S. Switchable Selectivity in the Pd-Catalyzed Alkylative Cross-Coupling of Esters. *Org. Lett.* **2018**, *20*, 4094–4098. [CrossRef]

56. Dardir, A.H.; Melvin, P.R.; Davis, R.M.; Hazari, N.; Beromi, M.M. Rapidly Activating Pd-Precatalyst for Suzuki-Miyaura and Buchwald-Hartwig Couplings of Aryl Esters. *J. Org. Chem.* **2017**, *83*, 469–477. [CrossRef]

57. Chatupheeraphat, A.; Liao, H.H.; Srimontree, W.; Guo, L.; Minenkov, Y.; Poater, A.; Cavallo, L.; Rueping, M. Ligand-Controlled Chemoselective C(acyl)-O Bond vs C(aryl)-C Bond Activation of Aromatic Esters in Nickel Catalyzed C(sp^2)-C(sp^3) Cross-Couplings. *J. Am. Chem. Soc.* **2018**, *140*, 3724–3735. [CrossRef]

58. Guo, L.; Rueping, M. Transition-Metal-Catalyzed Decarbonylative Coupling Reactions: Concepts, Classifications, and Applications. *Chem. Eur. J.* **2018**, *24*, 7794–7809. [CrossRef]

59. Pu, X.; Hu, J.; Zhao, Y.; Shi, Z. Nickel-Catalyzed Decarbonylative Borylation and Silylation of Esters. *ACS Catal.* **2016**, *6*, 6692–6698. [CrossRef]

60. Lei, P.; Meng, G.; Shi, S.; Ling, Y.; An, J.; Szostak, R.; Szostak, M. Suzuki-Miyaura Cross-Coupling of Amides and Esters at Room Temperature: Correlation with Barriers to Rotation around C–N and C–O Bonds. *Chem. Sci.* **2017**, *8*, 6525–6530. [CrossRef] [PubMed]

61. Shi, S.; Lei, P.; Szostak, M. Pd-PEPPSI: A General Pd-NHC Precatalyst for Suzuki-Miyaura Cross- Coupling of Esters by C–O Cleavage. *Organometallics* **2017**, *36*, 3784–3789. [CrossRef]

62. Li, G.; Shi, S.; Szostak, M. Pd-PEPPSI: Water-Assisted Suzuki-Miyaura Cross-Coupling of Aryl Esters at Room Temperature using a Practical Palladium-NHC (NHC = N-Heterocyclic Carbene) Precatalyst. *Adv. Synth. Catal.* **2018**, *360*, 1538–1543. [CrossRef]

63. Shi, S.; Szostak, M. Pd–PEPPSI: A General Pd–NHC Precatalyst for Buchwald–Hartwig Cross-Coupling of Esters and Amides (Transamidation) under the Same Reaction Conditions. *Chem. Commun.* **2017**, *53*, 10584–10587. [CrossRef] [PubMed]

64. Buchspies, J.; Pyle, D.J.; He, H.; Szostak, M. Pd-Catalyzed Suzuki-Miyaura Cross-Coupling of Pentafluorophenyl Esters. *Molecules.* **2018**, *23*, 3134–3144. [CrossRef]

65. Liebman, J.; Greenberg, A. The Origin of Rotational Barriers in Amides and Esters. *Biophys. Chem.* **1974**, *1*, 222–226. [CrossRef]

66. Lee, S.H.; Nikonov, G.I. Transfer Hydrogenation of Ketones, Nitriles, and Esters Catalyzed by a Half-Sandwich Complex of Ruthenium. *ChemCatChem.* **2015**, *7*, 107–113. [CrossRef]

67. Zhang, L.; Zhang, G.; Zhang, M.; Cheng, J. Cu(OTf)$_2$-Mediated Chan-Lam Reaction of Carboxylic Acids to Access Phenolic Esters. *J. Org. Chem.* **2010**, *75*, 7472–7474. [CrossRef]

68. Neuvonen, H.; Neuvonen, K.; Pasanen, P. Substituent Influences on the Stability of the Ring and Chain Tautomers in 1,3-O,N-Heterocyclic Systems: Characterization by ^{13}C-NMR Chemical Shifts, PM3 Charge Densities, and Isodesmic Reactions. *J. Org. Chem.* **2004**, *69*, 3794–3800. [CrossRef]

69. Kaplan, J.P.; Raizon, B.M.; Desarmeinen, M.; Feltz, P.; Headley, P.M.; Worms, P.; Lioyd, K.G.; Bartholini, G. New anticonvulsants: Schiff bases of γ-aminobutyric acid and γ-aminobutyramide. *J. Med. Chem.* **1980**, *23*, 702–704. [CrossRef] [PubMed]

70. Qin, C.; Wu, H.; Chen, J.; Liu, M.; Cheng, J.; Su, W.; Ding, J. Palladium-Catalyzed Aromatic Esterification of Aldehydes with Organoboronic Acids and Molecular Oxygen. *Org. Lett.* **2008**, *10*, 1537–1540. [CrossRef] [PubMed]

Sample Availability: Samples of the compounds are not available from the authors.

Article

A DFT Study on $Fe^{I}/Fe^{II}/Fe^{III}$ Mechanism of the Cross-Coupling between Haloalkane and Aryl Grignard Reagent Catalyzed by Iron-SciOPP Complexes

Akhilesh K. Sharma [1] and Masaharu Nakamura [1,2,*]

[1] International Research Center for Elements Science (IRCELS), Institute for Chemical Research, Kyoto University, Uji, Kyoto 611-0011, Japan

[2] Department of Energy and Hydrocarbon Chemistry, Graduate School of Engineering, Kyoto University, Kyoto 615-8510, Japan

* Correspondence: masaharu@scl.kyoto-u.ac.jp; Tel.: +81-774-38-3180

Academic Editor: Hans-Joachim Knölker

Received: 1 July 2020; Accepted: 5 August 2020; Published: 8 August 2020

Abstract: To explore plausible reaction pathways of the cross-coupling reaction between a haloalkane and an aryl metal reagent catalyzed by an iron–phosphine complex, we examine the reaction of FeBrPh(SciOPP) **1** and bromocycloheptane employing density functional theory (DFT) calculations. Besides the cross-coupling, we also examined the competitive pathways of β-hydrogen elimination to give the corresponding alkene byproduct. The DFT study on the reaction pathways explains the cross-coupling selectivity over the elimination in terms of $Fe^{I}/Fe^{II}/Fe^{III}$ mechanism which involves the generation of alkyl radical intermediates and their propagation in a chain reaction manner. The present study gives insight into the detailed molecular mechanic of the cross-coupling reaction and revises the Fe^{II}/Fe^{II} mechanisms previously proposed by us and others.

Keywords: iron catalysis; haloalkane coupling; Grignard reagent; $Fe^{I}/Fe^{II}/Fe^{III}$ mechanism; density functional theory

---

1. Introduction

Transition-metal-catalyzed cross-coupling reaction is one of the most versatile synthetic tools for constructing carbon frameworks of various functional molecules, e.g., pharmaceuticals, agrochemicals, and organic electronic materials. Iron catalysts have emerged as a sustainable alternative for conventional palladium and nickel catalysts and have attracted significant attention due to their practical merits of cost-effectiveness, low toxicity, and the abundance on the earth [1–4]. Despite the remarkable progress of iron-catalyzed cross-coupling reactions in organic synthesis [5–14], their reaction mechanisms remain elusive and attract considerable interests to their underlying molecular mechanisms [15–19].

More specifically, iron catalysts have proven useful for cross-coupling reactions of non-activated haloalkanes, a synthetically valuable yet challenging substrate, which is prone to undergo β-hydrogen elimination in the conventional palladium-catalyzed cross-couplings to form undesired alkene byproducts [20–23]. Numerous synthetic and mechanistic studies have thus appeared on this class of iron-catalyzed cross-coupling. The use of the controlled-addition technique of Grignard reagent in Kumada–Tamao-Corriu (KTC) coupling or using less reactive boron reagents in Suzuki–Miyaura (SM) coupling have proven, in combination with the use of bulky bisphosphine ligands, to be the successful strategy for the iron-catalyzed haloalkane coupling reactions [24–27].

There is a consensus that the haloalkane substrates undergo halogen abstraction by an (organo)iron species to generate the corresponding alkyl radical intermediates. A variety of radical mechanisms have been discussed thus far by Hu [28], Norrby [29], Bedford [30], Tonzetich [31], Fürstner [22], Neidig [32,33], and us [19,34]. Our group and the Neidig group have independently crystalized iron-bisphosphine complexes, such as $Fe^{II}X_2(SciOPP)$, $Fe^{II}XAr(SciOPP)$, $Fe^{II}Ar_2(SciOPP)$, where X = halide and Ar = Ph or Mes [32–34]. Furthermore, we identified the presence of $Fe^{II}BrMes(SciOPP)$, $Fe^{II}Mes_2(SciOPP)$ in the reaction mixture of the cross-coupling reaction by using solution-phase synchrotron X-ray absorption spectroscopic (XAS) studies [34]. According to this study, we proposed an Fe^{II}/Fe^{II} mechanism based on identification of $Fe^{II}Ar_2(SciOPP)$ (Figure 1a, Ar = Mes and X = Br). The Neidig group also identified the presence of iron(II) complexes in the solution phase through intensive mechanistic studies, consisting of Mössbauer and magnetic circular dichroism (MCD) spectroscopy, and synthetic approaches [32,33]. They identified that $Fe^{II}XAr(SciOPP)$ is the active iron species of the cross-coupling reaction suppressing undesired β-hydrogen elimination reactions (Figure 1a, Ar = Ph). On the other hand, the Bedford group crystalized iron(I) species, $[Fe^{I}Br(dpbz)_2]$ which is formed from iron(II)-precatalyst and they proposed a low-coordinate iron(I)-phosphine species is involved in the cross-coupling reactions [35]. Identification of such low-coordinate iron(I)-complexes in the reaction mixture was not fully successful, and their involvement in the reaction mechanism remains elusive. The active iron-species that generate alkyl radical intermediates and the iron-species that react with the radical intermediate to effect the carbon–carbon bond formation have also remained unknown.

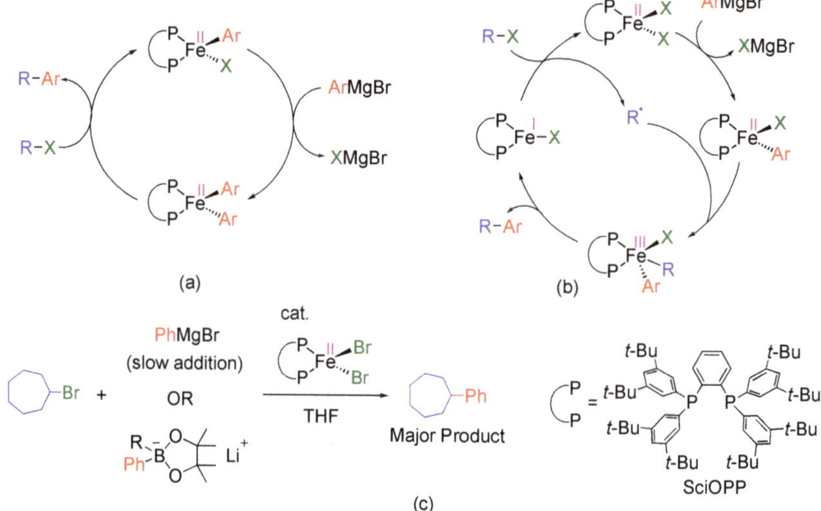

Figure 1. (a) Proposed catalytic cycle for cross-coupling between an alkyl halide and Grignard reagent catalyzed by iron-SciOPP complexes. (b) Proposed catalytic cycle based on a recent study involving iron-BenzP* catalyzed cross-coupling between α-chloroesters and Grignard reagent. (c) The iron-SciOPP-catalyzed cross-coupling reaction chosen for the current density functional theory (DFT) mechanistic study [26,27].

Recently, our group developed enantioselective cross-coupling reactions using chiral bisphosphine ligands and proposed that the radical species may not react with the iron(III)-species which takes part in the generation of the radical (in-cage mechanism) [36,37]. We further performed the detailed DFT analysis of the reaction of α-chloroesters with the Grignard reagent in presence of chiral bisphosphine ligand and found that $Fe^{I}/Fe^{II}/Fe^{III}$ mechanism operates in the reaction (Figure 1b) [38]. The iron(I) species activates the carbon–halogen bond of halo-ester substrates leading to the generation of α-ester-radical. In the next step, another molecule of $Fe^{II}XAr$ (active species) traps the generated radical (out-of-cage or chain mechanism) and lead to product formation through iron(III) species. Parallelly, the Gutierrez group reported DFT mechanistic study of the same reaction leading to the virtually same conclusion [39]. Despite these recent mechanistic insights, the use of electronically and sterically different iron-bisphosphine complexes, different substrates, and reagents in cross-coupling reactions make it difficult to generalize the mechanism understanding.

The iron-SciOPP catalysts have been known as a superior catalyst for the cross-coupling of alkyl halides and Grignard reagent [26]. Due to bulky nature and strong chelation ability of the ligand, it avoids the formation of multinuclear or multiligand iron–phosphine complexes and also suppresses the formation of undesired ate complexes, which presumably responsible for increased reactivity and selectivity by minimizing side reactions. The mechanistic studies by Neidig showed that $Fe^{II}BrPh(SciOPP)$ complex reacts with a haloalkane to give the corresponding cross-coupling product selectively, but the detailed mechanism has remained unclear [33]. In our pursuit to identify the iron species involved in the elementary steps of the iron-catalyzed haloalkane coupling, DFT is used to study the reaction between $Fe^{II}BrPh(SciOPP)$ and bromocycloheptane (Figure 1c), which has been believed as the key for the selective cross-coupling of a haloalkane with an aryl metal reagent in the presence of an iron-bisphosphine catalyst. The plausible reaction pathways, Fe^{II}/Fe^{II}, and $Fe^{I}/Fe^{II}/Fe^{III}$ are examined and compared in this study.

2. Results and Discussion

2.1. Reaction of Fe^{II}BrPh(SciOPP) with Bromocycloheptane

2.1.1. Cycloheptyl Radical Generation via C–Br Activation by Iron(II)-Species

The reaction mechanism from $Fe^{II}BrPh(SciOPP)$ is studied as it is predicted to be the starting species in cross-coupling reactions. The TSs with lowest energy spin states as found in our previous study were only considered, due to large system size [38]. The first step is the C–Br activation of bromocycloheptane by quintet $Fe^{II}BrPh(SciOPP)$ ($^5\mathbf{2_{PhBr}}$), through an atom transfer mechanism (Figure 2). To a tetrahedral geometry of $Fe^{II}BrPh(SciOPP)$, $^5\mathbf{2_{PhBr}}$, bromine-atom of substrate (**S**) approaches in the plane the same as P–Fe–P to form pre-reaction complex $^5\mathbf{2_{PhBr-RBr}}$, which is 6.7 kcal/mol higher in energy than the separated molecules. The $^5\mathbf{TS1}$ has a distorted square-pyramidal geometry, with the incoming bromine-atom positions in the square plane and the other bromine in the pyramidal position. The activation barrier of this process is 24.1 kcal/mol. The Mulliken spin densities on the iron center of the catalyst, and C_1 and bromine of substrate are $\varrho(Fe) = 3.316$, $\varrho(Br) = -0.055$, and $\varrho(C_1) = 0.672$, respectively (Figure S1), indicating that during this step, quintet iron(II) is converted to quartet-iron(III) species with the generation of doublet cycloheptyl radical, $^2\mathbf{R\cdot}$. The formed $^5\mathbf{3_{PhBrBr}R}$ and $^4\mathbf{3_{PhBrBr}}$ species are respectively 21.7 and 14.9 kcal/mol higher in energy than $^5\mathbf{2_{PhBr}}$, suggesting the reversible nature of the C–Br activation and will lead to a low concentration of radical $^2\mathbf{R\cdot}$.

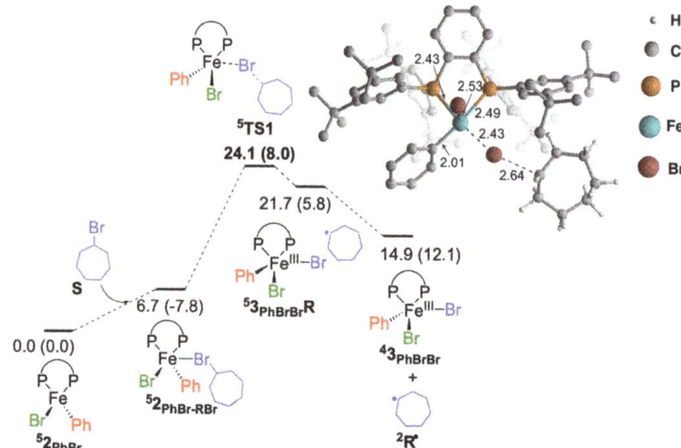

Figure 2. Free energy profile for the reaction of iron(II) species with bromocycloheptane. The relative free energies with respect to $^5\mathbf{2_{PhBr}}$ and bromocycloheptane are given in kcal/mol with total electronic energies including ZPE (zero-point energy) correction (ΔE_{ZPE}) are given in parenthesis.

2.1.2. Reaction of Cycloheptyl Radical with Iron(II)-Species: C–C Bond Formation and Generation of Iron(I) Species

The generated radical intermediate **R·** can react with either $^4\mathbf{3_{PhBrBr}}$ (iron(III)-species, in-cage mechanism) or $^5\mathbf{2_{PhBr}}$ (iron(II)-species, out-of-cage mechanism) present in the reaction mixture [36,38]. For the reaction of cycloheptyl radical (**R·**) with iron(II) species, the first step involves the out-of-cage movement of $^2\mathbf{R·}$ from the intermediate $^4\mathbf{3_{PhBrBr}R}$. Then the $^2\mathbf{R·}$ coordinates to the iron center through $^4\mathbf{TS2}$ with a lower barrier than $^5\mathbf{TS1}$ (20.7 kcal/mol from starting $^5\mathbf{2_{PhBr}}$ and 5.8 kcal/mol from the second $^5\mathbf{2_{PhBr}}$, Figure 3). It leads to the formation of lower energy iron(III) species $^4\mathbf{3_{PhBrR}}$, which is 11.7 kcal/mol lower in energy. The consequent reductive elimination forms C–C coupling product **P1**, which proceeds with the activation barrier of 13.1 kcal/mol ($^4\mathbf{TS3}$). The reductive elimination process is highly exergonic and the corresponding iron(I)-product complex ($^4\mathbf{1_{Br-PhR}}$) is 16.6 kcal/mol lower in energy than $^4\mathbf{3_{PhBrR}}$. Afterwards, the product will decoordinate from $^4\mathbf{1_{Br-PhR}}$ leading to the formation of 5.7 kcal/mol higher energy species $^4\mathbf{1_{Br}}$, probably through a barrierless process. The transient species $^4\mathbf{1_{Br}}$ readily reacts with a THF molecule (solvent) leading to the formation of lower energy species $^4\mathbf{1_{Br-THF}}$ (0.9 kcal/mol lower energy than $^4\mathbf{1_{Br-PhR}}$). The whole process for the reaction of $Fe^{II}BrPh(SciOPP)$ with bromocycloheptane leading to the formation of product and $^4\mathbf{1_{Br-THF}}$ is exergonic by 13.3 kcal/mol.

2.1.3. Reaction of Cycloheptyl Radical with $Fe^{III}Br_2Ph(SciOPP)$ (In-Cage Mechanism)

The reaction of cycloheptyl radical with $^4\mathbf{3_{PhBrBr}}$ leading to a cross-coupling product turns out to have a higher free energy barrier of 26.9 kcal/mol (Figure S2). This process occurs through an outer-sphere fashion ($^5\mathbf{TS4}$), where **R·** attacks directly the *ipso*-carbon of the phenyl group and does not coordinate with the iron center. The reaction leading to byproduct alkene formation occurs by the simultaneous transfer of bromine and hydrogen from bromocycloheptane to iron and the *ipso*-carbon of the phenyl group of $^5\mathbf{2_{PhBr}}$, respectively, via $^5\mathbf{TS5}$ with a 25.2 kcal/mol energy barrier. Both processes for the formation of the cross-coupling and the alkene products are highly exothermic (ca. 35–45 kcal/mol). The formation of alkene is favorable kinetically over the cross-coupling. However, it does not agree with experimental observation.

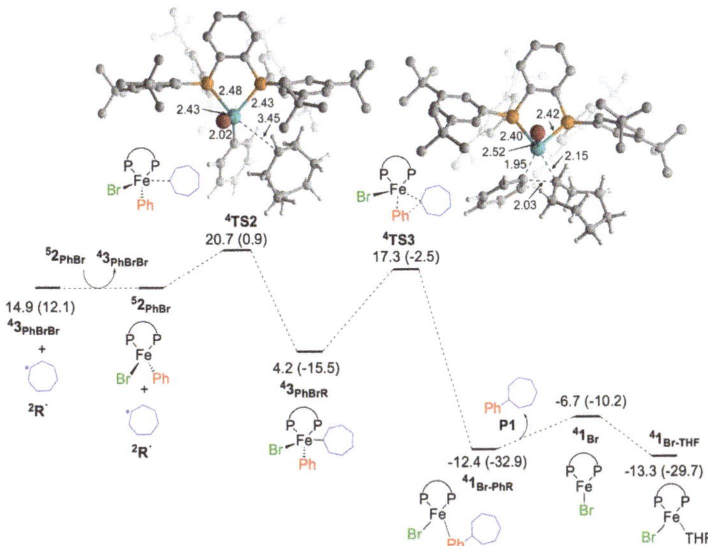

Figure 3. Free energy profile for cross-coupling by the reaction of iron(II) species with cycloheptyl radical, where radical is generated by the reaction of iron(II) species and bromocycloheptane. Free energies with respect to ⁵2PhBr and bromocycloheptane are given in kcal/mol (ΔE_{ZPE} are given in parenthesis).

Although the reaction of cycloheptyl radical with $Fe^{III}Br_2Ph(SciOPP)$, ⁴3PhBrBr leading to cross-coupling product and alkene formation is highly exothermic, it is unlikely as ⁵TS4 and ⁵TS5 are 6.2 and 4.5 kcal/mol higher in energy than ⁴TS2. Hence, the cross-coupling reaction proceeds by the reaction of alkyl radical with iron(II)-species. Similar to our previous study, iron-SciOPP complexes also do not favor Fe^{II}/Fe^{III} mechanism.

2.1.4. Reaction of Cycloheptyl Radical with Iron(II)-Species: Alkene Byproduct Formation

From complex ⁴3PhBrR the possibility of formation of alkene is also considered (Figure 4). It involves the transfer of β-hydrogen from a coordinated cycloheptyl group to phenyl ligand (⁴TS6). The activation barrier for this process is apparently high and ⁴TS6 is 14.1 kcal/mol higher in energy than ⁴TS3 and the formation of alkene is unlikely from ⁴3PhBrR. The TSs for coupling and alkene formation by the outer-sphere reaction of radical with phenyl of iron(II) species, ⁵2PhBr could not be located.

Further, we checked the possibility of alkene formation by H-atom transfer (⁴TS7) from cycloheptyl radical to iron-atom of the iron(II) complex, ⁵2PhBr. The activation barrier for this process is 12.9 kcal/mol. It leads to the formation of the iron(III)-hydride complex, ⁴3PhBrH. The following reductive elimination between the hydride and phenyl groups through ⁴TS8 lead to the formation of benzene. This pathway is unfavorable as ⁴TS7 is 7.7 kcal/mol higher in energy than ⁴TS2. Hence, the reaction of the alkyl radical with $Fe^{II}BrPh(SciOPP)$ does not give the alkene byproduct, which is in concert with the experimental observations.

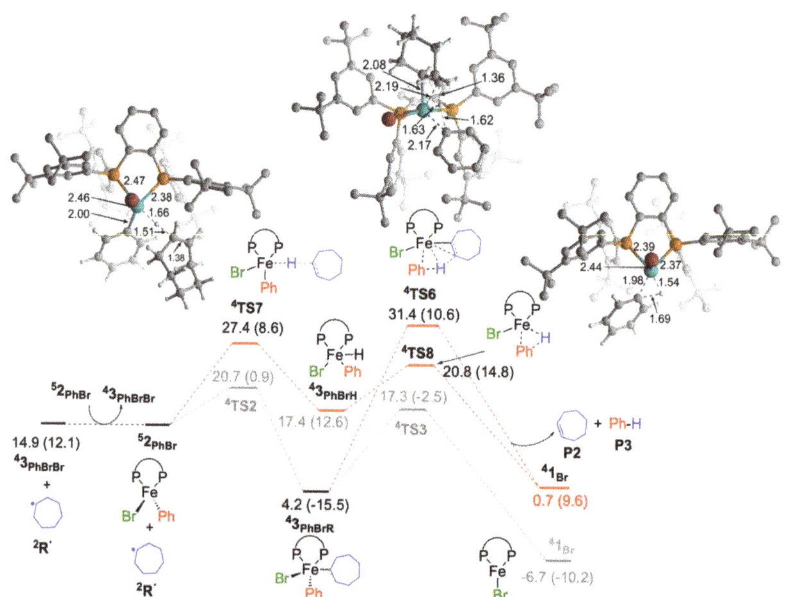

Figure 4. Free energy profile for olefin formation by reaction of Fe^{II} species with cycloheptyl radical, where radical is generated by the reaction of iron(II) species and bromocycloheptane. Free energies with respect to $^{5}2_{PhBr}$ and bromocycloheptane are given in kcal/mol (ΔE_{ZPE} are given in parenthesis).

2.1.5. Reaction of Iron(I) Species with Bromocycloheptane: Regeneration of $Fe^{II}Br_2$(SciOPP) and Cycloheptyl Radical

The iron(I) species formed during the reaction can react either with another iron(III) species to undergo a comproportionation to generate two iron(II) species, or with haloalkane substrate (of which concentration is larger than iron(III) species) to generate one iron(II) species and an alkyl radical intermediate. For the reaction to proceed the first step is the removal of THF from $^{4}1_{Br-THF}$ to form $^{4}1_{Br}$ and this process is endergonic by 6.6 kcal/mol (Figure 5). This step is required for both comproportionation and reaction with the haloalkane substrate. The coordination of the substrate through bromide to $^{4}1_{Br}$, leading to the formation of $^{4}1_{Br-BrR}$ is exergonic by 2.3 kcal/mol. The activation barrier for the Br-atom transfer ($^{4}TS9$) from $^{4}1_{Br-BrR}$ species is 6.8 kcal/mol (barrier from $^{4}1_{Br-THF}$). The process is highly exergonic and leads to formation of cycloheptyl radical and iron(II) species $^{5}2_{BrBr}$. The generated $^{5}2_{BrBr}$ species will undergo transmetalation and the radical species will react with the available $^{5}2_{PhBr}$ in the reaction mixture, propagating the radical chain reaction [40]. Hence, the reaction will proceed through the *out-of-cage* mechanism, and similar to our previous study, the current reaction also follows the $Fe^{I}/Fe^{II}/Fe^{III}$ pathway [38].

The comproportionation between $^{4}1_{Br}$ and $^{4}3_{PhBrBr}$ (generated by substrate reaction with iron(II) species $^{5}2_{PhBr}$) may be a barrierless process and is a highly exergonic process (37.2 kcal/mol). Since the concentration of the haloalkane substrate is high enough under the catalytic coupling condition, the reaction between the iron(I) species and haloalkane is more likely. Hence, iron(I) species is the active species after its generation during the reaction, which reduces the activation barrier for C–Cl activation from 24.1 to 6.8 kcal/mol.

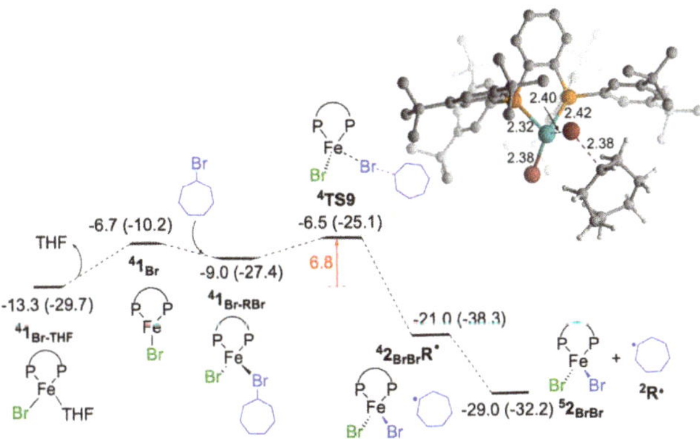

Figure 5. Free energy profile for probable reaction path of generated iron(I) species. Free energies with respect to ^{5}**2$_{PhBr}$** and bromocycloheptane are given in kcal/mol (ΔE_{ZPE} are given in parenthesis).

2.2. Proposed Catalytic Cycle Based on Fe^I/Fe^{II}/Fe^{III} Mechanism

Based on the above theoretical studies, we propose the reaction mechanism depicted in Figure 6. The Fe^{II}XAr(SciOPP) (^{5}**2$_{PhBr}$**) is first converted to Fe^{I}XTHF(SciOPP) (^{4}**1$_{Br-THF}$**) with the formation of the cross-coupling product. Afterwards, iron(I) species (^{4}**1$_{Br-THF}$**) mainly activate the alkly halide. Transmetalation of ^{4}**1$_{Br-THF}$** is less likely due to the low concentration of Grignard reagent, which excludes the possibility of involvement of Fe^{I}ArTHF(SciOPP) species. Hence, the catalytic cycle starts with the halogen atom abstraction from the alkyl halide (Alkyl-X) by iron(I) species (^{4}**1$_{Br}$**) affording a dihaloiron species Fe^{II}X$_2$(SciOPP) (^{5}**2$_{BrBr}$**) along with an alkyl radical intermediate (^{2}**R·**) (Figure 6 and Figure S3). The resultant alkyl radical ^{2}**R·** escapes from the solvent cage and is further trapped by iron(II) species, Fe^{II}XAr(SciOPP) (^{5}**2$_{PhBr}$**) (which have a higher concentration) via the out-of-cage pathway. It led to the formation of iron(III) species Fe^{III}XArR(SciOPP) (^{4}**3$_{PhBrR}$**). The reductive elimination from ^{4}**3$_{PhBrR}$** led to the formation of a cross-coupling product and regenerating iron(I) species (^{4}**1$_{Br-THF}$**). Based on this mechanism, the detection of both iron(II) and iron(I) species, and their roles in the cross-coupling reaction can be reasonably explained.

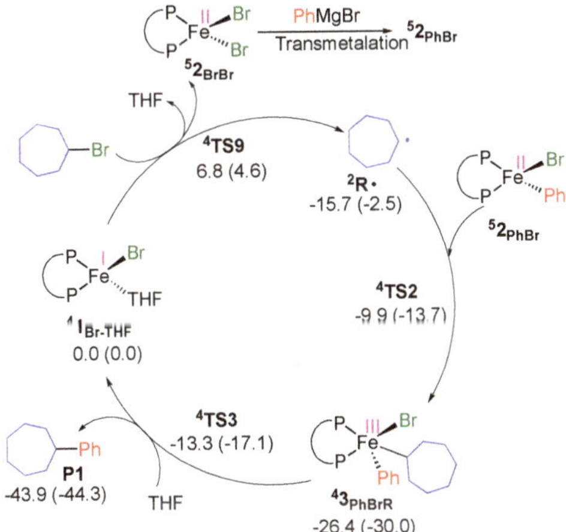

Figure 6. Revised catalytic cycle of iron-SciOPP catalyzed-KTC coupling of a haloalkane based on the current DFT study. Free energies with respect to ^{4}**1$_{Br-THF}$** and bromocycloheptane are given in kcal/mol (ΔE_{ZPE} are given in parenthesis).

3. Computational Methods

The Gaussian09 program was used for quantum mechanical calculations reported in the study [41]. The DFT method B3LYP [42,43] was used for geometry optimization and energy calculations with the inclusion of solvent effects by the PCM method [44]. Grimme's D3 dispersion correction method was used to include dispersion correction [45]. For geometry optimization, the 6-31G* basis set [46] was used for all atoms as it gave good agreement with the crystal structure of *Fe^{II}BrPh(SciOPP)* and *Fe^{II}Br$_2$(SciOPP)* (Figure S4). The geometry optimization with SDD [47] basis set for Fe, Br and 6-31G* for other atoms lead to relatively longer metal ligand distances. Hence, the 6-31G* basis set was used. The hessian calculation was performed to confirm minima (no imaginary frequency) and TS (one imaginary frequency). All TS were confirmed by visual inspection of imaginary frequency. The connectivity between TSs and minima were confirmed by IRC [48] calculations for 20 steps and subsequent optimization of the geometry at PCM/B3LYP/6-31G* level of theory.

The energies were further computed using the 6-311G** basis set [49] for all atoms and with Douglas–Kroll–Hess second-order scalar relativistic effects. The E$_{ZPE}$ reported in the paper is calculated by adding zero-point-energy obtained at the 6-31G* basis set to electronic energy of higher basis set (E$_{ZPE}$ = E$_{elec(6-311G^{**})}$ + ZPE$_{(6-31G^*)}$). The free energies were calculated by adding thermal free energy correction obtained at 6-31G* to electronic energy of the 6-311G** basis set (G = E$_{elec(6-311G^{**})}$ + G$_{Corr(6-31G^*)}$). The free energies were calculated at 298.15 K temperature and 1 atm pressure. Concisely, the energies reported in the paper were calculated at PCM$_{THF}$/B3LYP-D3/6-311G**//PCM$_{THF}$/B3LYP-D3/6-31G* level of theory.

The spin-state energies of iron complexes computed by a DFT method depend on the choice of the exchange-correlation functional [50,51]. Various functionals were recommended in the literature, while no consensus has been made for the optimal choice of the DFT method. Hence, care should be taken in choosing functional, especially for iron complexes with the small spin-state energy gaps. The DFT method (B3LYP-D3 Functional) used in this work predicted quintet as the ground state for iron(II) complexes, Fe^{II}BrPh(SciOPP) and Fe^{II}Br$_2$(SciOPP), which is in agreement with the experimentally known spin state [33,34]. Additionally, the spin state for related iron(I) complex (Fe(η^2-[TIPS-CC-H])Br(SciOPP)) was also predicted correctly [52]. Further, the OBPE-D3 functional (which is known to provide reliable spin state gap) [53] gave consistent results with B3LYP-D3 for the ground spin state (quartet) of iron(I) and iron(III) complexes (Table S1). The OPBE-D3 functional failed to give reliable results for the spin state of the Fe^{II}BrPh(SciOPP) complex. Hence, the B3LYP-method reliably predicts the spin state of the current system and is used in this study. Further, we have also checked the effect of long-range-corrected (LC) exchange functional on barrier height [54] using CAM-B3LYP-D3 functional. The CAM-B3LYP-D3 functional led to a similar conclusion as of B3LYP-D3, and some TSs showed a higher energy barrier (details are given in Table S2).

4. Conclusions

We have studied the mechanism of the reaction between Fe^{II}BrPh(SciOPP) and bromocycloheptane by using DFT methods and identified iron species of different oxidation states involved in a catalytic cycle of the iron-bisphosphine-catalyzed haloalkane couplings. The reaction mechanism follows the Fe^I/Fe^{II}/Fe^{III} pathway, where iron(I) activates the C–Br bond of the haloalkane substrate, another molecule of Fe^{II}BrPh(SciOPP) traps the generated alkyl radical species to form iron(III) species leading to the formation of the cross-coupling product and iron(I) species. The iron(I) species then again react with haloalkane generating alkyl radical and the reaction propagates through a radical chain mechanism. The current mechanism mirrors our previous mechanistic study on a chiral bisphosphine-catalyzed enantioselective cross-coupling despite the change of bisphosphine ligand from BenzP* to SciOPP.

The reaction of haloalkane with iron(II) species and the reaction of the alkyl radical with iron(III) species have a higher energy barrier than those of Iron(I) species and iron(II) species, respectively. The obtained energetics exclude the possibility of Fe^{II}/Fe^{II} and Fe^{II}/Fe^{III} pathways proposed previously.

Furthermore, we have found that an alkyl radical prefers to coordinate to the $Fe^{II}BrPh(SciOPP)$ species, instead of the β-hydrogen-atom abstraction ending with the alkene byproduct. The generated $Fe^{III}BrPh(alkyl)(SciOPP)$ species favors reductive elimination to give the corresponding coupling product over the alkene byproduct. These results deepen our understanding of the iron-SciOPP catalysis and provide insights into the probable root for the side product formations. Further understanding of the mechanism of alkene formation with $Fe^{II}Ph_2(SciOPP)$ catalyst will help design better catalysts and synthetic methodology. The theoretical and experimental research in this line is now actively pursued in our laboratory and will be reported in due course.

Supplementary Materials: The following are available online. Cartesian coordinates of optimized geometries and their energies, Figure S1: Mulliken spin densities of the stationary points, Figure S2: Free energy profile for the reaction of Fe^{II}/Fe^{III} mechanism, Figure S3: Energy profile for $Fe^{I}/Fe^{II}/Fe^{III}$ pathway starting the reaction from iron(I) species, Figure S4: Comparison of geometrical parameters of X-ray crystal structure of 2_{PhBr} and 2_{BrBr} with the optimized geometry in quintet spin state, Table S1: The comparison of energies for the iron(I), iron(II) and iron(III) complexes in different spin states at B3LYP-D3 and OPBE-D3 functionals. Table S2: Comparison of free energy barriers of catalytic cycle at B3LYP-D3 and CAM-B3LYP-D3 functionals.

Author Contributions: Conceptualization, M.N. and A.K.S.; methodology, A.K.S.; validation, A.K.S. and M.N.; writing—original draft preparation, A.K.S. and M.N.; writing—review and editing, M.N.; visualization, A.K.S.; supervision, M.N.; project administration, M.N.; funding acquisition, M.N. All authors have read and agreed to the published version of the manuscript.

Funding: This work was supported by a Grant-in-Aid for Scientific Research (20H02740) and Grant-in-Aid for Scientific Research on Innovative Areas "Synergistic Effects for Creation of Functional Molecules" (JP18064006) from the Ministry of Education, Culture, Sports, Science and Technology (MEXT), and through the "Funding Program for Next generation World-Leading Researchers (Next Program)" initiated by the Council for Science and Technology Policy (CSTP). This work was supported by Integrated Research Consortium on Chemical Sciences. M.N. thanks for the financial support from Nissan Chemical Corporation and Tosoh Finechem Corporation.

Acknowledgments: Supercomputing facility at the Institute of Molecular Science in Japan and the Academic Center for Computing at Media Studies at Kyoto University in Japan are acknowledged for computational resources. Computation time was also provided by the Super Computer System, Institute for Chemical Research, Kyoto University. WMC Sameera and Miho Isegawa are also acknowledged for useful discussion and contributions in computational design. This work was also supported by the Integrated Research Consortium on Chemical Sciences (IRCCS).

Conflicts of Interest: The authors declare no conflicts of interest.

References

1. Guðmundsson, A.; Bäckvall, J.-E. On the use of iron in organic chemistry. *Molecules* **2020**, *25*, 1349. [CrossRef] [PubMed]
2. Fürstner, A. Iron catalysis in organic synthesis: A critical assessment of what it takes to make this base metal a multitasking champion. *ACS Cent. Sci.* **2016**, *2*, 778–789. [CrossRef] [PubMed]
3. Bauer, I.; Knölker, H.-J. Iron catalysis in organic synthesis. *Chem. Rev.* **2015**, *115*, 3170–3387. [CrossRef]
4. Bolm, C.; Legros, J.; Le Paih, J.; Zani, L. Iron-catalyzed reactions in organic synthesis. *Chem. Rev.* **2004**, *104*, 6217–6254. [CrossRef] [PubMed]
5. Piontek, A.; Bisz, E.; Szostak, M. Iron-catalyzed cross-couplings in the synthesis of pharmaceuticals: In pursuit of sustainability. *Angew. Chem. Int. Ed.* **2018**, *57*, 11116–11128. [CrossRef]
6. Legros, J.; Figadere, B. Iron-promoted C–C bond formation in the total synthesis of natural products and drugs. *Nat. Prod. Rep.* **2015**, *32*, 1541–1555. [CrossRef]
7. Cahiez, G.; Knochel, P.; Kuzmina, O.; Steib, A.; Moyeux, A. Recent advances in iron-catalyzed Csp2–Csp2 cross-couplings. *Synthesis* **2015**, *47*, 1696–1705. [CrossRef]
8. Ilies, L.; Nakamura, E. Iron-catalyzed cross-coupling reactions. *PATAI'S Chem. Funct. Groups* **2012**, *83*, 1–209. [CrossRef]
9. Guérinot, A.; Cossy, J. Iron-catalyzed C–C cross-couplings using organometallics. *Top. Curr. Chem.* **2016**, *374*, 49. [CrossRef]
10. Tamura, M.; Kochi, J.K. Vinylation of grignard reagents. Catalysis by iron. *J. Am. Chem. Soc.* **1971**, *93*, 1487–1489. [CrossRef]

11. Reddy, C.K.; Knochel, P. New cobalt and iron-catalyzed reactions of organozinc compounds. *Angew. Chem. Int. Ed.* **1996**, *35*, 1700–1701. [CrossRef]

12. Cahiez, G.; Avedissian, H. Highly stereo and chemoselective iron-catalyzed alkenylation of organomagnesium compounds. *Synthesis* **1998**, *1998*, 1199–1205. [CrossRef]

13. Fürstner, A.; Leitner, A.; Méndez, M.; Krause, H. Iron-catalyzed cross-coupling reactions. *J. Am. Chem. Soc.* **2002**, *124*, 13856–13863. [CrossRef] [PubMed]

14. Hatakeyama, T.; Nakamura, M. Iron-catalyzed selective biaryl coupling: Remarkable suppression of homocoupling by the fluoride anion. *J. Am. Chem. Soc.* **2007**, *129*, 9844–9845. [CrossRef] [PubMed]

15. Bedford, R.B. How low does iron go? Chasing the active species in Fe-catalyzed cross-coupling reactions. *Acc. Chem. Res.* **2015**, *48*, 1485–1493. [CrossRef] [PubMed]

16. Neidig, M.L.; Carpenter, S.H.; Curran, D.J.; DeMuth, J.C.; Fleischauer, V.E.; Iannuzzi, T.E.; Neate, P.G.N.; Sears, J.D.; Wolford, N.J. Development and evolution of mechanistic understanding in iron-catalyzed cross-coupling. *Acc. Chem. Res.* **2018**, *52*, 140–150. [CrossRef]

17. Parchomyk, T.; Koszinowski, K. Iron-catalyzed cross-coupling: Mechanistic insight for rational applications in synthesis. *Synthesis* **2017**, *49*, 3269–3280. [CrossRef]

18. Mako, T.L.; Byers, J.A. Recent advances in iron-catalysed cross coupling reactions and their mechanistic underpinning. *Inorg. Chem. Front.* **2016**, *3*, 766–790. [CrossRef]

19. Noda, D.; Sunada, Y.; Hatakeyama, T.; Nakamura, M.; Nagashima, H. Effect of TMEDA on iron-catalyzed coupling reactions of ArMgX with Alkyl halides. *J. Am. Chem. Soc.* **2009**, *131*, 6078–6079. [CrossRef]

20. Nakamura, M.; Matsuo, K.; Ito, S.; Nakamura, E. Iron-catalyzed cross-coupling of primary and secondary Alkyl halides with Aryl grignard reagents. *J. Am. Chem. Soc.* **2004**, *126*, 3686–3687. [CrossRef]

21. Nagano, T.; Hayashi, T. Iron-catalyzed grignard cross-coupling with Alkyl halides possessing β-hydrogens. *Org. Lett.* **2004**, *6*, 1297–1299. [CrossRef] [PubMed]

22. Martin, R.; Fürstner, A. Cross-coupling of Alkyl halides with Aryl grignard reagents catalyzed by a low-valent iron complex. *Angew. Chem. Int. Ed.* **2004**, *43*, 3955–3957. [CrossRef] [PubMed]

23. Bedford, R.B.; Bruce, D.W.; Frost, R.M.; Goodby, J.W.; Hird, M. Iron(III) salen-type catalysts for the cross-coupling of aryl grignards with alkyl halides bearing β-hydrogens. *Chem. Commun.* **2004**, 2822. [CrossRef] [PubMed]

24. Bedford, R.B.; Huwe, M.; Wilkinson, M.C. Iron-catalysed Negishi coupling of benzylhalides and phosphates. *Chem. Commun.* **2009**, 600–602. [CrossRef]

25. Hatakeyama, T.; Kondo, Y.; Fujiwara, Y.-I.; Takaya, H.; Ito, S.; Nakamura, E.; Nakamura, M. Iron-catalysed fluoroaromatic coupling reactions under catalytic modulation with 1,2-bis (diphenylphosphino) benzene. *Chem. Commun.* **2009**, 1216–1218. [CrossRef]

26. Hatakeyama, T.; Hashimoto, T.; Kondo, Y.; Fujiwara, Y.; Seike, H.; Takaya, H.; Tamada, Y.; Ono, T.; Nakamura, M. Iron-catalyzed Suzuki–Miyaura coupling of Alkyl halides. *J. Am. Chem. Soc.* **2010**, *132*, 10674–10676. [CrossRef]

27. Hatakeyama, T.; Fujiwara, Y.-I.; Okada, Y.; Itoh, T.; Hashimoto, T.; Kawamura, S.; Ogata, K.; Takaya, H.; Nakamura, M. Kumada–Tamao–Corriu coupling of Alkyl halides catalyzed by an iron–bisphosphine complex. *Chem. Lett.* **2011**, *40*, 1030–1032. [CrossRef]

28. Bauer, G.; Wodrich, M.D.; Scopelliti, R.; Hu, X. Iron pincer complexes as catalysts and intermediates in Alkyl–Aryl Kumada coupling reactions. *Organometallics* **2014**, *34*, 289–298. [CrossRef]

29. Hedström, A.; Izakian, Z.; Vreto, I.; Wallentin, C.-J.; Norrby, P.-O. On the radical nature of iron-catalyzed cross-coupling reactions. *Chem. A Eur. J.* **2015**, *21*, 5946–5953. [CrossRef]

30. Bedford, R.B.; Brenner, P.B.; Richards, E.; Carvell, T.W.; Cogswell, P.M.; Gallagher, T.; Harvey, J.; Murphy, D.; Neeve, E.C.; Nunn, J.; et al. Expedient iron-catalyzed coupling of Alkyl, Benzyl and Allyl halides with Arylboronic esters. *Chem. A Eur. J.* **2014**, *20*, 7935–7938. [CrossRef]

31. Przyojski, J.A.; Veggeberg, K.P.; Arman, H.D.; Tonzetich, Z.J. Mechanistic studies of catalytic carbon–carbon cross-coupling by well-defined iron NHC complexes. *ACS Catal.* **2015**, *5*, 5938–5946. [CrossRef]

32. Daifuku, S.L.; Al-Afyouni, M.H.; Snyder, B.E.R.; Kneebone, J.L.; Neidig, M.L. A combined Moössbauer, magnetic circular dichroism, and density functional theory approach for iron cross-coupling catalysis: Electronic structure, In Situ Formation, and reactivity of iron-Mesityl-Bisphosphines. *J. Am. Chem. Soc.* **2014**, *136*, 9132–9143. [CrossRef] [PubMed]

33. Daifuku, S.L.; Kneebone, J.L.; Snyder, B.E.R.; Neidig, M.L. Iron (II) active species in iron–bisphosphine catalyzed Kumada and Suzuki–Miyaura cross-couplings of Phenyl Nucleophiles and secondary Alkyl halides. *J. Am. Chem. Soc.* **2015**, *137*, 11432–11444. [CrossRef] [PubMed]

34. Takaya, H.; Nakajima, S.; Nakagawa, N.; Isozaki, K.; Iwamoto, T.; Imayoshi, R.; Gower, N.J.; Adak, L.; Hatakeyama, T.; Honma, T.; et al. Investigation of organoiron catalysis in Kumada–Tamao–Corriu-type cross-coupling reaction assisted by solution-phase X-ray absorption spectroscopy. *Bull. Chem. Soc. Jpn.* **2015**, *88*, 410–418. [CrossRef]

35. Adams, C.J.; Bedford, R.B.; Carter, E.; Gower, N.J.; Haddow, M.; Harvey, J.; Huwe, M.; Cartes, M.Á.; Mansell, S.M.; Mendoza, C.; et al. Iron(I) in Negishi cross-coupling reactions. *J. Am. Chem. Soc.* **2012**, *134*, 10333–10336. [CrossRef]

36. Jin, M.; Adak, L.; Nakamura, M. Iron-catalyzed enantioselective cross-coupling reactions of α-chloroesters with Aryl grignard reagents. *J. Am. Chem. Soc.* **2015**, *137*, 7128–7134. [CrossRef]

37. Iwamoto, T.; Okuzono, C.; Adak, L.; Jin, M.; Nakamura, M. Iron-catalysed enantioselective Suzuki–Miyaura coupling of racemic alkyl bromides. *Chem. Commun.* **2019**, *55*, 1128–1131. [CrossRef]

38. Sharma, A.K.; Sameera, W.M.C.; Jin, M.; Adak, L.; Okuzono, C.; Iwamoto, T.; Kato, M.; Nakamura, M.; Morokuma, K. DFT and AFIR study on the mechanism and the origin of enantioselectivity in iron-catalyzed cross-coupling reactions. *J. Am. Chem. Soc.* **2017**, *139*, 16117–16125. [CrossRef]

39. Lee, W.; Zhou, J.; Gutierrez, O. Mechanism of Nakamura's Bisphosphine-iron-catalyzed asymmetric C(sp2)–C(sp3) cross-coupling reaction: The role of spin in controlling Arylation pathways. *J. Am. Chem. Soc.* **2017**, *139*, 16126–16133. [CrossRef]

40. Schley, N.D.; Fu, G.C. Nickel-catalyzed Negishi Arylations of Propargylic bromides: A mechanistic investigation. *J. Am. Chem. Soc.* **2014**, *136*, 16588–16593. [CrossRef]

41. Frisch, M.J.; Trucks, G.W.; Schlegel, H.B.; Scuseria, G.E.; Robb, M.A.; Cheeseman, J.R.; Scalmani, G.; Barone, V.; Mennucci, B.; Petersson, G.A.; et al. *Gaussian 09, Revision D.01/E.01*; Gaussian, Inc.: Wallingford, CT, USA, 2009.

42. Becke, A.D. Density-functional exchange-energy approximation with correct asymptotic behavior. *Phys. Rev. A* **1988**, *38*, 3098–3100. [CrossRef] [PubMed]

43. Lee, C.; Yang, W.; Parr, R.G. Development of the Colle-Salvetti correlation-energy formula into a functional of the electron density. *Phys. Rev. B* **1988**, *37*, 785–789. [CrossRef] [PubMed]

44. Tomasi, J.; Mennucci, B.; Cammi, R. Quantum mechanical continuum solvation models. *Chem. Rev.* **2005**, *105*, 2999–3094. [CrossRef] [PubMed]

45. Grimme, S.; Antony, J.; Ehrlich, S.; Krieg, H. A consistent and accurate ab initio parametrization of density functional dispersion correction (DFT-D) for the 94 elements H-Pu. *J. Chem. Phys.* **2010**, *132*, 154104. [CrossRef] [PubMed]

46. Ditchfield, R. Self-consistent molecular-orbital methods. IX. An extended Gaussian-type basis for molecular-orbital studies of organic molecules. *J. Chem. Phys.* **1971**, *54*, 724. [CrossRef]

47. Dolg, M.; Wedig, U.; Stoll, H.; Preuss, H. Energy-adjusted ab initio pseudopotentials for the first row transition elements. *J. Chem. Phys.* **1987**, *86*, 866–872. [CrossRef]

48. Fukui, K. The path of chemical reactions—The IRC approach. *Acc. Chem. Res.* **1981**, *14*, 363–368. [CrossRef]

49. Krishnan, R.; Binkley, J.S.; Seeger, R.; Pople, J.A. Self-consistent molecular orbital methods. XX. A basis set for correlated wave functions. *J. Chem. Phys.* **1980**, *72*, 650–654. [CrossRef]

50. Radoń, M. Benchmarking quantum chemistry methods for spin-state energetics of iron complexes against quantitative experimental data. *Phys. Chem. Chem. Phys.* **2019**, *21*, 4854–4870. [CrossRef]

51. Siig, O.S.; Kepp, K.P. Iron(II) and Iron(III) spin crossover: Toward an optimal density functional. *J. Phys. Chem. A* **2018**, *122*, 4208–4217. [CrossRef]

52. Kneebone, J.L.; Brennessel, W.W.; Neidig, M.L. Intermediates and reactivity in iron-catalyzed cross-couplings of Alkynyl grignards with Alkyl Halides. *J. Am. Chem. Soc.* **2017**, *139*, 6988–7003. [CrossRef] [PubMed]

53. Swart, M. Accurate spin-state energies for iron complexes. *J. Chem. Theory Comput.* **2008**, *4*, 2057–2066. [CrossRef] [PubMed]
54. Song, J.-W.; Hirosawa, T.; Tsuneda, T.; Hirao, K. Long-range corrected density functional calculations of chemical reactions: Redetermination of parameter. *J. Chem. Phys.* **2007**, *126*, 154105. [CrossRef] [PubMed]

Sample Availability: Samples of the compounds are not available from the authors.

MDPI

St. Alban-Anlage 66

4052 Basel

Switzerland

Tel. +41 61 683 77 34

Fax +41 61 302 89 18

www.mdpi.com

Molecules Editorial Office

E-mail: molecules@mdpi.com

www.mdpi.com/journal/molecules